T0297065

Introduction to aerospace materials

Related titles:

Welding and joining of aerospace materials
(ISBN 978-1-84569-532-3)
As the demands on aircraft and the materials from which they are manufactured increase, so do the demands on the techniques used to join them. *Welding and joining of aerospace materials* reviews welding techniques such as inertia friction, laser and hybrid laser-arc welding. It also discusses other joining techniques such as riveting, bonding and brazing.

Failure mechanisms in polymer matrix composites
(ISBN 978-1-84569-750-1)
Polymer matrix composites are replacing materials such as metals in industries such as aerospace, automotive and civil engineering. As composites are relatively new materials, more information on the potential risk of failure is needed to ensure safe design. This book focuses on three main types of failure: impact damage, delamination and fatigue. Chapters in Parts I to IV describe the main types of failure mechanism and discuss testing methods for predicting failure in composites. Chapters in Parts V and VI discuss typical kinds of in-service failure and their implications for industry.

Aerodynamic measurements
(ISBN 978-1-84569-992-5)
Aerodynamic measurements presents a comprehensive review of the theoretical bases for experimental techniques used in aerodynamics. Limitations of each method in terms of accuracy, response time and complexity are addressed. This book serves as a guide to choosing the most pertinent technique for each type of flow field including: 1D, 2D, 3D, steady or unsteady, subsonic, supersonic or hypersonic.

Details of these books and a complete list of titles from Woodhead Publishing can be obtained by:

- visiting our web site at www.woodheadpublishing.com
- contacting Customer Services (e-mail: sales@woodheadpublishing.com; fax: +44 (0) 1223 832819; tel.: +44 (0) 1223 499140 ext. 130; address: Woodhead Publishing Limited, 80 High Street, Sawston, Cambridge CB22 3HJ, UK)
- in North America, contacting our US office (e-mail: usmarketing@woodheadpublishing.com; tel.: (215) 928 9112; address: Woodhead Publishing, 1518 Walnut Street, Suite 1100, Philadelphia, PA 19102-3406, USA)

If you would like e-versions of our content, please visit our online platform: www.woodheadpublishingonline.com. Please recommend it to your librarian so that everyone in your institution can benefit from the wealth of content on the site.

Introduction
to aerospace
materials

Adrian P. Mouritz

WOODHEAD
PUBLISHING

Oxford   Cambridge   Philadelphia   New Delhi

Published by Woodhead Publishing Limited,
80 High Street, Sawston, Cambridge CB22 3HJ, UK
www.woodheadpublishing.com
www.woodheadpublishingonline.com

Woodhead Publishing, 1518 Walnut Street, Suite 1100, Philadelphia, PA 19102-3406, USA

Woodhead Publishing India Private Limited, G-2, Vardaan House, 7/28 Ansari Road,
Daryaganj, New Delhi – 110002, India
www.woodheadpublishingindia.com

First published 2012, Woodhead Publishing Limited
© Woodhead Publishing Limited, 2012
The author has asserted his moral rights.

British Library Cataloguing in Publication Data
A catalogue record for this book is available from the British Library.

ISBN 978-1-85573-946-8 (print)
ISBN 978-0-85709-515-2 (online)

The publisher's policy is to use permanent paper from mills that operate a sustainable forestry policy, and which has been manufactured from pulp which is processed using acid-free and elemental chlorine-free practices. Furthermore, the publisher ensures that the text paper and cover board used have met acceptable environmental accreditation standards.

Typeset by Replika Press Pvt Ltd, India
Printed by Lightning Source

Cover image © Christopher Weyer

Contents

Preface

The purpose of this book is to give the reader an introduction to the science and engineering of the materials used in aircraft, helicopters and spacecraft. The topic of aerospace materials is core to aerospace engineering, and sits alongside the other key disciplines of aircraft technology: design, aerodynamics, flight control systems, avionics, propulsion technology, airframe structures and so on. The focus of this book is on the structural materials used in the airframe and propulsion systems. The book examines the materials used in the main structures (e.g. fuselage, wings, landing gear, control surfaces) and the propulsion systems (e.g. jet engines, helicopter rotor blades). The reason for the focus on structural materials is simple: they have a major influence on the cost, performance and safety of aircraft. The other applications of materials on aircraft, such as cabin equipment (e.g. seating, flooring) and electronic equipment (e.g. flight control computers, communication systems, avionics) are outside the scope of this book.

The objective of this book is to describe the science and technology of aerospace materials for college-level students and practising engineers. The reader does not need to have already completed an introductory course in materials engineering to understand this book. The information contained in this book is sufficient for the reader to understand the topics without needing an in-depth knowledge of materials. The book attempts to provide a balance between the *science* and *engineering* of materials so that the reader may understand the underpinning science that determines the behaviour of materials and enough engineering to prepare students for professional practice.

The book is divided into the following topics:

- Introduction to materials for aerospace structures and engines (chapters 1–3)
- Engineering science and properties of aerospace materials (chapters 4 and 5)
- Production, metallurgy and properties of aerospace metal alloys (chapters 6–12)

- Production and properties of composite materials (including polymers) (chapters 13–16)
- Wood (chapter 17)
- Performance issues with aerospace materials (including damage detection) (chapters 18–23)
- Recycling of aerospace materials (chapter 24)
- Materials selection for aerospace structures and engines (chapter 25).

The challenge for any textbook is to provide the proper balance of breadth and depth of the subject. The chapters contain sufficient information to provide an introduction to the topic. Most chapters give applications, case studies and other examples to illustrate the practical aspects of aerospace materials and their performance. It is not the intent of this book to provide in-depth information on every topic, and references are added at the end of each chapter for further reading and research. The books and articles suggested as references are not the only sources of information, although they provide a useful starting point to deepen the reader's understanding of each topic beyond the introductory information provided in the chapters. General references to Internet sites are not provided because they change or disappear without warning; however, the Internet has a wealth of information and many case studies. A glossary of terminology is also found at the end of most chapters so that the reader does not have to wade through the text to find definitions.

Adrian P. Mouritz

1
Introduction to aerospace materials

1.1   The importance of aerospace materials

The importance of materials science and technology in aerospace engineering cannot be overstated. The materials used in airframe structures and in jet engine components are critical to the successful design, construction, certification, operation and maintenance of aircraft. Materials have an impact through the entire life cycle of aircraft, from the initial design phase through to manufacture and certification of the aircraft, to flight operations and maintenance and, finally, to disposal at the end-of-life.

Materials affect virtually every aspect of the aircraft, including the:

- purchase cost of new aircraft;
- cost of structural upgrades to existing aircraft;
- design options for the airframe, structural components and engines;
- fuel consumption of the aircraft (light-weighting);
- operational performance of the aircraft (speed, range and payload);
- power and fuel efficiency of the engines;
- in-service maintenance (inspection and repair) of the airframe and engines;
- safety, reliability and operational life of the airframe and engines; and
- disposal and recycling of the aircraft at the end-of-life.

Aerospace materials are defined in this book as structural materials that carry the loads exerted on the airframe during flight operations (including taxiing, take-off, cruising and landing). Structural materials are used in safety-critical airframe components such as the wings, fuselage, empennage and landing gear of aircraft; the fuselage, tail boom and rotor blades of helicopters; and the airframe, skins and thermal insulation tiles of spacecraft such as the space shuttle. Aerospace materials are also defined as jet engine structural materials that carry forces in order to generate thrust to propel the aircraft. The materials used in the main components of jet engines, such as the turbine blades, are important to the safety and performance of aircraft and therefore are considered as structural materials in this book.

An understanding of the science and technology of aerospace materials is critical to the success of aircraft, helicopters and spacecraft. This book provides the key information about aerospace materials used in airframe

1

structures and jet engines needed by engineers working in aircraft design, aircraft manufacturing and aircraft operations.

1.2   Understanding aerospace materials

Advanced materials have an important role in improving the structural efficiency of aircraft and the propulsion efficiency of jet engines. The properties of materials that are important to aircraft include their physical properties (e.g. density), mechanical properties (e.g. stiffness, strength and toughness), chemical properties (e.g. corrosion and oxidation), thermal properties (e.g. heat capacity, thermal conductivity) and electrical properties (e.g. electrical conductivity). Understanding these properties and why they are important has been essential for the advancement of aircraft technology over the past century.

Understanding the properties of materials is reliant on understanding the relationship between the science and technology of materials, as shown in Fig. 1.1. Materials science and technology is an interdisciplinary field that involves chemistry, solid-state physics, metallurgy, polymer science, fibre technology, mechanical engineering, and other fields of science and engineering.

Materials science involves understanding the composition and structure of materials, and how they control the properties. The term composition means the chemical make-up of the material, such as the types and concentrations of alloying elements in metals or the chemical composition of polymers. The structure of materials must be understood from the atomic to final component levels, which covers a length scale of many orders of magnitude (more than 10^{12}). The important structural details at the different length scales from the atomic to macrostructure for metals and fibre-polymer composites, which are the two most important groups of structural materials used in aircraft, are shown in Fig. 1.2. At the smallest scale the atomic and molecular structure of materials, which includes the bonding between atoms, has a large influence on properties such as stiffness and strength. The crystal structure and nanoscopic-sized crystal defects in metals and the molecular structures of the fibres and polymer in composites also affect the properties. The microstructure of materials typically covers the length scale

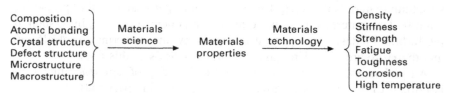

1.1 Relationship between materials science and materials technology.

(a)

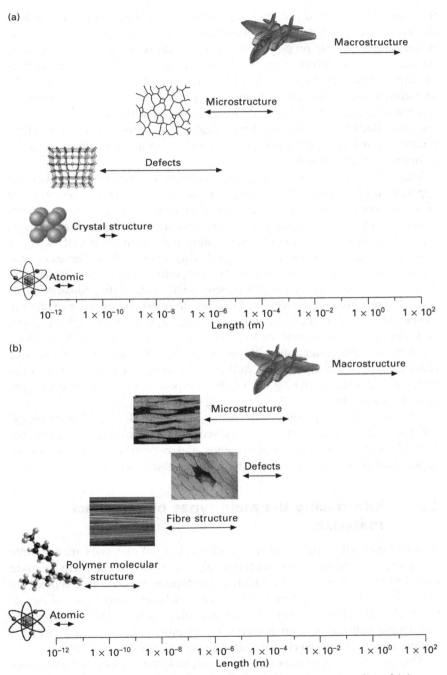

1.2 Structural factors at different sizes affect the properties of (a) metals and (b) fibre–polymer composites.

from around 1 to 1000 μm, and microstructural features in metals such as the grain size, grain structure, precipitates and defects (e.g. voids, brittle inclusions) affect the properties. Microstructural features such as the fibre arrangement and defects (e.g. voids, delaminations) affect the properties of composites. The macrostructural features of materials, such as its shape and dimensions, may also influence the properties. The aim of materials science is to understand how the physical, mechanical and other properties are controlled over the different length scales. From this knowledge it is then possible to manipulate the composition and structure of materials in order to improve their properties.

Materials technology (also called materials engineering) involves the application of the material properties to achieve the service performance of a component. Put another way, materials technology aims to transform materials into useful structures or components, such as converting soft aluminium into a high strength metal alloy for use in an aircraft wing or making a ceramic composite with high thermal insulation properties needed for the heat shields of a spacecraft. The properties needed by materials are dependent on the type of the component, such as its ability to carry stress without deforming excessively or breaking; to resist corrosion or oxidation; to operate at high temperature without softening; to provide high structural performance at low weight or low cost; and so on. Materials technology involves selecting materials with the properties that best meet the service requirements of a component as well as maintaining the performance of the materials over the operating life of the component by resisting corrosion, fatigue, temperature and other damaging events.

Most aerospace engineering work occurs in the field of materials technology, but it is essential to understand the science of materials. This book examines the interplay between materials science and materials technology in the application of materials for aircraft structures and jet engines.

1.3   Introducing the main types of aerospace materials

An extraordinarily large number and wide variety of materials are available to aerospace engineers to construct aircraft. It is estimated that there are more than 120 000 materials from which an aerospace engineer can choose the materials for the airframe and engine. This includes many types of metals (over 65 000), plastics (over 15 000), ceramics (over 10 000), composites, and natural substances such as wood. The number is growing at a fast pace as new materials are developed with unique or improved properties.

The great majority of materials, however, lack one or more of the essential properties required for aerospace structural or engine applications. Most materials are too expensive, heavy or soft or they lack sufficient corrosion resistance, fracture toughness or some other important property. Materials used in aerospace structures and engines must have a combination of essential properties that few materials possess. Aerospace materials must be light, stiff, strong, damage tolerant and durable; and most materials lack one or more of the essential properties needed to meet the demanding requirements of aircraft. Only a tiny percentage of materials, less than 0.05%, are suitable to use in the airframe and engine components of aircraft, helicopters and spacecraft.

It is estimated that less than about one hundred types of metal alloys, composites, polymers and ceramics have the combination of essential properties needed for aerospace applications. The demand on materials to be lightweight, structurally efficient, damage tolerant, and durable while being cost-effective and easy to manufacture rules out the great majority for aerospace applications. Other demands on aerospace materials are emerging as important future issues. These demands include the use of renewable materials produced with environmentally friendly processes and materials that can be fully recycled at the end of the aircraft life. Sustainable materials that have little or no impact on the environment when produced, and also reduce the environmental impact of the aircraft by lowering fuel burn (usually through reduced weight), will become more important in the future.

The main groups of materials used in aerospace structures are aluminium alloys, titanium alloys, steels and composites. In addition to these materials, nickel-based alloys are important structural materials for jet engines. These materials are the main focus of this book. Other materials have specific applications for certain types of aircraft, but are not mainstream materials used in large quantities. Examples include magnesium alloys, fibre–metal laminates, metal matrix composites, woods, ceramics for heat insulation tiles for rockets and spacecraft, and radar absorbing materials for stealth military aircraft.

Many other materials are also used in aircraft: copper for electrical wiring; semiconductors for electronic devices; synthetic fabrics for seating and other furnishing. However, none of these materials are required to carry structural loads. In this book, the focus is on the materials used in aircraft structures and jet engines, and not the nonstructural materials which, although important to aircraft operations, are not required to support loads.

Seldom is a single material able to provide all the properties needed by an aircraft structure and engine. Instead, combinations of materials are used to achieve the best balance between cost, performance and safety. Table 1.1 gives an approximate grading of the common aerospace materials for several key factors and properties for airframes and engines. There are large differences between the performance properties and cost of materials. For example, aluminium and steel are the least expensive; composites are the lightest; steels have the highest stiffness and strength; and nickel alloys have

Table 1.1 Grading of aerospace materials on key design factors

Property	Aluminium	Titanium	Magnesium	High-strength steel	Nickel superalloy	Carbon fibre composite
Cost	Cheap	Expensive	Medium	Medium	Expensive	Expensive
Weight (density)	Light	Medium	Very light	Heavy	Heavy	Very light
Stiffness (elastic modulus)	Low/medium	Medium	Low	Very high	Medium	High
Strength (yield stress)	Medium	Medium/high	Low	Very high	Medium	High
Fracture toughness	Medium	High	Low/medium	Low/medium	Medium	Low
Fatigue	Low/medium	High	Low	Medium/high	Medium	High
Corrosion resistance	Medium	High	Low	Low/medium	High	Very high
High-temperature creep strength	Low	Medium	Low	High	Very high	Low
Ease of recycling	High	Medium	Medium	High	Medium	Very low

the best mechanical properties at high temperature. As a result, aircraft are constructed using a variety of materials which are best suited for the specific structure or engine component.

Figure 1.3 shows the types and amounts of structural materials in various types of modern civil and military aircraft. A common feature of the different aircraft types is the use of the same materials: aluminium, titanium, steel and composites. Although the weight percentages of these materials differ between aircraft types, the same four materials are common to the different aircraft and their combined weight is usually more than 80–90% of the structural mass. The small percentage of 'other materials' that are used may include magnesium, plastics, ceramics or some other material.

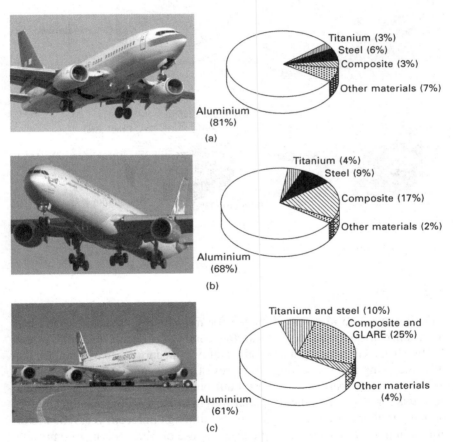

1.3 Structural materials and their weight percentage used in the airframes of civilian and military aircraft. (a) Boeing 737, (b) Airbus 340-330, (c) Airbus A380, (d) Boeing 787, (e) F-18 *Hornet* (C/D), (f) F-22 *Raptor*. Photographs supplied courtesy of (a) K. Boydston, (b) S. Brimley, (c) F. Olivares, (d) C. Weyer, (e) J. Seppela and (f) J. Amann.

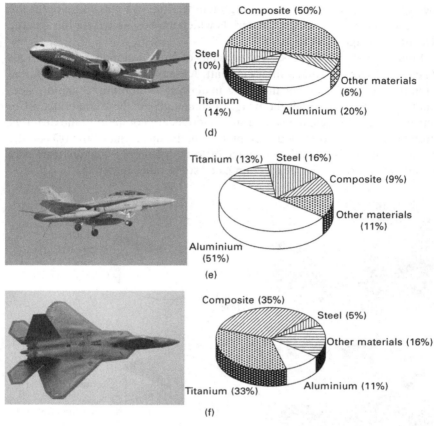

1.3 Continued

1.3.1 Aluminium

Aluminium is the material of choice for most aircraft structures, and has been since it superseded wood as the common airframe material in the 1920s/1930s. High-strength aluminium alloy is the most used material for the fuselage, wing and supporting structures of many commercial airliners and military aircraft, particularly those built before the year 2000. Aluminium accounts for 70–80% of the structural weight of most airliners and over 50% of many military aircraft and helicopters, although in recent years the use of aluminium has fallen owing to the growing use of fibre–polymer composite materials. The competition between the use of aluminium and composite is intense, although aluminium will remain an important aerospace structural material.

Aluminium is used extensively for several reasons, including its moderately low cost; ease of fabrication which allows it to be shaped and machined into structural components with complex shapes; light weight; and good stiffness, strength and fracture toughness. Similarly to any other aerospace material, there are several problems with using aluminium alloys, and these include susceptibility to damage by corrosion and fatigue.

There are many types of aluminium used in aircraft whose properties are controlled by their alloy composition and heat treatment. The properties of aluminium are tailored for specific structural applications; for example, high-strength aluminium alloys are used in the upper wing skins to support high bending loads during flight whereas other types of aluminium are used on the lower wing skins to provide high fatigue resistance.

1.3.2 Titanium

Titanium alloys are used in both airframe structures and jet engine components because of their moderate weight, high structural properties (e.g. stiffness, strength, toughness, fatigue), excellent corrosion resistance, and the ability to retain their mechanical properties at high temperature. Various types of titanium alloys with different compositions are used, although the most common is Ti–6Al–4V which is used in both aircraft structures and engines.

The structural properties of titanium are better than aluminium, although it is also more expensive and heavier. Titanium is generally used in the most heavily-loaded structures that must occupy minimum space, such as the landing gear and wing–fuselage connections. The structural weight of titanium in most commercial airliners is typically under 10%, with slightly higher amounts used in modern aircraft such as the Boeing 787 and Airbus A350. The use of titanium is greater in fighter aircraft owing to their need for higher strength materials than airliners. For instance, titanium accounts for 25% of the structural mass of the F-15 *Eagle* and F-16 *Fighting Falcon* and about 35% of the F-35 *Lightning II*. Titanium alloys account for 25–30% of the weight of modern jet engines, and are used in components required to operate to 400–500 °C. Engine components made of titanium include fan blades, low-pressure compressor parts, and plug and nozzle assemblies in the exhaust section.

1.3.3 Magnesium

Magnesium is one of the lightest metals, and for this reason was a popular material for lightweight aircraft structures. Magnesium was used extensively in aircraft built during the 1940s and 1950s to reduce weight, but since then the usage has declined as it has been replaced by aluminium alloys and composites. The use of magnesium in modern aircraft and helicopters is typically less than 2% of the total structural weight. The demise of magnesium as an important structural material has been caused by several factors, most notably higher cost and lower stiffness and strength compared with aluminium alloys. Magnesium is highly susceptible to corrosion which leads to increased requirements for maintenance and repair. The use of magnesium alloys is now largely confined to non-gas turbine engine parts, and applications include gearboxes and gearbox housings of piston-engine aircraft and the main transmission housing of helicopters.

1.3.4   Steel

Steel is the most commonly used metal in structural engineering, however its use as a structural material in aircraft is small (under 5–10% by weight). The steels used in aircraft are alloyed and heat-treated for very high strength, and are about three times stronger than aluminium and twice as strong as titanium. Steels also have high elastic modulus (three times stiffer than aluminium) together with good fatigue resistance and fracture toughness. This combination of properties makes steel a material of choice for safety-critical structural components that require very high strength and where space is limited, such as the landing gear and wing box components. However, steel is not used in large quantities for several reasons, with the most important being its high density, nearly three times as dense as aluminium and over 50% denser than titanium. Other problems include the susceptibility of some grades of high-strength steel to corrosion and embrittlement which can cause cracking.

1.3.5   Superalloys

Superalloys are a group of nickel, iron–nickel and cobalt alloys used in jet engines. These metals have excellent heat resistant properties and retain their stiffness, strength, toughness and dimensional stability at temperatures much higher than the other aerospace structural materials. Superalloys also have good resistance against corrosion and oxidation when used at high temperatures in jet engines. The most important type of superalloy is the nickel-based material that contains a high concentration of chromium, iron, titanium, cobalt and other alloying elements. Nickel superalloys can operate for long periods of time at temperatures of 800–1000 °C, which makes them suitable for the hottest sections of gas turbine engines. Superalloys are used in engine components such as the high-pressure turbine blades, discs, combustion chamber, afterburners and thrust reversers.

1.3.6   Fibre–polymer composites

Composites are lightweight materials with high stiffness, strength and fatigue performance that are made of continuous fibres (usually carbon) in a polymer matrix (usually epoxy). Along with aluminium, carbon fibre composite is the most commonly used structural material for the airframe of aircraft and helicopters. Composites are lighter and stronger than aluminium alloys, but they are also more expensive and susceptible to impact damage.

Carbon fibre composites are used in the major structures of aircraft, including the wings, fuselage, empennage and control surfaces (e.g. rudder, elevators, ailerons). Composites are also used in the cooler sections of jet engines, such as the inlet fan blades, to reduce weight. In addition to carbon fibre composites, composites containing glass fibres are used in radomes and semistructural components such as fairings and composites containing aramid fibres are used in components requiring high impact resistance.

1.3.7   Fibre–metal laminates

Fibre–metal laminates (FML) are lightweight structural materials consisting of thin bonded sheets of metal and fibre–polymer composite. This combination creates a material which is lighter, higher in strength, and more fatigue resistant than the monolithic metal and has better impact strength and damage tolerance than the composite on its own. The most common FML is GLARE® (a name derived from glass reinforced aluminium) which consists of thin layers of aluminium alloy bonded to thin layers of fibreglass composite. FMLs are not widely used structural materials for aircraft; the only aircraft at present that use GLARE® are the Airbus 380 (in the fuselage) and C17 GlobeMaster III (in the cargo doors).

1.4   What makes for a good aerospace material?

Selecting the best material for an aircraft structure or engine component is an important task for the aerospace engineer. The success or failure of any new aircraft is partly dependent on using the most suitable materials. The cost, flight performance, safety, operating life and environmental impact from engine emissions of aircraft is dependent on the types of materials that aerospace engineers choose to use in the airframe and engines. It is essential that aerospace engineers understand the science and technology of materials in order to select the best materials. The selection of materials for aircraft is not guesswork, but is a systematic and quantitative approach that considers a multitude of diverse (and in some instances conflicting) requirements. The selection of materials is performed during the early design phase of aircraft, and has a lasting influence which remains until the aircraft is retired from service.

The key requirements and factors that aerospace engineers must consider in the selection of materials are listed below and in Table 1.2.

Table 1.2 Selection factors for aerospace structural materials

Costs	Purchase cost. Processing costs, including machining, forming, shaping and heat treatment costs. In-service maintenance costs, including inspection and repair costs. Recycling and disposal costs.
Availability	Plentiful, consistent and long-term supply of materials.
Manufacturing	Ease of manufacturing. Low-cost and rapid manufacturing processes.
Density	Low specific gravity for lightweight structures.
Static mechanical properties	Stiffness (elastic modulus). Strength (yield and ultimate strength).
Fatigue durability	Resistance against initiation and growth of cracks from various sources of fatigue (e.g. stress, stress-corrosion, thermal, acoustic).
Damage tolerance	Fracture toughness and ductility to resist crack growth and failure under load. Notch sensitivity owing to cut-outs (e.g. windows), holes (e.g. fasteners) and changes in structural shape. Damage resistance against bird strike, maintenance accidents (e.g. dropped tools on aircraft), impact from runway debris, hail impact.
Environmental durability	Corrosion resistance. Oxidation resistance. Moisture absorption resistance. Wear and erosion resistance. Space environment (e.g. micrometeoroid impact, ionizing radiation).
Thermal properties	Thermally stable at high temperatures. High softening temperatures. Cryogenic properties. Low thermal expansion properties. Non/low flammability. Low-toxicity smoke.
Electrical and magnetic properties	High electrical conductivity for lightning strikes. High radar (electromagnetic) transparency for radar domes. Radar absorbing properties for stealth military aircraft.

Cost. The whole-of-life cost of aerospace materials must be acceptable to the aircraft operator, and obviously should be kept as low as possible. Whole-of-life costs include the cost of the raw material; cost of processing and assembling the material into a structural or engine component; cost of in-service maintenance and repair; and cost of disposal and recycling at the end of the aircraft life.

Availability. There must be a plentiful, reliable and consistent source of materials to avoid delays in aircraft production and large fluctuations in purchase cost.

Manufacturing. It must be possible to process, shape, machine and join the materials into aircraft components using cost-effective and time-efficient manufacturing methods.

Weight. Materials must be lightweight for aircraft to have good manoeuvrability, range and speed together with low fuel consumption.

Mechanical properties. Aerospace materials must have high stiffness, strength and fracture toughness to ensure that structures can withstand the aircraft loads without deforming excessively (changing shape) or breaking.

Fatigue durability. Aerospace materials must resist cracking, damage and failure when subjected to fluctuating (fatigue) loads during flight.

Damage tolerance. Aerospace materials must support the ultimate design load without breaking after being damaged (cracks, delaminations, corrosion) from bird strike, lightning strike, hail impact, dropped tools, and the many other damaging events experienced during routine operations.

Thermal properties. Aerospace materials must have thermal, dimensional and mechanical stability for high temperature applications, such as jet engines and heat shields. Materials must also have low flammability in the event of aircraft fire.

Electrical properties. Aerospace materials must be electrically conductive to dissipate the charge in the event of lightning strike.

Electromagnetic properties. Aerospace materials must have low electromagnetic properties to avoid interfering with the electronic devices used to control and navigate the aircraft.

Radar absorption properties. Materials used in the skin of stealth military aircraft must have the ability to absorb radar waves to avoid detection.

Environmental durability. Aerospace materials must be durable and resistant to degradation in the aviation environment. This includes resistance against corrosion, oxidation, wear, moisture absorption and other types of damage caused by the environment which can degrade the performance, functionality and safety of the material.

1.5   Summary

The materials used in aircraft have a major influence on the design, manufacture, in-service performance and maintainability. Materials impact on virtually every aspect of the aircraft, including cost, design options, weight, flight performance, engine power and fuel efficiency, in-service maintenance and repair, and recycling and disposal at the end-of-life.

Understanding the materials used in aircraft relies on understanding both the science and technology of materials. Materials science involves studying the effects of structure and composition on the properties. Materials technology involves understanding how the material properties can be used to achieve the in-service performance requirements of a component.

Although there are over 120 000 materials, less than about 100 different materials are used in the airframe and engines of aircraft. The four major types of structural materials are aluminium alloys, fibre–polymer composites (particularly carbon fibre–epoxy), titanium alloys and high-strength steels; these materials account for more than 80% of the airframe mass in most commercial and military aircraft. An important high temperature material for jet engines is nickel-based superalloy. Other materials are used in the airframe or engines in small amounts, and include fibre–metal laminates, ceramic matrix composites, magnesium alloys and, in older and light aircraft, wood.

Selection of the best material to meet the property requirements of an aircraft component is critical in aerospace engineering. Many factors are considered in materials selection, including whole-of-life cost; ease of manufacturing; weight; structural efficiency; fatigue and damage tolerance; thermal, electrical, electromagnetic and radar absorption properties; and durability against corrosion, oxidation and other damaging processes.

1.6   Further reading and research

Askeland, D. R. and Phulé, P. P., *The science and engineering of materials*, Thomson, 2006.

Ashby, M. F. and Jones, D. R. H., *Engineering materials 1: an introduction to their properties and applications*, Butterworth–Heinemann, 1996.

Barrington, N. and Black, M., 'Aerospace materials and manufacturing processes at the millennium', in *Aerospace materials*, edited B. Cantor, H. Assender and P. Grant, Institute of Physics Publishing, Bristol, 2001, pp. 3–14.

Peel, C. J. and Gregson, P. J., 'Design requirements for aerospace structural materials', in *High performance materials in aerospace*, edited H. M. Flower, Chapman and Hall, London, 1995, pp. 1–48.

Aerospace materials: past, present and future

2.1   Introduction

The development of new materials and better utilisation of existing materials has been central to the advancement of aerospace engineering. Advances in the structural performance, safety, fuel economy, speed, range and operating life of aircraft has been reliant on improvements to the airframe and engine materials. Aircraft materials have changed greatly in terms of mechanical performance, durability, functionality and quality since the first powered flight by the Wright Brothers in 1903. Furthermore, the criteria which are used to select materials for aircraft have also changed over the past 100 years. Figure 2.1 presents a timeline for the approximate years when new criteria were introduced into the selection of aircraft materials.

The main criteria for materials selection for the earliest aircraft (c. 1903–1920) was minimum weight and maximum strength. The earliest aircraft were designed to be light and strong; other design criteria such as cost, toughness and durability were given less importance in the quest for high strength-to-weight. Many of the criteria which are now critical in the choice of materials were not recognised as important by the first generation of aircraft designers, and their goal was simply to use materials that provided high strength for little weight. At the time the best material to achieve the strength-to-weight requirement was wood.

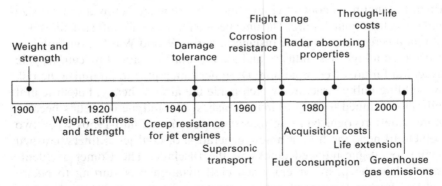

2.1 Historical timeline indicating when key criteria for materials selection were introduced into aircraft design.

15

The situation changed during the 1920s/1930s when the criteria for materials selection widened to consider a greater number of factors affecting aircraft performance and capability. The design of aircraft changed considerably as commercial and commuter aviation became more popular and the military began to recognise the tactical advantages of fast fighters and heavy bombers. Improved performance from the 1930s led to aircraft capable of flying at fast speeds over long distances while carrying heavy payloads. The requirement for high strength-to-weight remained central to the choice of material, as it had with earlier aircraft, but other criteria such as high stiffness and durability also became important. Higher stiffness allowed sleeker and more compact designs, and hence improved performance. These new criteria not only required new materials but also the development of new production methods for transforming these materials into aircraft components. Aluminium alloys processed using new heat treatments and shaped using new metal-forming processes were developed to meet the expanding number of selection criteria. The importance of availability emerged as a critical issue in the selection of materials during the Second World War. For example, the supply of aluminium to Japan was cut off in the late 1930s/early 1940s, which forced their military to use magnesium in the construction of many fighter aircraft.

Major advances in aerospace technology, particularly jet aircraft, first-generation helicopters and rockets/missiles, occurred shortly after World War II. These advances placed greater demands on the performance requirements of the the airframe and engine materials. Another significant milestone was the introduction of pressurised cabin aircraft for high altitude flight during the 1940s. The increased pressure loads exerted on the fuselage led to the development of stressed skin panels made using high-strength material.

Around the same time, the need for materials with fatigue and fracture properties emerged as a critical safety issue, and represents the introduction of the damage tolerance criterion. Damage tolerance is the capability of an aircraft structure to contain cracks and other damage below a critical size without catastrophic failure. The unexpected failure of aircraft structures was common before and, in some instances, during World War II. Aviation was considered a high risk industry and aircraft crashes caused by catastrophic structural failures were common. Designers attempted to minimise the risk by building bulky structures which made the aircraft heavy, but structural failures continued leading to many crashes. The fatigue of metals became more widely recognised as an important issue in the mid-1950s when two Comet airliners, the first of a new generation of civil jet airliners, crashed owing to fatigue-induced cracks in the fuselage. The Comet accidents occurred in the post-war era when civil aviation was starting to boom, and the crashes threatened public confidence in aviation safety. Fracture toughness and fatigue resistance joined other important properties such as weight, stiffness and strength as essential properties in the choice of aircraft materials.

The development of supersonic aircraft together with advances in rocket technology during the 1960s prompted the need for high-temperature materials. The aerospace industry invested heavily in the development of new materials for supersonic airliners such as Concorde, high-speed fighters and surveillance aircraft for the Cold War, and spacecraft and satellites for the Space Race. The investments led to the development of heat-resistant airframe materials such as titanium alloys and special aluminium alloys that were capable of withstanding frictional heating effects during supersonic flight without softening. The need for more powerful engines for aircraft and rockets also drove the development of high-temperature materials capable of operating above 800 °C. New types of nickel-based alloys and other heat-resistant materials were developed to survive within the hottest sections of jet engines.

The need for damage-tolerant materials became more intense in the late 1970s when unexpected failures occurred in ultra-high-strength steel components in United States Air Force (USAF) aircraft. It became clear that the failures involved manufacturing defects and fatigue cracks so small that they could not be found reliably. The USAF introduced a damage tolerant design philosophy which accepted the presence of cracks in aircraft and managed this by achieving an acceptable life by a combination of design and inspection. Achieiving this required the use of materials that were resistant to fatigue cracking and failure.

The certification of new commercial aircraft required manufacturers to demonstrate that fatigue cracks could be detected before reaching the critical length associated with catastrophic failure. Aviation safety authorities such as the FAA introduced stringent regulations on the damage tolerance of safety-critical structures. New commercial aircraft would not be certified and permitted to fly unless new criteria on damage tolerance were met. This change in the certification requirements further increased the need for damage-tolerant materials with excellent fracture toughness and fatigue properties for both airframe and engine applications.

Although always important, weight reduction of civil aircraft became critical during the 1970s owing to rising fuel costs and the revenue opportunities associated with increased range and heavier payload. The OPEC fuel crisis of the 1970s, when the price of Avgas jumped by more than 500%, threatened the financial viability of the global aviation industry and sent many airline companies broke. The aerospace industry implemented new measures to minimise weight and maximise structural performance, and this included the greater use of higher-strength aluminium alloys and the introduction of carbon–epoxy fibre composite materials into secondary structures such as engine cowlings and undercarriage doors.

The realisation of the financial benefits of extending the life of ageing aircraft during the 1980s and 1990s provided greater focus on improved damage tolerance and corrosion resistance. With the rising cost of new aircraft and greater competition among airline companies, including the introduction of low-cost carriers into the aviation market, the need to prolong the operating life of aircraft became critical. New aluminium alloys with improved corrosion resistance and composites materials which are completely resistant to corrosion were used in greater quantities. The 1990s was an era when factors such as the costs of manufacturing and maintenance became increasingly important in the choice of materials. The 1990s was also an era when novel structural materials with radar absorbing properties and low thermal emissions were used in large quantities on stealth military aircraft. Although aircraft with limited stealth capability had been in operation since the 1970s, the need for extremely low radar visibility became a critical requirement that drove the development of radar-absorbing materials.

All of the factors for materials selection outlined above apply today in the choice of materials for modern aircraft: weight, stiffness, strength, damage tolerance, fracture toughness, fatigue, corrosion resistance, heat resistance and so on. The first decade of the 21st century is characterised by an emphasis on materials that reduce the manufacturing cost (by cheaper processing and assembly using fewer parts) and lower through-life operating cost (through longer life with fewer inspections and less maintenance). Reductions in greenhouse gas emissions by reducing aircraft weight and improving engine fuel efficiency are also contemporary issues in materials selection. There is also growing interest in producing materials with environmentally friendly manufacturing processes and using sustainable materials that are easily recycled.

The evolution of aircraft technology and the associated drivers in materials selection has meant that the airframe and engine materials are constantly changing. The approximate year of introduction of the main aerospace materials is shown in Fig. 2.2. Many materials have been introduced, with most being developed specifically for aerospace but later finding applications in other sectors such as rail, automotive or engineering infrastructure. It is important to recognise that continuous improvements have occurred with each type of material since their introduction into aircraft. For example, on-going developments in aluminium alloys have occurred since the 1920s to improve properties such as strength, toughness and corrosion resistance. Similarly, advances with composite materials since the 1970s have reduced costs while increasing mechanical properties and impact toughness. Aircraft designers now have the choice of dozens of aluminium alloys with properties tailored to specific applications and operating conditions.

In this chapter, we study the historical development of the major types of aerospace materials: wood, aluminium, magnesium, titanium, nickel

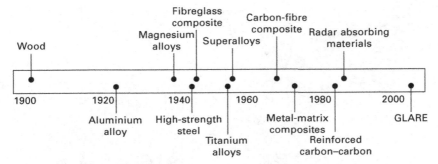

2.2 Historical timeline indicating the approximate year when the main types of materials were first used in aircraft.

superalloys and composites. The introduction of these materials into aircraft structures or engines, and how their usage and properties have changed over time is discussed. Also, the current status and future growth in aircraft production and how this may impact on the use of materials is examined. The on-going advances in materials technology for next-generation aircraft, helicopters and space-craft are also reviewed.

2.2   Brief history of aerospace materials

2.2.1   Wood

The era of aerospace materials arguably started with the first powered flight of *Kitty Hawk* by Orville and Wilbur Wright. The principal criterion used in the selection of materials for the first generation of aircraft (1903–1930) was maximum strength for minimum weight. Every other consideration in materials selection, including stiffness, toughness and durability, were secondary compared with the main consideration of high strength-to-weight. Weight had to be kept to an absolute minimum because of the low power (below 150 hp) of early aircraft engines. The airframes in the earliest aircraft were constructed almost entirely of wood because there were no other suitable materials that combined strength and lightness. The high-strength materials of the early 1900s, such as steel and cast iron, were about 10 times denser than wood and, therefore, too heavy for the airframe.

Wood was the material of choice in early aircraft because of its light weight, stiffness and strength (Fig. 2.3). Wood was also used because it was plentiful, inexpensive, and its properties were well understood through use in other structural applications such as buildings and bridges. Another important reason for using wood was the craftsmen who handbuilt the earliest aircraft were able to easily shape and carve timber into lightweight frames, beams and other structural components. However, wood is not the

(a)

(b)

2.3 Wooden aircraft. (a) Sopwith *Camel*. Photograph supplied courtesy of the National Museum of the US Air Force. (b) de Havilland *Mosquito*. Photograph supplied courtesy of M. J. Freer.

ideal material and has many inherent problems. The mechanical properties are variable and anisotropic which meant aircraft had to be over-designed to avoid structural damage. Many early aircraft experienced structural failures owing to inconsistent strength properties as the result of 'soft' or 'weak' spots in the wood. Furthermore, wood absorbs moisture, warps and decays over time, which meant that aircraft required continuous maintenance and on-going repairs.

The first generation of aircraft builders evaluated many types of timber, and found that fir, spruce and several other softwoods were best suited for making structural components with a high ratio of strength-to-weight. The aircraft industry later discovered that laminated plywood construction provided greater strength and toughness than single-piece wood. Laminated plywood consists of thin bonded sheets of timber orientated with the wood grain at different angles. The use of plywood reduced the weight penalty experienced with one-piece timber construction that had to be over-designed. As a result plywood rapidly gained popularity as a structural material in the period between the two world wars. Even during the World War II some fighters and light bombers were constructed from wood and plywood. Probably the most famous wooden aircraft during the war was the de Havilland *Mosquito*, which, for its time, was a highly advanced fighter/bomber capable of flying at 650 km h^{-1}.

The large-scale production of fighters, bombers and heavy load transport aircraft during World War II led to the demise of wood as an important structural material. Abundant supplies of high quality timber were not available to many countries during the war, which forced the greater use of alternative materials such as aluminium. Also, wood lacks the stiffness and strength required for many military aircraft, particularly bombers, cargo transporters and other heavy lift aircraft that have high loading on their wings and airframe. The use of wood continued to decline in the post-war era with the development of pressurised cabins for high-altitude flying. Today, few aircraft are constructed using wood, except for some gliders, ultra-lights and piston-driven aircraft, because cheaper, lighter and more structurally efficient materials are available.

2.2.2   Aluminium

The development of aircraft with greater engine power during the 1920s placed increased demands on wood construction that it struggled to meet. The loads on the wings and airframe increased as aircraft became larger and heavier. The wing loading on aircraft built during the 1910s was 30–40 kg m^{-2}, which could be supported using wooden frames. However, the construction of larger, heavier aircraft in the following decades increased pressure loading on the wings to 500–1000 kg m^{-2}. Figure 2.4 shows the general trend towards higher wing loads for military aircraft and passenger airliners over the past century. Loads on other parts of the aircraft, particularly the fuselage and tailplane, have also increased. Wood lacks the stiffness, strength and toughness to withstand high loads, and aircraft builders sought other lightweight materials with better structural properties.

With steel being too heavy, the aircraft industry in the 1920s turned to aluminium alloys as a replacement for wood. Aluminium is one of the lightest metals; being about 2.5 times lighter than steel. It is stiffer, stronger, tougher

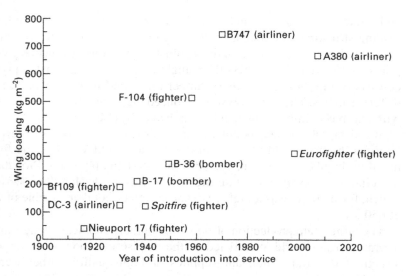

2.4 Plot of wing pressure load against year of introduction for several fighters, bombers and airliners.

and more durable than timber. Also, aluminium can be easily fabricated into thin skin panels and readily machined into spars, stiffeners and beams for the fuselage and wings.

Aluminium had been available in commercial quantities to aircraft manufacturers since the early 1900s, but it was too soft. Aluminium was first used in the airframe of Zeppelin airships during World War I, but it lacked the strength to be used in fixed-wing aircraft that are more heavily loaded. Metallurgists during the early decades of the twentieth century improved the strength properties of aluminium by the addition of alloying elements and development of heat-treatment processes. Various types and amounts of alloying elements were added to aluminium using a trial and error approach to assess the effect on strength. The metals industry experimented with many alloying elements to increase strength and hardness. The industry also tested different heat treatments and metal-forming processes to improve the mechanical properties. A major breakthrough occurred when the addition of a few percent of copper and other alloying elements was found to increase the strength by several hundred percent. The development of the aluminium alloy 'Duralumin' in 1906 was largely responsible for the uptake of aluminium by the aircraft industry from the 1920s. Duralumin is an aluminium alloy containing copper (4.4%), magnesium (1.5%) and manganese (0.6%) which is strengthened by heat treatment. Duralumin sparked an explosion in the use of aluminium in highly-stressed aircraft structures, such as the skins, ribs and stiffeners in the wings and fuselage. The use of aluminium also provided the capability to increase the speed, range and size (payload) of aircraft over that possible with wood.

Following the initial success of Duralumin, the mechanical properties of aluminium alloys improved dramatically in the era between the two world wars owing to on-going research and development. From the 1950s, there was a much better understanding of the effects of alloy composition, impurity control, processing conditions and heat treatment on the properties of aluminium. Figure 2.5 shows the sustained improvement in the strength of aluminium alloys since the 1920s. Similar improvements have been achieved with other important properties, including longer fatigue life, greater fracture toughness and damage tolerance, and better corrosion resistance. These developments have been driven largely by the demands of the aerospace industry for more structurally efficient materials. Other major advances in aluminium technology occurred in the 1960s/70s when Al–Li alloys, which have higher stiffness and lower weight than conventional alloys, were developed. New heat-treatment processes developed in the 1970s/80s resulted in better toughness, damage tolerance and corrosion resistance.

Aluminium is the material of choice for most aircraft structures, and typically accounts for 70–80% of the structural weight of most commercial airliners and over 50% of military aircraft and helicopters. In recent years, however, the percentage of airframe weight consisting of aluminium has declined owing to greater use of carbon fibre–polymer composites. Figure 2.6 shows the recent decline in the use of aluminium in airliners owing to greater usage of composite materials in the fuselage, wings and other major

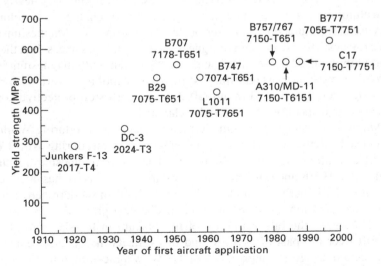

2.5 Yield strength of aluminium alloys and the year of introduction into service.

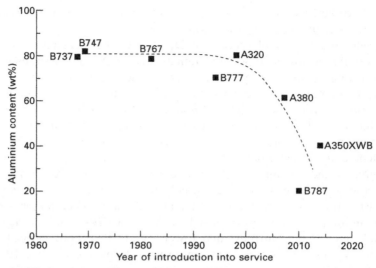

2.6 Decline in use of aluminium in passenger aircraft.

structures. Despite this drop, aluminium will remain an important structural material for both aircraft and helicopters.

2.2.3   Magnesium

Like aluminium, magnesium has been used for many years as an airframe material because of its low weight. Magnesium is lighter (by nearly 40%) than aluminium, although it has never been a serious challenger to aluminium because of its higher cost and inferior structural properties. Magnesium alloys have lower stiffness, strength, fatigue resistance and toughness than the types of aluminium used in aircraft. The greatest problem with magnesium is poor corrosion resistance. Magnesium is highly susceptible to various forms of corrosion, and when used in aircraft requires corrosion protective coatings and regular inspections for corrosion damage.

Magnesium was first used in German military aircraft during World War I and used extensively in German and Japanese aircraft during World War II owing to limited supplies of aluminium. The use of magnesium reached its peak in the 1950s and 60s, and, since the early 1970s, usage has declined and now it is used sparingly (under 1–2% by weight) in modern aerostructures owing to corrosion problems and low mechanical properties. For example, Fig. 2.7 shows the fall in the use of magnesium in Russian-built Tupolev aircraft over the past fifty years, and this reflects the general reduction in magnesium usage in many aircraft types. Magnesium remains a useful material in aircraft and helicopters even though its usage is low, and it is unlikely to be completely eliminated from aircraft.

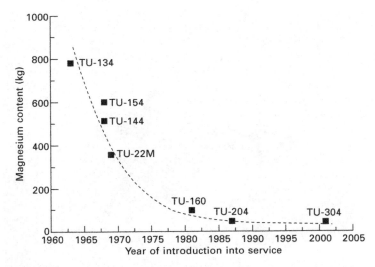

2.7 Decline in use of magnesium with successive versions of Tupolev aircraft since the 1960s, reflecting its general decline as an aerospace structural material.

2.2.4   Titanium

Titanium was first used in military and commercial aircraft during the 1950s. The original need for titanium arose from the development of supersonic aircraft capable of speeds in excess of Mach 2. The skins of these aircraft require heat-resistant materials that do not soften owing to frictional heating effects at supersonic speeds. Conventional aluminium alloys soften when the aircraft speed exceeds about Mach 1.5 whereas titanium remains unaffected until Mach 4–5. The USAF developed the SR-71 *Blackbird* with an all-titanium skin construction in the mid-1960s (Fig. 2.8). The SR-71 was a high-altitude reconnaissance aircraft with a maximum speed in excess of Mach 3, and, at the time, was one of the world's most sophisticated and fastest aircraft. The SR-71 demonstrated the application of titanium in airframe structures. Titanium is stiffer, stronger and more fatigue resistant than aluminium, and for these reasons it has been used increasingly in heavily loaded structures such as pressure bulkheads and landing gear components.

The use of titanium in commercial aircraft has increased over recent decades, albeit slowly owing to the high cost of titanium metal and the high costs of manufacturing and machining titanium components. Although the mechanical properties of titanium are better than those of aluminium, the material and manufacturing costs are much higher and it is uneconomical to use in structural components unless they need to be designed for high loads. For this reason, the structural weight of titanium in airliners is typically under

2.8 SR-71 supersonic aircraft contain large amounts of titanium in the airframe.

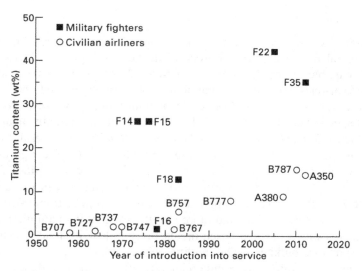

2.9 Amount of titanium used in aircraft.

10%, although higher amounts are used in new aircraft types such as the Airbus 350 and Boeing 787 as shown in Fig. 2.9. Titanium usage is greater in fighter aircraft because the loads on the wings and fuselage are higher and the cost is less critical in materials selection. Titanium is also used in gas turbine engine components required to operate at temperatures of 450–500 °C. Titanium has high static, fatigue and creep strengths as well as excellent corrosion resistance at elevated temperatures, which makes it suitable for jet engines. Titanium engine components include fan blades, guide vanes, shafts and casings in the inlet region; low-pressure compressor; and plug and nozzle assemblies in the exhaust section. Titanium alloys account for 25–30% of the weight of many modern jet engines.

2.2.5   Superalloys

The development of aircraft, helicopters and rockets is reliant on the development of materials for gas turbine engines or rocket motors that can operate at high temperatures for a long time without softening or degrading. Superalloys are an important group of high-temperature materials used in the hottest sections of jet and rocket engines where temperatures reach 1200–1400 °C. Superalloys are based on nickel, cobalt or iron with large additions of alloying elements to provide strength, toughness and durability at high temperature.

Since the introduction of jet engines in the post-World War II period the aerospace industry has invested heavily in alloy development, metal-casting processes, and metal-forming technologies to raise the maximum operating temperature of superalloys. The need to improve the efficiency and thrust of engines has resulted in an enormous increase in the temperature at the entry of the high-pressure turbine section. This temperature has risen over the past sixty years from 800 to 1600 °C, and future engines will probably be required to operate at about 1800 °C. These increases are only possible with the development of materials capable of operating for long periods at extremely high temperatures. Figure 2.10 shows the general trend for improvement in the operating temperature limit (creep strength) of nickel-based superalloys used in high-pressure turbine blades since the late-1960s. Advances in alloy composition, impurity control and casting technology (including the development of directional solidification and single crystal casting methods) together with the development of thermal ceramic coatings for superalloys have increased greatly the maximum operating temperature, thus resulting in increased engine performance by raising the power-to-weight ratio and fuel economy. Fuel consumption is an important metric in evaluating the operational efficiency of an airliner, and Fig. 2.11 shows the improvement in the fuel economy since the late 1950s. Over the fifty-year period, fuel consumption of engines has dropped by 50% and the aircraft fuel burn per seat has fallen over 80%. Continuous advances in the main factors that affect fuel burn rate, viz, airframe design, engine design, flight control and navigation, and advanced materials, have led to large improvements in fuel economy. Not only does this cut the operating cost of aircraft, but it

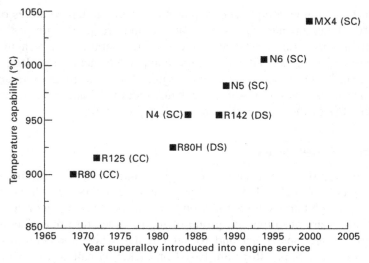

2.10 Improvement in the temperature capability (creep strength) of nickel superalloys used in jet engines since the late 1960s. The alloy type and casting methods (CC = chill casting; DS = directional solidification casting; SC = single-crystal casting) are given.

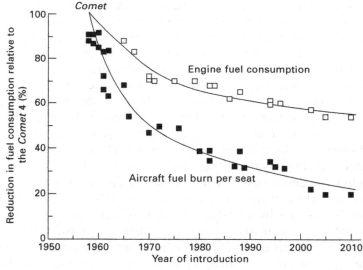

2.11 Reductions in the engine fuel consumption and aircraft fuel burn per seat based on the *Comet* 4 aircraft.

also reduces greenhouse gas emissions and other pollutants owing to lower fuel consumption.

The durability and operating life of engine components have also improved dramatically in recent decades owing to advances in their materials. For example, when the Boeing 707 entered service in 1958, the engines were removed for maintenance after about 500 h of operation. Most of the maintenance related to deterioration of the high-pressure turbine blades that were made of early versions of superalloys. Today, a Boeing 747 class engine can operate for 20 000 h without major maintenance. This remarkable improvement is in part the result of advances in the metallurgy of nickel-based superalloys and other high-temperature materials, including ceramic coatings.

2.2.6   Composite materials

Fibre-reinforced polymer composites are another important group of aerospace materials that have a long history of usage. Composites were first used in the 1940s for their high strength-to-weight ratio and corrosion resistance. The first generation of composite material consisted of glass fibres in a low-strength polymer matrix. The potential application of this material was demonstrated during the late 1940s and 1950s in various prototype aircraft components and filament-wound rocket motor cases. However, the aerospace industry was initially reluctant to use composites in large quantities because the original fibreglass materials were expensive to produce; difficult to manufacture with a high degree of quality control; their mechanical properties were inconsistent and variable owing to inadequate processing methods; and they were prone to delamination cracking when subjected to impact events such as bird strike. Furthermore, the elastic modulus of fibreglass composite is low and therefore it is not suitable in structural applications that require high stiffness. Composites were gradually introduced into semistructural aircraft components during the 1950s and 1960s, such as engine cowlings and undercarriage landing gear doors, to reduce weight and avoid corrosion.

The evolution of primary aircraft structures from aluminium to composite has been slow owing to the commercial risk involved with making the change. The aerospace industry, particularly those companies producing civil aircraft, is conservative owing to the financial and safety risk of changing structural materials. The industry recognised that the changeover from aluminium to composite may possibly provide benefits such as increased airframe life and reduced production costs and weight, but none of these were guaranteed. Furthermore, the transition from metals to composites requires a complete change of production facilities, which comes at huge expense. Aluminium remains a satisfactory material, despite problems with fatigue and corrosion, and the strong incentive to replace this material with composite was lacking for many years. Furthermore, the large weight savings attributed to composites are not always achieved, and often the reduction in mass achieved by replacing aluminium with composite has been modest.

A major change in the use of composite material occurred in the 1960s with the commercial production of carbon fibres. Carbon-fibre composites are lightweight, stiff, strong, fatigue resistant and corrosion resistant, and for these reasons their potential application in both airframes and engines was immediately recognised by the aerospace industry. However, the high cost of carbon fibres, poor understanding of the design rules, structural properties and durability together with technical challenges in certification meant that the initial use of carbon-fibre composites was small. Until the 1970s, the use of carbon-fibre composites was limited to semistructural components which accounted for less than 5% of the airframe weight. Corrosion problems with aluminium and the OPEC energy crisis in the 1970s were incentives for the aerospace industry to expand the use of carbon-fibre composites in both fighter aircraft and commercial airliners. As design methods and manufacturing processes improved and the cost of carbon fibre dropped the amount of composite material used in aircraft increased during the 1980s and 1990s, as shown in Fig. 2.12.

Major milestones in the use of carbon-fibre composites were applications in primary structures of fighter aircraft such as the *Harrier* (AV-8B) and *Hornet* (F-18) and in the tail section of the Boeing 777 in the 1990s. The use of composites in the fuselage and wings of modern airliners such as the A380, A350 and B787 are recent major events. The use of carbon-fibre composites in helicopter components, such as the body, tail boom and rotor blades, has also increased dramatically since the 1990s. Composites are the first material since the 1930s to seriously challenge the long-held dominant

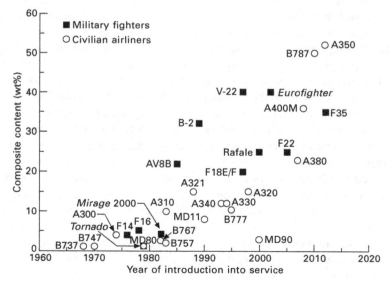

2.12 Amount of composite materials used in aircraft.

position of aluminium in airframe construction, and the competition between these two materials is likely to be intense in coming years.

Composite materials are increasingly being used in low-temperature components in jet engines because of their light weight. Carbon-fibre composites were first used in gas turbine engines in the 1960s for their light weight and high mechanical properties. Because of their low softening temperature, however, composites can only be used in low-temperature engine components such as the air inlet fan where the temperature remains below 180–200 °C. An early engine application for carbon-fibre composites was in fan blades in the Rolls-Royce RB-211 high-bypass turbofan, which was designed in the late 1960s for aircraft such as the Lockheed L-1011 (*TriStar*). Unfortunately, Rolls-Royce pushed the state-of-the-art with carbon-fibre composites too far because the blades were vulnerable to damage from bird impact and there were manufacturing problems. This combined with other technical problems and major cost blow-outs in the development of the RB-211 caused Rolls-Royce to become insolvent in 1971 and the company was nationalised by the UK Government. This incident demonstrates the serious problems that can occur when new materials are introduced into critical aircraft components before their capabilities are fully characterised and certified. Despite this initial set-back, major aircraft engine manufacturers continued to develop composite components during the 1970s/1980s, and in recent years carbon-fibre materials have been used reliably in fan blades and inlet casings. The use of composites is expected to increase further in coming years with the development of higher-temperature polymers and improvements in impact damage tolerance.

Other types of composites have been developed for airframe and engine applications, although their usage is much less than that of polymer matrix materials. Metal matrix composites (MMC) were first developed in the 1950s/60s to improve the structural efficiency of monolithic metals such as aluminium. MMCs are composed of a hard reinforcing phase dispersed in a continuous metal matrix phase. The reinforcement is often a ceramic or man-made fibre (boron, carbon) in the form of small particles, whiskers or continuous filaments. The development of MMCs resulted in materials that are stiffer, harder, stronger and, in some cases, lighter and more fatigue resistant than the base metal. Early applications of MMCs included fuselage struts in the space shuttle orbiter, ventral fins and fuel access doors in the F-16 *Fighting Falcon*, and main rotor blade sleeves in some helicopters. The use of MMCs in jet engines has been evaluated, but, to date, the applications are limited to components such as fan exit guide vanes in specific engine types and they are not widely used in large-scale commercial production of engine parts. MMCs have largely failed to make a major impact in structural or engine applications. During the development of MMCs it became obvious that these materials are expensive to produce; difficult to forge, machine and join; and have low ductility and toughness; and for these reasons they are not often used.

Ceramic matrix composites (CMCs) are another class of composite material introduced into aircraft and spacecraft in the 1970s. CMCs consist of ceramic reinforcement embedded in a ceramic matrix. CMCs were developed for high-temperature applications which require materials with higher strength and toughness than conventional monolithic ceramics. The most famous CMC is reinforced carbon–carbon which gained fame through its use in heat shields on the space shuttle, brake discs for aircraft, and engine nozzle liners for rockets and missiles.

Fibre-metal laminates (FML) were developed as damage-tolerant composite materials for aircraft structures during the 1980s. The original FML was called ARALL, which consists of thin layers of aramid fibre composite sandwiched between layers of aluminium alloy. Difficulties with manufacturing and problems with moisture absorption lead to the development of an alternative FML known as GLARE, which comprises alternating layers of fibreglass composite and aluminium. GLARE has higher strength, fatigue resistance, damage tolerance and corrosion durability than monolithic aluminium, and was first used widely in the upper fuselage of the Airbus 380 and later in cargo doors for the C-17 *Globemaster III* heavy-lift transporter. The future of FMLs in other large aircraft is uncertain owing to high production and manufacturing costs.

2.3   Materials for the global aerospace industry

On-going advances in materials technology are essential to the success of the aerospace industry in the design, construction and in-service operation of aircraft. The aerospace industry is broadly defined as an industry network that designs, builds and provides in-service support to aircraft, helicopters, guided missiles, space vehicles, aircraft engines, and related parts. The industry includes small to medium-sized enterprises that design, manufacture or service specific aerospace items for large global companies such as Boeing, EADS and Lockheed-Martin who design, assemble, sell and provide in-service support to the entire aircraft system.

Improvements in materials, whether by making them cheaper, lighter, stronger, tougher or more durable, have been crucial to the development of better and safer aircraft. It is worthwhile examining the current state and future growth of the aerospace industry to understand the need for on-going advances in aerospace materials. The global aerospace industry in 2008 had an annual turnover of about $275 billion. This makes aerospace one of the most valuable industries in the world. The largest national players are the USA (which has about 50% of the market), European Union, Japan, Canada and, increasingly, China. The aerospace industry employs world-wide about

1.5 million people, with many engaged in highly skilled professions. The aerospace industry is not only important because of the economic wealth and employment it generates, but also from the generation of knowledge. Globally, aerospace drives innovation and skills in the engineering and information technology sectors which are then used by other (non-aerospace) industries. This includes the development of new materials and their manufacturing processes.

Figure 2.13 shows the approximate value of the different market segments in the aerospace industry. About 50% (or ~$140 billion) of the annual turnover is derived from the manufacture and sales of aircraft, of which the greatest share (23%) is from large commercial airliner sales. In comparison, military aircraft (7%), business jets (6%) and helicopters (4%) account for a relatively small, but still valuable, share of the market. The other 50% of industry turnover is related to the in-service support and maintenance of aircraft.

Aerospace is not a stable, constant and predictable industry; but is a volatile industry subject to fluctuations in the growth and recession of the global economy. Not surprisingly, the growth of the aerospace industry closely tracks the demand for air travel, which over the past thirty years has grown at an average rate of 8%. Figure 2.14 shows the number of new large commercial aircraft sales in the period between the mid-1970s and mid-2000s. The number of aircraft sales grew during prolonged periods of global economic growth and dropped during economic recessions or major

A: Regional/commuter aircraft (5%)
B: Business jets (6%)
C: General aviation (1%)
D: Large commercial jets (23%)
E: Helicopters (4%)
F: Unmanned aerial vehicles (2%)
G: Military aircraft (7%)
H: Support and maintenance (52%)

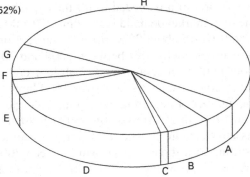

2.13 Global aerospace industry by market segment.

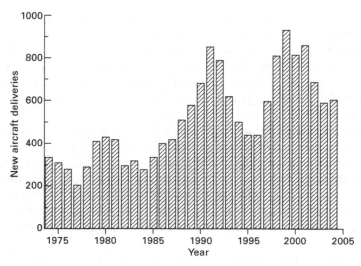

2.14 New deliveries of large commercial aircraft between 1974 and 2004.

terrorist attacks that affected public confidence in aviation safety. This trend is expected to continue, with faster growth occurring during periods of economic prosperity and slower (or negative) growth during recessions and when the public lacks confidence in aviation safety.

Despite the fluctuations in new aircraft sales, the major aerospace companies predict an era of sustained growth. Several key statistics illustrate the projected growth in civil aviation between 2004 and 2024:

- The world-wide fleet of large passenger aircraft is expected to more than double by 2024, growing to over 35 000. About 57% of the fleet operating today (9600 airliners) is forecast to still be in operation in 2024 and the remainder (7200 airliners) to be retired. An additional 18 500 aircraft are needed to fill the capacity demand.
- The total market demand is for almost 26 000 new passenger and freight aircraft worth about $2.3 trillion at current price list.
- New aircraft deliveries between 2004 and 2025 are forecast to include about 11 000 single-aisle and small jet freighters, 2000 small twin-aisle and regional freighters, and over 1600 large twin-aisle aircraft.
- The number of new freighter aircraft sales per year is forecast to double from 1760 (in 2004) to about 3500 (in 2024).
- Passenger traffic is forecast to grow at an average of 4.8% per year, resulting in a three-fold growth between 2004 and 2023. Airfreight is forecast to grow even faster, with freight tonnes kilometres increasing at an average of 6.2% per year.

In addition to a growing number of new aircraft, the age of existing aircraft has risen considerably over the past decade and is likely to continue rising. The average age of military aircraft is also increasing, as shown in Fig. 2.15.

Growth of the aviation industry must be supported by the construction of new aircraft and the life extension of existing aircraft. The materials used in new aircraft have a major economic impact on the profitability of manufacturers and airline companies. Similarly, the materials used in existing aircraft have a large influence on the airframe life and maintenance costs. The future of the aerospace industry is reliant on advances in materials technology. Development of new materials and research into the life extension and durability of existing materials is essential for the on-going success of the industry.

2.4   Future advances in aerospace materials

The future success of the aerospace industry both in terms of the cost-effective manufacture of new aircraft and the cost-effective extension of the operating life of existing aircraft is reliant on on-going improvements to existing materials and the development of new materials. Advances in materials technology is classified as evolutionary or revolutionary. Evolutionary advances mean that small, incremental improvements are made to existing materials, such as a new alloy composition, processing method or heat treatment. Examples include the addition of new alloying elements to nickel superalloys to increase the creep resistance and maximum operating temperature or the development

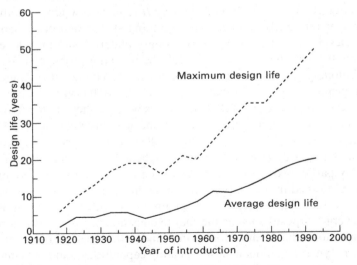

2.15 Trends in design life for US military aircraft.

of a new thermal ageing treatment for aluminium alloys to increase their resistance to stress corrosion. The evolutionary approach is often preferred by the aerospace industry since past experience has shown that almost every new material has some initial problems. The aerospace industry is more comfortable with incremental improvements on conventional materials, for which they have good knowledge of the design, manufacture, maintenance and repair issues.

Revolutionary advances are the application of new materials to structures or engines that are different to previously used materials. An example was the first-time that carbon-fibre composite was used to fabricate a primary load-bearing structure (the tail section) on a commercial airliner (B777) in the mid-1990s. Another example was the use of GLARE in the A380 in 2005, which was the first application of a fibre-metal laminate in an aircraft fuselage. Revolutionary materials usually have had limited success in being directly incorporated into aircraft owing to the high costs of manufacturing, qualification and certification. The cost and time associated with developing a new material and then testing and certifying its use in safety-critical components can cost an aerospace company hundreds of millions of dollars and take 5–10 years or longer. The introduction of new material can require major changes to the production infrastructure of aircraft manufacturing plants as well as to the in-service maintenance and repair facilities. For example, some suppliers of structural components to the Boeing 787 had to make major changes to their production facilities from metal to composite manufacturing, which required new design methods, manufacturing processes and quality control procedures as well as reskilling and retraining of production staff. As another example, the introduction of new radar absorbing materials on stealth aircraft such as the F-35 *Lightning II* require new repair methods in the event of bird strike, hail impact, lightning strikes and other damaging events. Despite the challenges, revolutionary materials are introduced into new aircraft when the benefits outweigh the potential problems and risks.

Evolutionary and revolutionary advances across a broad range of materials technologies are on-going for next-generation aircraft. It is virtually impossible to give a complete account of these advances because they are too numerous. Some important examples are: development of high-temperature polymers for composites capable of operating at 400 °C or higher; new polymer composites that are strengthened and toughened by the addition of nano-sized clay particles or carbon nanotubes; new damage tolerant composites that are reinforced in the through-thickness direction by techniques such as stitching, orthogonal weaving or z-pinning; multifunctional materials that serve several purposes such as thermal management, load-bearing strength, self-assessment and health monitoring, and self-actuation functions; bio-inspired polymer materials with self-healing capabilities; sandwich materials that contain high-performance metal-foam cores, truss or periodic open-cell cores; new tough ceramic materials with improved structural capabilities; rapidly solidified amorphous metals with improved mechanical properties and corrosion resistance; and new welding and joining processes for dissimilar materials. The aerospace materials for this century are sure to be just as ground-breaking and innovative as the materials used in the past century. On-going improvement in structural and engine materials is essential for the advancement of aerospace engineering. After many years of commercial service it might be expected that structural and engine material technology would be approaching a plateau, and the pace of innovation would decline. However, customer demands for higher performance, lower operating costs and more 'environmentally friendly' propulsion systems continue to drive materials research to ever more challenging goals.

2.5   Summary

The materials used in aircraft structures and engines have changed dramatically over the past century or so to meet the advances in aircraft technology. As aircraft have become faster, larger and more technically advanced, the demands on the materials have become more intense. The evolution of aerospace materials has been controlled by the evolution of the factors used in their selection. Originally materials were selected based on their strength and weight, but as aircraft have become more advanced the selection has become based on a multitude of structural performance, durability, damage tolerance, economic, environmental and other factors. This has forced improvements to the properties of aerospace materials over the past 100 years, and these improvements will continue with on-going research and development of new materials and processing methods.

The most used aircraft structural material since the 1930s is aluminium alloy. Over the past 80 years or so there have been continuous improvements in the strength, corrosion resistance and other properties of aluminium through research that has led to better alloy compositions, control of impurities, superior heat treatments and forming processes. Aluminium will remain an important structural material for both civil and military aircraft despite the increasing use of composite materials.

The use of carbon-fibre composites has increased greatly since the mid-1990s, and this material is now competing 'head-to-head' with aluminium as the dominant aerospace structural material. The increased use of composites is the result of many factors, including improvements in the properties of the fibres and polymer matrix, better design techniques and manufacturing methods, and the requirements for greater structural efficiency, reduced operating costs, and better fatigue and corrosion resistance for aircraft.

The development of jet engine technology since the mid-1940s has relied heavily on advances in high-temperature materials, particularly nickel-based superalloys. Improvements to alloy composition and casting methods have resulted in large increases in the temperature limit of engines over the past 60 years or so, resulting in greater thrust and power. There have also been large improvements in the fuel economy and emissions for jet engines, resulting, in part, from advances in materials technology.

The general approach to implementing new materials into aircraft is through evolutionary advances, which simply means incremental improvements to the properties of the materials already in use. Revolutionary advances which involve the application of new material are less common, although they do occur when the benefits outweigh the cost and risk.

2.6   Further reading and research

Schatzberg, E. M., 'Materials and the development of aircraft: wood–aluminium–composites', in *Around Glare*, edited C. Vermeeren, Kluwer Academic Publishers, 2002, pp. 43–72.

Vermeeren, C. A. J. R., 'An historic overview of the development of fibre metal laminates', *Applied Composite Materials*, **10** (2003), 189–205.

Williams, J. C. and Starke, E. A., 'Progress in structural materials for aerospace systems', *Acta Materialia*, **51** (2003), 5775–5799.

3
Materials and material requirements for aerospace structures and engines

3.1   Introduction

It is not possible to fully understand the topic of aerospace materials without also understanding the aerospace structures in which they are used. The choice of aerospace materials is governed by the design, function, loads and environmental service conditions of the structure. Understanding aircraft structures allows aerospace engineers to select the most appropriate material. The materials used in safety-critical structures, which are structures that can result in loss of the aircraft when they fail, require high mechanical properties and excellent durability in the aviation environment. Examples of safety-critical structures are the fuselage, wings, landing gear and empennage. The main components in gas turbine engines are also safety-critical, such as the turbine blades and discs.

The materials used in aircraft structures require a combination of high stiffness, strength, fracture toughness, fatigue endurance and corrosion resistance. Aerospace materials must carry the structural and aerodynamic loads while being inexpensive and easy to fabricate. The materials must also be damage tolerant and provide durability over the aircraft design life, which for military fighter aircraft is typically in the range of 8000–14 000 flight hours (over a period of 15 to 40 years) and for large commercial airliners is 30 000–60 000 flight hours (25–30 years). During this period, the aircraft structures should not crack, corrode, oxidise or suffer other forms of damage while operating under adverse conditions that involve high loads, freezing and high temperatures, lightning strikes and hail impact, and exposure to potentially corrosive fluids such as jet fuel, lubricants and paint strippers.

In addition to high mechanical properties and long-term durability, it is essential that the materials are light. The airframe accounts for a large percentage (typically between 20 and 40%) of the all-up weight of most aircraft, and any savings in weight by using light materials which are structurally efficient result in less fuel burn, greater range and speed, and smaller engine requirements. Table 3.1 presents a breakdown in the mass of the main structures of several aircraft types as a percentage of the total weight. The other mass of the aircraft is comprised of the power plant, fuel, instruments, equipment, cabin fittings and payload, as shown in Fig. 3.1.

39

Table 3.1 Breakdown of structural weight for various aircraft types as a percentage of total weight

Structural component	Sailplane	Light and executive aircraft	Subsonic airliner	Military fighter
Fuselage	25	11	7	12
Wing	30	14	8	12
Stabilisers	3	2	4	4
Undercarriage	2	4	4	4
Total %	60	31	23	32

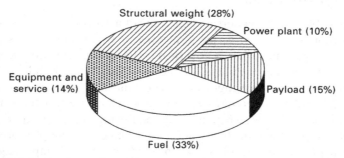

3.1 Approximate breakdown of take-off weight of large passenger aircraft.

This chapter presents an overview of the safety-critical structures for aircraft, helicopters and spacecraft. The main structures in fixed-wing aircraft, which are the fuselage, wings, empennage, control surfaces, undercarriage, and gas turbine engines are examined. The loads and environmental service conditions for these structures are briefly explained. The key mechanical and durability properties required by the materials used in the different structures are discussed, and the types of materials that are used are identified. The chapter also presents a short overview of the main structures in helicopters, and describes the types of materials which are used in the main body, tail boom and rotor blades. The main structures in re-entry spacecraft such as the space shuttle are explained, and again the materials which are used are identified.

3.2   Fixed-wing aircraft structures

Figure 3.2 shows the main structural components in a modern military aircraft, which are the fuselage, wings, empennage, landing gear and control surfaces such as flaps, elevators and ailerons. Aircraft structures must be lightweight and structurally efficient, and this is achieved by the combination of optimised design and high-performance materials. Many major aircraft sections, including the fuselage, wing and empennage, are shell-like structures

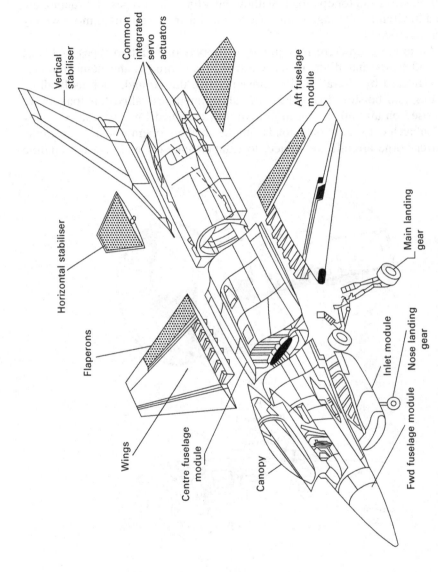

Vertical stabiliser

Common integrated servo actuators

Horizontal stabiliser

Aft fuselage module

Flaperons

Main landing gear

Wings

Inlet module

Nose landing gear

Centre fuselage module

Canopy

Fwd fuselage module

3.2 Main structural components of a modern military aircraft.

known as monocoque or semimonocoque (Fig. 3.3). A monocoque structure is an unreinforced shell that must be thick to avoid buckling under an applied load. A semimonocoque structure consists of a thin shell supported by longitudinal stiffening members and transverse frames to resist bending, compression and torsion loads without buckling. Both types of structure are used in aircraft, although semimonocoque construction is used more widely than monocoque.

An aircraft structure is required to support one or two types of loads: ground loads and flight loads. Grounds loads are, as the name implies, encountered by aircraft during movement on the ground, such as landing, taxiing and hoisting loads. Flight loads (sometimes called 'air loads') are imposed on aircraft during flight. Aircraft designed for a special role may be subjected to additional loads unique to their operation. For example, carrier-borne aircraft are subject to catapult take-off and arrested landing

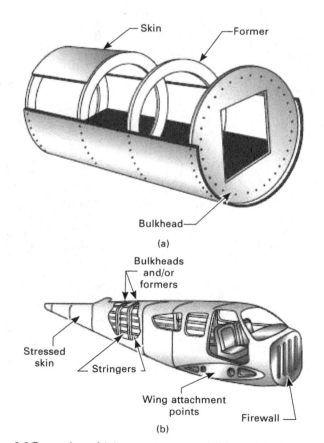

3.3 Examples of (a) monocoque and (b) semimonocoque fuselage structures.

loads, as well as high vertical descent loads on landing. Ground and flight loads can be further divided into surface loads and body loads. Surface loads such as aerodynamic and hydrostatic pressures act over the skin of an aircraft. Aerodynamic pressures acting on the underside of aircraft wings or helicopter rotor blades are examples of surface loads. Body forces act over the volume of the structure and are produced by gravitational forces and inertia loads. For example, the force applied on the undercarriage during landing is a body force. Forces exerted on the aircraft wing-box from the wings and fuselage are another instance of body loads. The surface and body loads applied to aircraft structures are resolved in the structure as tension, compression, bending, shear, torsion or a combination of these load types.

The loads exerted on the airframe, which includes the fuselage, wings and empennage, are carried by both the skins and internal frames which includes stringers, spars, circumferential frames, pressure bulkheads and various other reinforcing members. The internal structure of an airframe is complex, but is designed with an optimum configuration of load-bearing members to provide high structural performance combined with light weight (Fig. 3.4). A critical aspect in the optimum design of an airframe is the selection of materials for each structural detail that possess the required combination of physical, mechanical and durability properties.

3.2.1   Fuselage

The fuselage is a long cylindrical shell, closed at its ends, which carries the internal payload. The dominant type of fuselage structure is semimonocoque construction. These structures provide better strength-to-weight ratios for the central portion of the body of an airplane than monocoque construction. A semimonocoque fuselage consists of a thin shell stiffened in the longitudinal direction with stringers and longerons and supported in the radial direction using transverse frames or rings (Fig. 3.5). The strength of a semimonocoque fuselage depends mainly on the longitudinal stringers (longerons), frames and pressure bulkhead. The skin carries the cabin pressure (tension) and shear loads, the longitudinal stringers carry the longitudinal tension and compression loads, and circumferential frames maintain the fuselage shape and redistribute loads into the airframe.

The primary loads on the fuselage are concentrated around the wing-box, wing connections, landing gear and payload. During flight the upward loading of wings coupled with the tailplane loads usually generates a bending stress along the fuselage. The lower part of the fuselage experiences a compressive stress whereas the upper fuselage (called the crown) is subject to tension. Shear loads are generated along the sides of the fuselage and torsion loads when the aircraft rolls and turns. Pressurisation of the cabin for high-attitude flying exerts an internal tensile (hoop) stress on the fuselage.

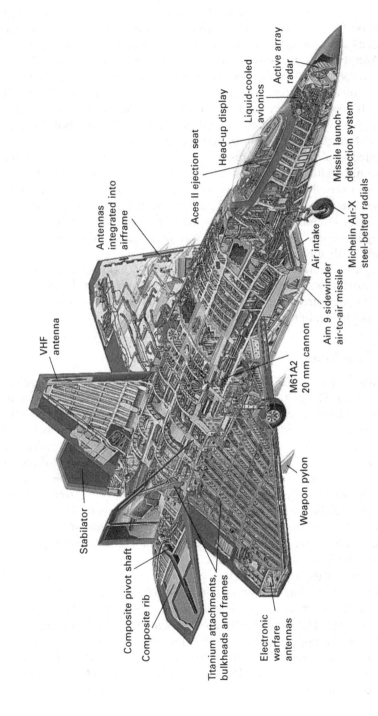

Active array
radar

Liquid-cooled
avionics

Head-up display

Aces II ejection seat

Missile launch-
detection system

Michelin Air-X
steel-belted radials

Aim 9 sidewinder
air-to-air missile

Air intake

Antennas
integrated into
airframe

M61A2
20 mm cannon

VHF
antenna

Weapon pylon

Stabilator

Composite pivot shaft

Composite rib

Titanium attachments,
bulkheads and frames

Electronic
warfare
antennas

3.4 Internal structural design of a modern aircraft.

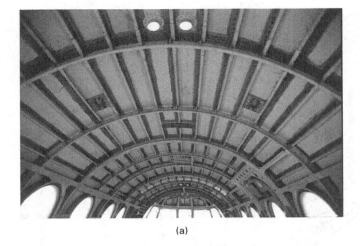

(a)

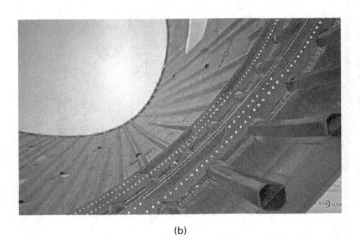

(b)

3.5 Semimonocoque fuselage structures made using (a) aluminium alloys and (b) carbon–epoxy composite ((a) reproduced with permission from R. Wilkinson, *Aircraft structures and systems*, Longman).

Figure 3.6 shows the property requirements for fuselage materials. Important properties for fuselage materials are stiffness, strength, fatigue resistance, corrosion resistance, and fracture toughness. Although all of these properties are important, fracture toughness is often the limiting design consideration in aluminium fuselages. Fuselage materials need good resistance against fatigue cracking owing to pressurisation and depressurisation of the fuselage with every flight. Aluminium alloy has been the most common fuselage

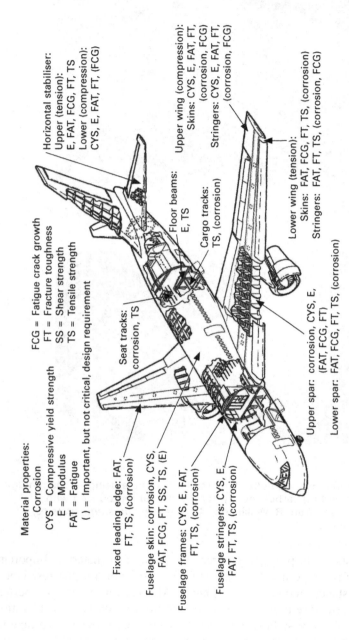

Material properties:
Corrosion
CYS = Compressive yield strength
E = Modulus
FAT = Fatigue
( ) = Important, but not critical, design requirement

FCG = Fatigue crack growth
FT = Fracture toughness
SS = Shear strength
TS = Tensile strength

Horizontal stabiliser:
Upper (tension):
E, FAT, FCG, FT, TS
Lower (compression):
CYS, E, FAT, FT, (FCG)

Upper wing (compression):
Skins: CYS, E, FAT, FT, (corrosion, FCG)
Stringers: CYS, E, FAT, FT, (corrosion, FCG)

Floor beams:
E, TS

Cargo tracks:
TS, (corrosion)

Seat tracks:
corrosion, TS

Lower wing (tension):
Skins: FAT, FCG, FT, TS, (corrosion)
Stringers: FAT, FT, TS, (corrosion)

Fixed leading edge: FAT, FT, TS, (corrosion)

Fuselage skin: corrosion, CYS, FAT, FCG, FT, SS, TS, (E)

Fuselage frames: CYS, E, FAT, FT, TS, (corrosion)

Fuselage stringers: CYS, E, FAT, FT, TS, (corrosion)

Upper spar: corrosion, CYS, E, (FAT, FCG, FT)

Lower spar: FAT, FCG, FT, TS, (corrosion)

3.6 Material property requirements for the main aircraft structures reproduced with permission from J. T. Staley and D. J. Lege, 'Advances in aluminium alloy products for structural applications in transportation', *Journal de Physique IV*, **3** (1993), 179–190).

material over the past eighty years, although carbon fibre–epoxy composite is regularly used in the fuselage of military fighters and increasingly in large passenger aircraft. For example, the Boeing 787 fuselage is constructed using carbon–epoxy composite. GLARE, which is a metallic laminate material, and carbon-epoxy are used extensively in the fuselage of the Airbus 380.

3.2.2   Wings

The main function of the wing is to pick up air loads to maintain flight and to transmit these loads to the fuselage (via the wing-box and wing connections). Additional loads on the wing are internal fuel pressure, landing gear forces, wing leading and trailing edge loads, and the engine weight (when wing mounted). The wing loading, which is simply the total flying weight divided by the wing area, varies between aircraft types. For example, the maximum wing loading with full take-off weight for a Piper Arrow is about 80 kg m^{-2}, Airbus 380 is 700 kg m^{-2}, Boeing 747 is 750 kg m^{-2} and Tornado is 1100 kg m^{-2}. The wings account for 20–25% of the structural weight of an aircraft and, therefore, it is imperative that lightweight materials are used.

Wings are constructed of thin skins supported on the inside by stringers and spars, and are designed to carry bending, shear and torsion loads. The bending load is a combination of tension and compression forces. When the aircraft is on the ground the wings hang down under their own weight, the weight of fuel stored inside them, and the weight of engines if these are wing-mounted. This creates a tension load along the upper wing surface and a compression load on the lower surface. During flight when the loads are much higher, however, the bending loads are reversed. The wing bends upwards in flight to support the weight of the aircraft, and this generates compression in the upper surface and tension in the lower surface as shown in Fig. 3.7.

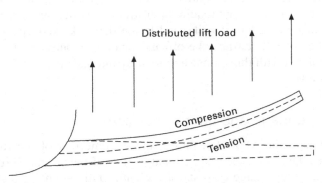

3.7 Bending action of an aircraft wing during flight.

The property requirements for the materials used in the wing are given in Fig. 3.6. The materials used in wings must have high stiffness and strength to withstand the bending, shear and torsion loads. Other requirements include light weight, damage resistance against bird strike at the leading edges, and durability. An important requirement is high fatigue strength to resist damage and failure from fluctuating loads owing to flight manoeuvring, turbulence and wind gusts. In military combat aircraft the fluctuations in stress are generally higher than commercial aircraft owing to the need for frequent and fast manoeuvring. These fluctuating loads can induce fatigue damage. Fatigue of metal structures is favoured by fluctuating tensile loads whereas fatigue damage does not occur in compression, and therefore the lower (tension) and upper (compression) wing surfaces have different material requirements. For this reason, several materials are used in a single aircraft wing. For example, subsonic aircraft wings have traditionally been made using two types of aluminium alloys: high compressive strength alloy (such as 2024 Al) for the upper wing surface and high tensile strength alloy (e.g. 7075 Al) for the lower surface. Wings are increasingly being constructed using carbon–epoxy composite materials owing to their combination of high strength and fatigue resistance. Wings can be constructed using both metals and composites, such as the skins consisting of carbon–epoxy composite and the stringers and spars made of high-strength aluminium or titanium alloys.

Supporting structures on the wing such as attachments to the fuselage and landing gear are designed for strength, fatigue and fracture toughness. The wing-box and wing connections are more highly loaded than the wing itself and therefore are made of materials with higher strength, fatigue life and fracture toughness than the aluminium alloys used in the main wing section. The wing-box and wing connections in modern aircraft are usually constructed with titanium alloy or carbon–epoxy composite.

The wings hold control surfaces such as flaps, elevators and ailerons. Flaps control the amount of lift needed for take-off and landing; elevators control the pitch; and ailerons control the roll. Control surfaces are, in the main, lightly loaded and do not require high strength. However, they require impact resistance against bird strike and flying debris kicked up from the runway by the aircraft during take-off and landing. Control surfaces are usually constructed with thin skins supported by internal stiffeners or foam/cellular materials.

3.2.3   Empennage and control surfaces

The empennage is the whole tail unit at the extreme rear of the fuselage and it provides the stability and directional control of the aircraft (Fig. 3.8). Structurally, the empennage consists of the entire tail assembly, including the vertical stabiliser, horizontal stabilisers, rudder, elevators, and the rear

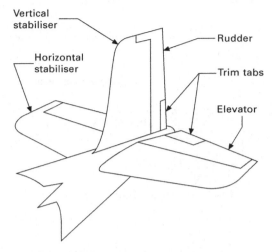

3.8 Typical empennage lay-out.

section of the fuselage to which they are attached. The stabilisers are fixed wing sections which provide stability for the aircraft to keep it flying straight. The horizontal stabiliser prevents the up-and-down, or pitching, motion of the aircraft nose. Important material properties are elastic modulus, strength, fatigue resistance and fracture toughness. The rudder is used to control yaw, which is the side-to-side movement of the aircraft nose. The elevator is the small moving section at the rear of the horizontal stabiliser used to generate and control the pitching motion. The loads on the rudder and elevator are smaller than those acting on the vertical and horizontal stabilisers, although properties such as stiffness, strength and toughness are still critically important.

The empennage in large aircraft also houses the auxiliary power unit (APU). An APU is a relatively small gas turbine used to generate power to start the main turbine engines and to provide electricity, hydraulic pressure and air conditioning while the aircraft is on the ground. The empennage in older versions of passenger aircraft also houses the main turbine engine. Aluminium alloy is the most common structural material used in the empennage and control surfaces, although fibre–polymer composites are increasingly being used for weight saving.

3.2.4   Landing gear

The landing gear, which is also called the undercarriage, is a complex system consisting of structural members, hydraulics, energy absorption components, brakes, wheels and tyres (Fig. 3.9). Additional components attached to and functioning with the landing gear may include steering devices and retracting mechanisms. Of the many components, it is the structural members that

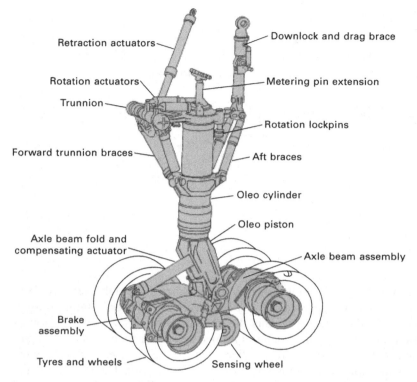

Retraction actuators

Downlock and drag brace

Rotation actuators

Metering pin extension

Trunnion

Rotation lockpins

Forward trunnion braces

Aft braces

Oleo cylinder

Oleo piston

Axle beam fold and
compensating actuator

Axle beam assembly

Brake
assembly

Tyres and wheels

Sensing wheel

3.9 Main-wheel bogie (from S. Pace, *North American Valkyrie XB-70A*, Aero Series vol. 30, Tab Books, 1984).

support the heavy landing loads and stop the landing gear from collapsing under the aircraft weight. The materials must be strong enough to support heavy take-off weight when an aircraft has a full load of fuel and the high impact loads on landing. Landing gear materials must therefore have high static strength, good fracture toughness and fatigue strength, and the most commonly used materials are high-strength steel and titanium alloy.

3.2.5   Jet engines

The materials used in jet engines are subjected to the most arduous working temperatures in an aircraft. Jet engines are gas turbines that compress air to high pressure and this air is then heated to extreme temperature by burning fuel to produce hot, high-pressure gases which are expelled from the engine exhaust thus propelling the aircraft forward. The engine materials must perform for long periods under high temperatures and stresses while exposed to hot corrosive and oxidising gases generated by the burning fuel. Jet engine materials must possess high tensile strength, toughness, fatigue strength and creep resistance together with excellent resistance against corrosion and oxidation at high temperature.

Most conventional materials cannot survive the severe conditions in the hottest section of jet engines, the combustion chamber, where temperatures reach ~1500 °C (2760 °F). A group of materials called superalloys, which includes nickel-based, cobalt-based and iron–nickel alloys, are used in the hot sections of jet engines. Ceramic materials with high heat insulating properties are coated on the superalloys to provide protection against the extreme heat. Titanium alloys and composites, which are lighter than superalloys but have lower temperature capacity, are used in cooler parts of the engine, such as the inlet section.

3.3   Helicopter structures

Figure 3.10 shows the main sections and internal structure for a typical modern helicopter. The main body (or airframe) of the helicopter is most heavily loaded at two points: the connection to the tail boom and the connection to the main rotor drive shaft or turbine engine. The tail boom applies torsion and bending loads to the body during flight whereas high tension and shear forces occur around the drive shaft connection. To carry these loads with a weight-efficient design, the main body is constructed with a truss frame network covered with a thin skin. Most of the load is carried by the frame which consists of longitudinal, transverse and inclined beams. The frame and skin of modern helicopters are constructed of aluminium alloy or fibre–polymer composite or some combination of the two. The aluminium alloys used in

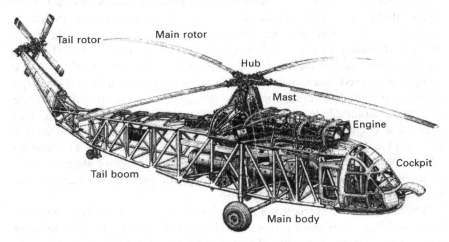

3.10 Structural design of a typical helicopter (from Flight International).

helicopters airframes are usually the 2000 (Al–Cu) and 7000 (Al–Cu–Zn) alloys whereas the composite is usually carbon fibre–epoxy. Composite material is also used in the helicopter body in preference to aluminium alloy when a high strength-to-weight ratio is required. Glass-fibre composite may be used in the more lightly loaded body components and aramid composites are used in structures where vibration damping is required (such as around the drive shaft) or high-energy absorption (such as the underfloor). Titanium alloy or stainless steel can be used in regions of high stress or heat.

The main rotor consists of a mast, hub and blades. Main rotor systems are classified according to how the blades are attached and move relative to the hub. There are three basic classifications: rigid, semirigid and fully articulated. Some modern rotor systems use a combination of these types. When the main rotor turns it generates both the aerodynamic lift force that supports the weight of the helicopter and the thrust that counteracts drag in forward flight. The main rotor also generates torque that tends to make the helicopter spin, and the tail rotor is used to counteract this effect. On twin-rotor helicopters the rotors spin in opposite directions, and this cancels the torque reaction.

The mast is a cylindrical shaft that extends upwards from, and is driven by, the transmission. The material properties for the mast include high elastic modulus, strength and fatigue resistance, and, therefore, it is usually made of high-strength steel or titanium alloy. At the top of the mast is an attachment point for the rotor blades called the hub, which can be made using a variety of high-strength materials such steel, titanium or composite. The blades are long, narrow airfoils with a high aspect ratio, and this design minimises the drag resistance from the tip vortices. Helicopters have between two and six blades attached to the hub, and each blade produces an equal share of the lifting force. Rotor blades are made from various metals, including aluminium, steel, titanium, and composites such as carbon–epoxy laminate or sandwich materials with a lightweight honeycomb core. Figure 3.11 shows the materials used in the blades of a *Sea King* helicopter. Carbon–epoxy composite is used extensively in blades because of their light weight, high strength, potential for multi-functional design, and, most critically, fatigue resistance. Rotor blades experience many tens of millions of load cycles over the average life of a helicopter, and composites can extend the service life by a factor of up to 200 compared with aluminium blades. The leading edge of the blade is covered with an erosion shield made of stainless steel or titanium to resist damage from impacting dirt particles kicked-up from the ground during take-off and landing.

The tail boom is constructed of hollow aluminium or carbon-fibre composite tubes and frames extending from the rear of the main body. On the other end is the tail rotor assembly. The boom houses the drive mechanism for the tail rotor. The tail rotor is a smaller, vertically mounted rotor whose

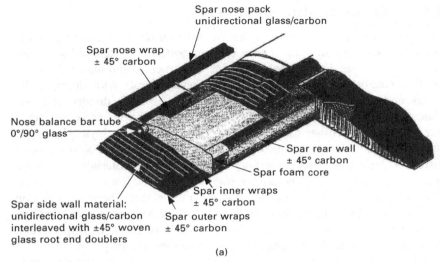

Spar nose pack
unidirectional glass/carbon

Spar nose wrap
± 45° carbon

Nose balance bar tube
0°/90° glass

Spar rear wall
± 45° carbon

Spar foam core

Spar inner wraps
± 45° carbon

Spar side wall material:
unidirectional glass/carbon
interleaved with ±45° woven
glass root end doublers

Spar outer wraps
± 45° carbon

(a)

(b)

3.11 Material structure of the main rotor blade for the *Sea King*
helicopter ((a) from M. R. Edwards, Materials for military helicopters,
*Proceedings of the Institution of Mechanical Engineers, Part G:
Journal of Aerospace Engineering*, **216** (2002), 77–88).

role is to control yaw, including acting counter to the torque reaction of
the main rotor. The tail rotor blades are similar in construction to the main
blades, and are often made using a combination of metals and composites.
The tail rotor can be enclosed within a metal or composite casing to protect
the blades from erosion and bird strike.

3.4   Space shuttle structures

The space shuttle is a complex system consisting of an external fuel tank, two solid rocket boosters and the Space Transportation System (STS) orbiter vehicle. In this section, we only examine the structure and materials of the orbiter. The orbiter resembles a conventional aircraft with double-delta wings, and uses many of the same materials. The orbiter is divided into nine major structural sections (Fig. 3.12). Most of the sections are constructed like a passenger airliner using aircraft-grade aluminium alloys. The major structural assemblies are connected and held together by rivets, bolts and other fasteners, again much like an airliner. However, some materials used in the space shuttle are unique, and are not found in fixed- or rotary-wing aircraft. One distinguishing feature of the orbiter is the reusable thermal insulation system. Over 25 000 ceramic and carbon–carbon composite tiles, that can withstand temperatures of about 1200 °C and above 2000 °C, respectively, are used to insulate the underlying structure during re-entry.

The forward fuselage section is robustly designed to carry the high body bending loads and nose gear landing loads. The body skin panels, stringers, frames and bulkheads in the forward section are made with the same aluminium alloy (2024 Al) found in conventional aircraft structures. The windows are made using the thickest ever pieces of optical quality glass. Each window consists of three individual panes: the innermost pane is 15.9 mm (0.625 in) thick tempered aluminosilicate glass whereas the centre and outer panes are 33 mm (1.3 in) and 15.9 mm fused silica glass. This design can withstand the extreme heat and thermal shock during re-entry when temperatures reach 600–700 °C.

The mid-fuselage section is the 18.3 m (60 ft) long structure that interfaces with the forward and aft fuselage sections and the wings. The mid-fuselage includes the wing carry-through structure, which is heavily loaded during re-entry, and the payload bay (including its doors). The fuselage is constructed

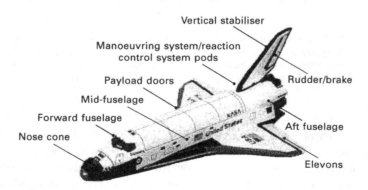

3.12 Main sections in the space shuttle orbiter.

with monolithic and honeycomb sandwich panels of aluminium, which are stiffened with load-bearing vertical and horizontal frames. The frame is constructed using 300 struts of metal matrix composite (boron fibre/aluminium tubes), which has exceptionally high stiffness and provides a weight saving of 45% compared with a conventional aluminium construction. The payload bay doors are a sandwich composite construction (carbon fibre–epoxy skins and Nomex core) with carbon-fibre composite stiffeners. This construction reduces the weight by over 400 kg (900 lb) or 23% compared with an aluminium honeycomb material.

The aft fuselage consists of an outer shell, thrust section and internal secondary structure, and it supports the manoeuvring/reaction control systems pods, main engines and vertical tail. The aft fuselage skins are made of aluminium alloy reinforced with boron fibre–epoxy composite struts. These struts transfer the main engine thrust loads to the mid-fuselage and external tanks during take-off. At take-off the two solid rocket boosters generate a combined thrust of 25 MN (5.6 million lb), which is over 200 times the twin engine thrust of a Boeing 737. Owing to the extreme thrust, titanium alloy strengthened with boron–epoxy struts is used near the engines.

The wing and vertical tail is constructed mostly with aircraft-grade aluminium alloy. The outboard wing section is made with high temperature nickel honeycomb sandwich composite and the inboard wing section of titanium honeycomb. The elevons, used for vehicle control during atmospheric flight, are constructed of aluminium honeycomb.

3.5   Summary

The weight of the wings (8–14%), fuselage (7–12%) and engines (5–7%) account for a large proportion of the maximum take-off weight of aircraft. Therefore, selection of weight-efficient structural materials is essential for flight performance, increased range and reduced fuel consumption.

Materials used in the pressurised fuselage of aircraft must carry tension loads generated by cabin pressure and shear loads. The stringers and circumferential frames provide strength to the fuselage shell to carry tension and compression loads. Fuselage materials require a combination of properties that include light weight (low density), high elastic modulus and yield strength, and resistance against fracture, fatigue and corrosion. High strength aluminium alloy and carbon–epoxy composite are the most common fuselage materials, although exceptions occur such as the use of the metal laminate GLARE in the upper fuselage of the Airbus 380.

Wings are designed to carry bending, shear and tension loads. During flight, the bending load is a combination of compression forces on the upper wing surface and tension loads on the lower surface. Materials used in the upper wing section must be lightweight with high elastic modulus and compression strength. The underside of the wing requires materials with low density, high stiffness and yield strength, and excellent tensile fatigue resistance. Wings are made using aluminium alloy and/or carbon–epoxy composite. The wing-box and wing connections are highly loaded structures built into the fuselage that are constructed of composite material or titanium alloy. The leading edges of the wings must be made using lightweight, damage-tolerant materials that resist bird strike and other impact events.

Landing gear in aircraft and helicopters must withstand heavy landing loads and support the take-off weight. Landing gear materials require high elastic modulus, strength and fracture toughness as well as fatigue resistance under repeated impact loads. The material most often used in landing gear is high strength steel.

Materials used in gas turbine engines are required to operate under high stress and temperature conditions for long periods of time. Engine materials require a combination of properties that includes high strength, toughness, fatigue resistance and creep strength at elevated temperature. Engine materials must also resist damage from oxidation and hot corrosive gases. The material of choice for the hottest engine components is nickel-based superalloys.

3.6   Further reading and research

Culter, J. and Liber, J., *Understanding aircraft structures*, Blackwell Publishing, Oxford, 2005.

Edwards, M. R., 'Materials for military helicopters', *Proceedings of the Institution of Mechanical Engineers, Part G: Journal of Aerospace Engineering*, **216** (2002), 77–88.

Megson, T. H. G., *Aircraft structures for engineering students*, Arnold, London, 1999.

Niu, M. C.-Y., *Airframe structural design: practical design information and data on aircraft structures*, Conmilit Press Ltd., Hong Kong, 2002.

Peel, C., 'Advances in aerospace materials and structures', in *Aerospace materials*, edited B. Cantor, H. Assender and P. Grant, Institute of Physics Publishing, Bristol, 2001, pp. 91–118.

Wilkinson, R., *Aircraft structures and systems*, MechAero Publishers, St Albans, 2001.

Strengthening of metal alloys

4.1 Introduction

Metal alloys used in aircraft structures and engines must have high mechanical properties to ensure weight-efficiency. It is essential that metals used in aircraft have sufficient strength to avoid permanent deformation and damage under the structural, aerodynamic and other loads experienced during flight. The mechanisms by which aerospace metals achieve their high strength are complex. Various mechanisms occur at the atomic, nanometre, microstructural and millimetre scales and these control the strength properties, as shown in Fig. 4.1 for the aluminium alloys used in an aircraft wing. Without these strengthening mechanisms occurring at different length scales, metals would be too soft and susceptible to plastic deformation to use in highly loaded structures and engine components. The mechanisms, when used in combination, provide aerospace engineers with weight-efficient metals with the capacity to withstand the extremely high loads experienced by modern aircraft.

The strength properties of aerospace metals, which include proof strength, ultimate strength, fatigue strength and creep strength, are controlled by a multitude of factors. The main factors are:

- alloy composition;
- arrangement and bonding of the atoms;
- type, size and concentration of precipitates and second-phase particles;
- types and concentration of imperfections and defects;
- metal casting processes, forming techniques and manufacturing methods.

This chapter describes the main factors that control the strength properties of metal alloys. The chapter explains the fundamental engineering science behind the development of high-strength metals for use in weight-efficient aircraft structural components. To understand the strengthening of metals, it is necessary to have a basic understanding of the arrangement of atoms in metals. A brief description of the crystal structures of the metals used in aircraft is provided. Following this is an overview of the various imperfections in the crystal structure which affect the strength of metals. The mechanisms by which these imperfections increase the strength of metals are outlined.

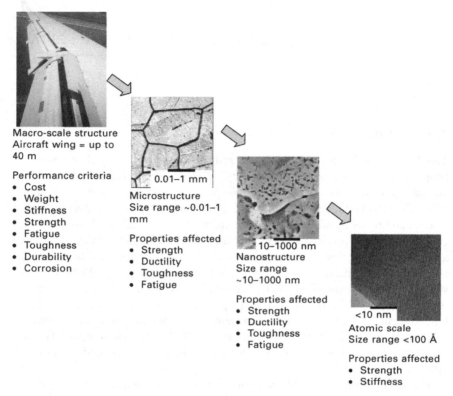

Macro-scale structure
Aircraft wing = up to
40 m

Performance criteria
- Cost
- Weight
- Stiffness
- Strength
- Fatigue
- Toughness
- Durability
- Corrosion

0.01–1 mm

Microstructure
Size range ~0.01–1
mm

Properties affected
- Strength
- Ductility
- Toughness
- Fatigue

10–1000 nm

Nanostructure
Size range
~10–1000 nm

Properties affected
- Strength
- Ductility
- Toughness
- Fatigue

<10 nm

Atomic scale
Size range <100 Å

Properties affected
- Strength
- Stiffness

4.1 Strengthening processes used to maximise the strength of
aluminium alloys in aircraft wings.

The strengthening of metals by dislocations is explained. Dislocations are
central to all the main strengthening mechanisms in metals, and what they
are and how they increase strength is described. The strengthening of metals
by precipitation hardening, intermetallic compounds, and control of the
grain structure is explained. The information in this chapter provides a basic
understanding of the various ways that aerospace engineers can increase the
strength of metal alloys for structural applications.

4.2   Crystal structure of metals

The atoms in solid metals are arranged in an ordered and repeating lattice
pattern called the crystal structure. A crystalline material consists of a
regular array of atoms that is repeated over a long distance compared
with the atomic size. A simple analogy is the stacking of oranges in a
grocery store, with each orange representing a single atom and each layer

of oranges being a lattice plane (Fig. 4.2). The arrangement pattern of the atoms is defined by the unit cell of the crystal. The unit cell is the basic building block, having the smallest repeatable structure of the crystal, and it contains a full description of the lattice structure. A crystalline material is constructed of many unit cells joined face-to-face in a three-dimensional repeating structure.

Most metals at room temperature are found in one of three crystalline patterns: body centred cubic (bcc), face centred cubic (fcc) or hexagonal close packed (hcp). The unit cell structures of these crystals are shown in Fig. 4.3. For the main types of metals used in aircraft structures and engines, aluminium and nickel have a fcc structure; magnesium is hcp; and titanium is bcc (called α-Ti) or hcp ($\beta\alpha$-Ti) depending on its alloy composition and heat treatment.

The atoms in crystals are packed close together and have a large number of nearest neighbour atoms (usually 8–12). The distance between the atoms within a unit cell is determined by the type of metal and its crystal structure.

4.2 Stacking of atoms within a crystalline material has similarities with the stacking of fruit.

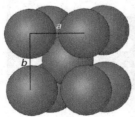

Body centred cubic   Face centred cubic   Hexagonal close packed

4.3 Unit cells for body centered cubic, face centered cubic, and hexagonal close packed crystals (*a* = width of unit cell; *b* = height of unit cell).

For example, the spacing between aluminium atoms is a/b = 0.40 nm, between magnesium atoms is a = 0.32 nm and b = 0.52 nm, and between titanium atoms (hcp) is a = 0.30 nm and b = 0.47 nm, where a is the width of the unit cell and b is the height of the unit cell. The tight packing is one reason why metals have high densities and high elastic stiffness compared with other solid materials such as polymers.

The atoms within the lattice crystal are connected by metallic bonding. Rather than the electrons being bound to the nucleus of a specific atom, the electrons are 'free' to move around the positively charged metal ions of the lattice. The electrons, which are called delocalised or conduction electrons, divide their density equally over all the atoms. For this reason, metal crystals are visualised as an array of positive ions in a sea of electrons, as shown in Fig. 4.4. The crystal structure is held together by the strong forces of attraction between the positive nuclei and delocalised electrons. Metallic bonding influences many of the properties of metals, such as their elastic modulus and electrical conductivity.

4.3   Defects in crystal structures

A perfect crystal structure exists when all the atoms are arranged in order through the entire material. When this occurs the metal has extraordinarily high strength. However, the crystal structure is rarely perfect, and instead contains imperfections that reduce the strength. For instance, the theoretical strength of pure aluminium is about 4600 MPa whereas the actual strength is only about 100 MPa owing to imperfections in the crystal structure. As another example, the strength of pure titanium is reduced from 7300 MPa to below 200 MPa owing to lattice imperfections. The most common imperfections are point defects, dislocations and grain boundaries.

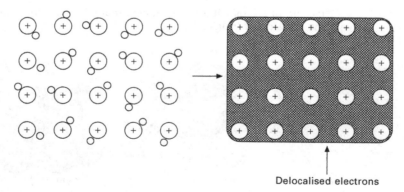

Delocalised electrons

4.4 Representation of metallic bonding. Single metal atoms have their electrons bound to the nucleus. In a metal, the electrons are delocalised and shared between the atoms.

4.3.1   Point defects

Point defects are localised disruptions in an otherwise perfect arrangement of atoms in a crystal lattice structure. A point defect involves a single atom or pair of atoms, and thus is different from extended defects such as dislocations and grain boundaries (which are described later). Point defects occur at one or two atomic sites, but their presence is felt over much larger distances in the lattice structure. There are three main types of point defects found in metals: vacancies, interstitial defects and substitutional defects.

Vacancies

A vacancy is a point defect where an atom is missing from its normal site in the crystal structure (Fig. 4.5). The vacancy disrupts the regular arrangement of atoms, and causes a local increase in the strain energy owing to the minor misalignment in the atom positions. Vacancies in aerospace alloys are formed during solidification and processing. During cooling and solidification, the atoms, which are randomly spaced when the metal is molten, arrange themselves to form an ordered crystalline pattern at the freezing point. Not all the atomic sites are filled during freezing, which occurs over a narrow temperature range (typically less than 1 °C) and, thereby, vacancies are created. Plastic forming and shaping of the solid metal also creates vacancies. During plastic forming, the atoms are permanently displaced from the original lattice position to change the shape of the metal component. Vacancies are created by the displacement of the atoms because lattice points in the crystal structure

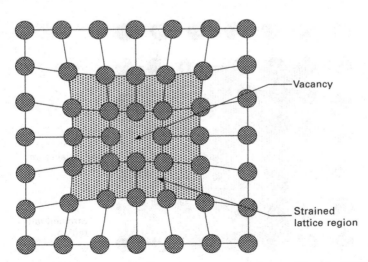

4.5 Point defect vacancy within a crystal structure. The shaded region indicates the region of high strain caused by the vacancy.

are left unoccupied by atoms of the base metal (or substitutional element). Vacancies are also created when materials operate at high temperature (e.g. jet engine components). The concentration of vacancies is determined by many factors, with 10^{12} to 10^{16} per cubic metre being the typical range for aerospace alloys.

Interstitial defects

An interstitial defect is formed when a foreign (solute) atom is positioned in the crystal structure at a point that is normally unoccupied (Fig. 4.6). The defect is formed when a solute atom such as an alloying or impurity element sits within a gap between the crystal lattice points of the base metal (solvent). An interstitial atom is usually smaller than the solvent atoms located at the lattice points, but is larger than the interstitial site it occupies. Consequently, the surrounding crystal structure is distorted. An interstitial defect is often formed in metal alloys when the size of the solute atom is less than about 85% of the size of the host metal atom. In many cases, the solute atoms are less than half as small as the base metal. Carbon in iron (steel) is one example of an element that is interstitial.

Substitutional defects

A substitutional defect is formed when an atom from the host metal is replaced with a solute atom. Substitutional defects are created by the deliberate addition of alloying elements or the unavoidable inclusion of impurity elements.

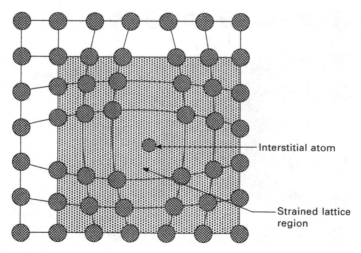

4.6 Interstitial defect within a crystal structure. The shaded region indicates the region of high strain caused by the interstitial defect.

Substitutional atoms may be larger than the host atom, thus reducing the interatomic spacing in the surrounding crystal structure or the substitutional atoms can be smaller which forces the surrounding host atoms to have a greater interatomic spacing (Fig. 4.7). In either case, substitutional defects disturb the surrounding crystal structure and cause a local increase in the lattice strain energy.

4.3.2   Dislocations

Dislocations are one of the most important types of lattice defects because of their strong influence on material properties such as strength, deformability, ductility and fracture toughness. Dislocations are an atomic disruption along a plane of the crystal lattice. Dislocations are known as line defects because they have a high aspect (length-to-width) ratio; being relatively long (typically several dozen to many thousand of atom spacings) and very thin (only 1 or 2 atom spacings wide). Dislocations are formed during solidification and plastic forming of metals.

Similar to point defects, dislocations are present in high concentrations in high-strength metals. Soft metals have a relatively low concentration of dislocations (typically under 10^6 cm of total dislocation length per cubic centimetre of material). This might seem a lot, but it is small compared with high-strength metals that contain anywhere between 10^{12} and 10^{14} cm dislocation line per cubic centimetre. If the dislocations were stacked end-to-end, then the dislocations in an aircraft panel (measuring 1.5 m long, 1.5 m wide and 3 mm thick) would extend over 1000 million km (or the distance much greater than from the sun to earth)! Such high dislocation line distances are only possible because they are packed close together (as shown in Fig. 4.8).

Edge dislocations

There are two types of dislocations: edge and screw. An edge dislocation can be visualised as being caused by the termination of a lattice plane in the crystal, as illustrated in Fig. 4.9. The surrounding planes are not straight, but instead bend around the edge of the terminating plane so that the crystal structure is perfectly ordered on both sides. The analogy with a stack of paper is apt. If a half sheet of paper is inserted in a stack of paper, the defect in the stack is only noticeable at the edge of the half sheet, and this is the location of the dislocation. The edge dislocation that is shown in Fig. 4.9 runs along the base of the terminated plane. Because of their structure and shape, dislocations create a local strain field in the lattice. Above the dislocation line there is a compressive strain (owing to compaction from the extra plane of atoms) and below is a tensile strain. This strain field, like the

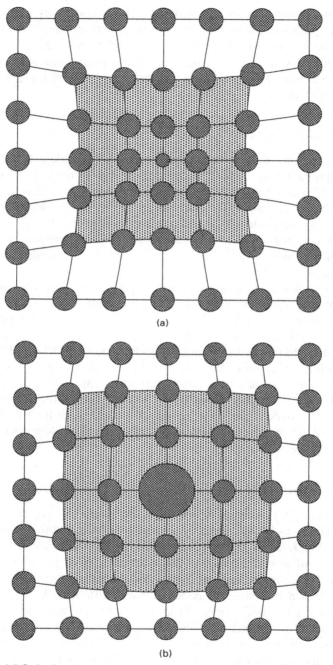

4.7 Substitutional atoms which are (a) smaller and (b) larger than the solvent atoms in the crystal structure. The shaded region indicates the region of high strain caused by the substitutional defect.

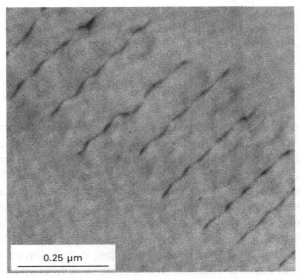

4.8 High magnification image of dislocations (from G. Zlateva and Z. Martinova, *Microstructure of metals and alloys*, CRC Press, Boca Raton, 2008).

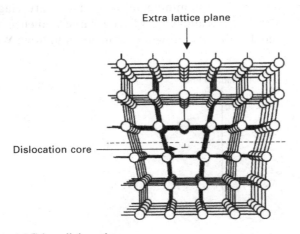

4.9 Edge dislocation.

strain field surrounding point defects, is important in the strengthening of metals, as is explained later in this chapter.

Screw dislocations

The screw dislocation can be visualised as cutting a crystal along a lattice plane and then sliding one half across the other half by an atomic spacing, as shown in Fig. 4.10. If the cut only goes part way through the crystal, and then slips, the boundary of the cut is a screw dislocation. The screw dislocation derives its name from the spiral stacking of crystal planes around the centre line.

Dislocation slip

Dislocations found in metals are a mixture of edge and screw dislocations, and both types have a large influence on the material properties when they move through the lattice structure under an external force. Dislocations are unable to move when the applied force is below the elastic limit of the metal. In elastic deformation, the shape change in the material is a result of stretching of interatomic bonds, but the force is too weak to cause dislocation motion. When the applied stress reaches the yield strength of the metal then the dislocations move by a process called slip. In fact, the yield strength is defined by the force needed to cause dislocation slip; weak forces are needed to cause dislocation slip in low-strength metals whereas stronger forces are required for higher strength materials. For instance, the applied force needed to cause dislocation slip in a high-strength steel used in aircraft landing gear (with a yield strength of about 2000 MPa) is about four times greater than for aluminium alloy used in the fuselage and wings (around 500 MPa). Dislocations also have a large influence on the plastic deformation and ductility by allowing the material to flow. Without

Dislocation line

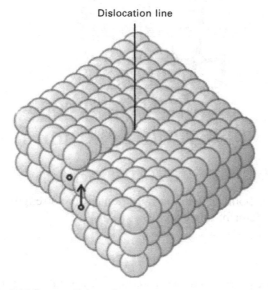

4.10 Screw dislocation.

the presence of dislocations, materials are brittle and plastic deformation is restricted.

The process of dislocation slip is represented in Fig. 4.11. When the applied force exceeds the elastic stress limit then sufficient energy exists to break the strained bonds between the metal atoms next to the dislocation. The terminated plane of atoms is forced sideways which breaks the bonds along one side of the dislocation. The ruptured bonds are immediately re-formed, but on the other side of the dislocation. This shift in bonding causes the dislocation to slip sideways by one atom spacing. When the process continues by maintaining the applied force then the dislocation is able to move through the crystal. The common analogy to dislocation slip is moving a carpet along a floor by using a fold or crease that is easily pushed along its length. Another way to visualise dislocation slip is to think about how a caterpillar moves forward. A caterpillar lifts some of its legs at any given time and uses that motion to move.

Dislocations move along specific lattice planes of a crystal, called slip planes. However, dislocations do not move with the same degree of ease

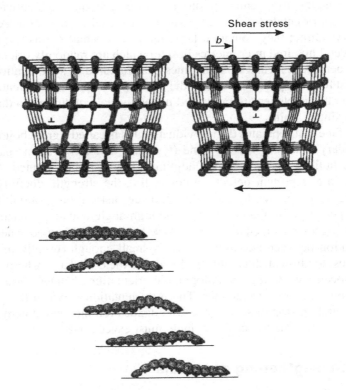

4.11 The process of dislocation slip is analogous to the motion of a caterpillar.

on all the crystallographic planes. Dislocations move most easily along the lattice planes with the closest packed atoms (i.e. most dense atomic packing). The number of slip planes is determined by the crystal structure. Bcc crystals have the highest number of slip planes (48), fcc an intermediate number (12) and hcp the smallest number (3). In general, the higher the number of slip planes then the greater the strengthening effect by strain hardening. For example, work-hardened β-titanium (bcc) has a tensile strength of about 400 MPa, aluminium (fcc) 150 MPa and magnesium (hcp) 110 MPa.

4.3.3   Grain boundaries

Very rarely do the atoms in a crystal structure extend in straight planes for the entire length of the metal structure. In most metals, the crystals are small and are orientated in many directions, as shown in Fig. 4.12. This is known as a polycrystalline metal. A grain is a volume of the polycrystalline material within which the arrangement of the atoms is nearly identical. However, the orientation of the atoms is different from each adjoining grain. That is, the arrangement of atoms within each grain is identical, but the grains are orientated differently. The metals used in aircraft structures are polycrystalline; they contain a large number of small grains (typically tens to several hundred micrometres in size) which are randomly orientated. However, aerospace metals are sometimes manufactured by solidification and mechanical forming processes so that the grains have a preferred orientation. Jet engine turbine blades are processed under special conditions so that the metal is a single grain.

The surface that separates the individual grains is called a grain boundary. The boundary is a very narrow zone (typically under two atom spacing) where the lattice structure of two adjoining grains is not aligned. Grain boundaries are important defects in controlling the strength, ductility and other material properties because they impede dislocation motion. There are two types of grain boundary, low- and high-angle boundaries, and both are found together in metals. A low-angle boundary is formed when the misorientation between adjoining grains is small enough (usually under a few degrees) to allow dislocations to slip across the boundary. A high-angle boundary occurs when the crystallographic orientation changes abruptly in passing from one grain to another. The misorientation between the grains is large enough to stop or severely restrict dislocation movement across the boundary, and for this to occur the angle must exceed 5–8°.

4.4   Strengthening of metals

An important property for any metal used in aircraft structures is high strength. Aerospace metals must be capable of withstanding high stress

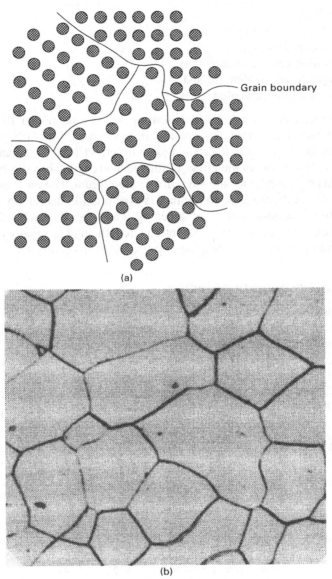

Grain boundary

(a)

(b)

4.12 Grain structure of: (a) polycrystalline material; (b) steel (from *Metals handbook, 8th ed. Vol. 7: Atlas of microstructures of industrial alloys*, American Society for Metals, Metals Park, Ohio, 1972).

without plastically deforming. The strengthening of metals is achieved by impeding the movement of dislocations under an externally applied load. Slowing or stopping dislocation slip increases the strength properties. There are various ways to increase the resistance against dislocation slip:

- work hardening;
- solid-solution hardening;
- dispersion hardening;
- precipitation (or age) hardening;
- grain refinement hardening; and
- duplex-structure hardening.

All these mechanisms hinder dislocation movement and render the metal stronger. Several mechanisms operate in each material, although the contribution of each to the total strengthening effect is not the same. For instance, Table 4.1 shows the effect of different strengthening mechanisms on the yield and ultimate tensile strengths of aluminium. The most effective strengthening mechanism for aluminium is precipitation hardening and the least effective is solid-solution strengthening. The situation can be different for other metals; for instance solid-solution strengthening is equally or more effective than precipitation strengthening for magnesium alloys.

4.4.1   Strain (work) hardening

Strain hardening (also called cold working) is an important strengthening process for aerospace alloys that involves plastically deforming the material during manufacturing to greatly increase the number of dislocations. During manufacture the metal is deformed into the final component shape (e.g. flat or curved skin panel, cylindrical landing gear strut) by forming processes such as rolling, forging, and extrusion (which are described in chapter 7). The metal must be plastically deformed to permanently change shape, and this deformation creates dislocations which increase the strength.

Table 4.1 Effect of different strengthening processes on aluminium

Material	Main strengthening mechanism	Yield strength (MPa)	Tensile strength (MPa)	Yield strength (alloy) / Yield strength (pure)
Pure Al		20	45	–
Pure Al–1.2% Mn	Solid solution strengthening	40	110	2
Pure Al (75% cold worked)	Strain hardening	150	170	7.5
Al (cold-worked and fine-grain structure)	Grain boundary hardening	190	200	9.5
Al–5% Mg	Dispersion hardening	150	290	7.5
Al–4% Cu–1.6% Mg	Precipitation hardening	440	480	22

Strain hardening is an effective strengthening process when many dislocations are created during plastic deformation. Dislocations are formed at grain boundaries, free surfaces, and at the interfaces between the lattice matrix and second phase particles. A large number of dislocations are also formed by homogenous nucleation via a mechanism called the Frank–Read source. which is shown in Fig. 4.13. The formation of dislocations requires a stress greater than the yield strength to be applied to the metal. This forces the dislocations to move along the slip planes until they encounter an obstacle that pins the ends of the dislocation line. The most common obstacle that pins dislocations is the presence of hard second phase particles within the grain.

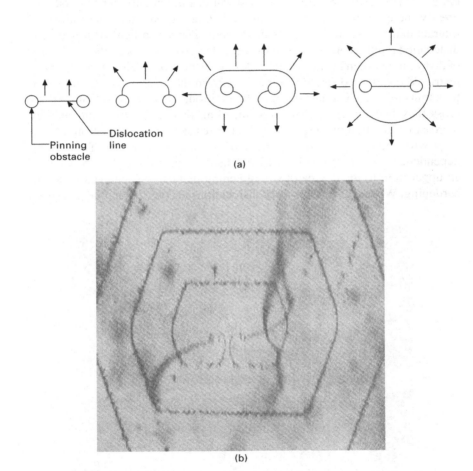

4.13 (a) Schematic representation of dislocation formation by the Frank–Read source. (b) High magnification photograph showing the formation of dislocations (from M. F. Ashby and D. R. H. Jones, *Engineering materials 1*, Butterworth–Heinemann, 1996).

The dislocation attempts to continue moving under the applied stress, but only the unpinned line section is capable of movement whereas the pinned ends remain stationary. This causes the dislocation to bend in the centre. The dislocation can bend so much that it loops around the particles which are causing the pinning. When the looping dislocation touches itself, a new dislocation is created. The new dislocation is free to move away whereas the pinned dislocation remains trapped, and this process of forming new dislocations is repeated many times.

Enormous numbers of dislocations can be created by plastic deformation: the dislocation line density can increase from 10^5–10^6 cm cm^{-3} to 10^{12}–10^{14} cm cm^{-3}. However, creating many dislocations is not sufficient by itself to greatly increase the strength. Strain hardening requires the dislocations to interact and impede each other's movement. The strain field surrounding a dislocation repulses other dislocations, and these forces impede the movement of dislocations when in close contact. As the dislocation motion becomes more restricted the applied stress required to deform the metal must be increased, thus translating into higher strength. In addition, when two dislocation lines cross they become entangled, thus impeding their movement and thereby increasing the strength (Fig. 4.14). As the resistance to dislocation motion rises with their density, the strength of the metal improves. This strong dependence of strength on dislocation density is shown in Fig. 4.15. There is an upper limit to the degree of strengthening that can be achieved by strain hardening. When a large number of dislocations become entangled in a small

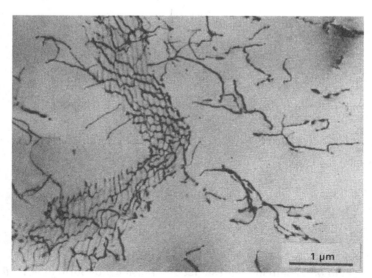

4.14 Tangles of dislocations within aluminium (from G. Zlateva and Z. Martinova, *Microstructure of metals and alloys*, CRC Press, Boca Raton, 2008).

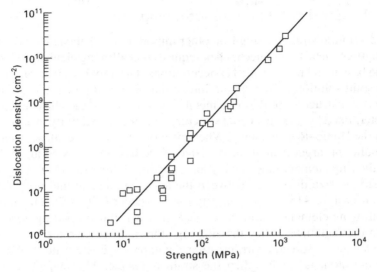

4.15 Effect of dislocation density on strength.

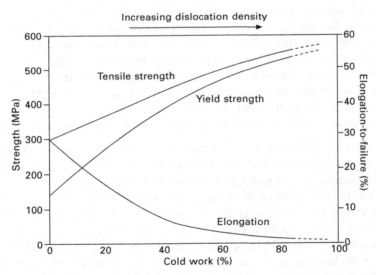

4.16 Effect of strain hardening (cold working) on the strength and ductility properties.

region, called a dislocation forest, then a small crack can develop and grow, causing the metal to break. For this reason, increasing the strength by strain hardening usually causes a corresponding loss in ductility and, at very high dislocation densities, the metal becomes brittle (Fig. 4.16).

4.4.2   Solid solution strengthening

Solid solution strengthening is another important mechanism for raising the strength of metals. The mechanism requires that alloying elements are added to the base metal in controlled concentrations. The alloying elements dissolve as a solid solution by occupying interstitial or substitutional lattice sites to the crystal structure of the base metal. The location of the alloying element is determined by its size, crystal structure and electronegativity in accordance with the Hume–Rothery rules. When the atomic size of the alloying element is smaller or larger than the atomic size of the base metal by about 15% then the alloying element may occupy an interstitial lattice site (Fig. 4.6). When the atomic size difference between the alloying element and host metal is similar (within ~15%) then substitution can occur (Fig. 4.7). The atoms of the alloying element distort the crystal structure, thus imparting a localised strain on the lattice. When the alloying atom is large then the atoms of the base metal are forced apart, creating a compressive strain field. When the alloying atom is smaller then the strain is tensile. The magnitude of the compressive or tensile strains in the distorted lattice increases with the atomic size difference between the alloying element and base metal.

Solid solution strengthening occurs because the lattice strain creates a barrier to dislocation movement. The degree of strengthening is determined by the ease with which dislocations move through the strained region of the crystal lattice. The local strain fields created by interstitial and substitutional atoms attract or repel the dislocation, thus impeding their motion. Greater stress must be applied to the material to force the dislocation to move against the strain field, and this increases the yield strength.

The amount of solid solution strengthening depends on two factors:

*   The size difference between the solute and base atoms. The greater the atomic size difference then the more severe is the lattice distortion, thus imparting greater resistance against dislocation slip.
*   The concentration of the solute atoms. The greater the amount of alloying element then the larger the strengthening effect (up to the solubility limit). Solid solution strengthening is an effective hardening process only when the alloying elements have a high degree of solid solubility in the host metal. If the alloying elements do not dissolve, but instead form second-phase particles then the solid solution strengthening effect is not as effective.

Figure 4.17 shows the importance of these two factors in the solid solution strengthening of iron (ferrite). The greater the difference between the atomic size of the alloying element and iron then the more powerful is the strengthening effect. Carbon and nitrogen are the most powerful strengthening elements because they are interstitial elements in iron due to

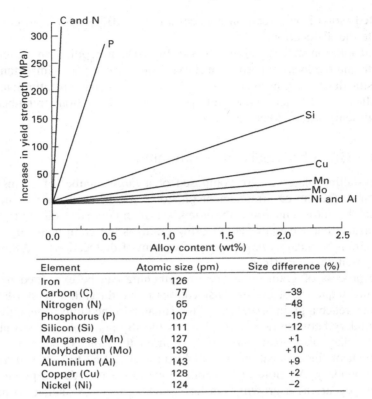

Element	Atomic size (pm)	Size difference (%)
Iron	126	
Carbon (C)	77	−39
Nitrogen (N)	65	−48
Phosphorus (P)	107	−15
Silicon (Si)	111	−12
Manganese (Mn)	127	+1
Molybdenum (Mo)	139	+10
Aluminium (Al)	143	+9
Copper (Cu)	128	+2
Nickel (Ni)	124	−2

4.17 Effects of alloying elements on the yield strength of iron by solid solution strengthening.

their smaller size. The other elements are substitutional in iron and have a smaller effect on yield strength. The strength also increases with the solute concentration. If too much of the alloying element is added, however, the solubility limit is exceeded and another strengthening mechanism operates (precipitation hardening).

Interstitial alloying and impurity elements can also strengthen metals by being trapped along the dislocations. For instance, when dislocations slip through iron (steel), the interstitial atoms such as carbon and nitrogen can be 'swept' along leading to the formation of interstitial concentrations or 'atmospheres' in the vicinity of dislocations. In an extreme case this can lead to lines of interstitial atoms along the dislocations (condensed atmospheres). Dislocations can be locked in position by strings of interstitial atoms, thus substantially raising the stress which would be necessary to cause dislocation movement. A very small concentration of interstitial atoms is needed to produce locking along the whole length of all dislocations in annealed iron. For a typical dislocation density of 10^8 lines cm^{-2} in annealed (soft) iron, a carbon concentration of 10^{-6} wt% is sufficient to saturate the dislocation.

Solid solution strengthening increases the yield strength, ultimate tensile strength and hardness of aerospace alloys. The process is also important for increasing the creep resistance (or loss of strength at high temperature); nickel superalloys used in jet engines rely partly on solid solution strengthening for high-temperature creep strength.

4.4.3   Grain boundary strengthening

Polycrystalline materials are composed of a large number of grains. As mentioned, the lattice arrangement of atoms within each grain is nearly identical, but the orientation of the atoms is different for each adjoining grain. The surface that separates neighbouring grains is the grain boundary (Fig. 4.12). Grain boundaries impede the movement of dislocations and thereby have a strengthening effect.

The process of grain boundary strengthening can be explained by the following sequence of events. Dislocations move through a crystal lattice until they reach a grain boundary. The mismatch in the lattice orientation at the boundary between two grains disrupts the slip plane of the dislocation. The boundary also creates a repulsive strain field that opposes the slip movement of the dislocation. The dislocation is forced to stop just ahead of the boundary. As more dislocations move to the boundary a process of 'pile-up' occurs as a growing cluster of dislocations are unable to move past the boundary (Fig. 4.18). Dislocations generate their own repulsive strain field and, with each successive pile-up of a dislocation at the grain boundary, the repulsive force acting on the dislocation nearest the boundary rises. Eventually, the repulsive stress is high enough to force the dislocation closest to the boundary to cross over to the adjoining grain. This process of restricting dislocation movement across grain boundaries is the basis of the strengthening mechanism.

The strength of polycrystalline materials is increased by reducing their grain size. The grain size of metals is often controlled by thermomechanical treatment, heat treatment or microalloying. Grains in aerospace alloys typically range from about 1 mm (coarse or large grains) to 1 μm (fine or small grains), and the strength increases as the grains become smaller. For example, Fig. 4.19 shows the effect of average grain size on the yield strength of steel. Reducing the grain size also has the additional beneficial effects of increasing the ductility, fracture toughness and fatigue life. Strength increases because the distance that dislocations must travel through a grain core to reach a grain boundary decreases with the grain size. Dislocations are more likely to reach a boundary, and thereby strengthen the metal, when the grain size is reduced.

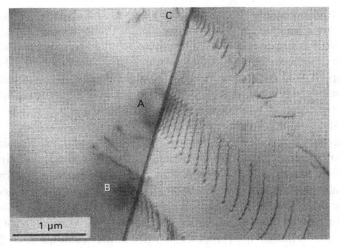

4.18 Dislocation pile-up at points A, B and C along a grain boundary (from G. Zlateva and Z. Martinova, *Microstructure of metals and alloys*, CRC Press, Boca Raton, 2008).

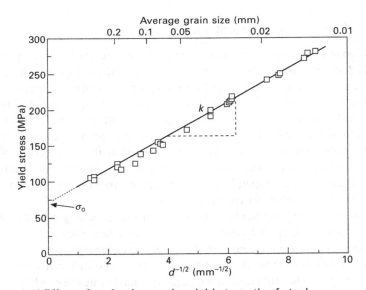

4.19 Effect of grain size on the yield strength of steel.

The grain size is related to the yield strength according to the Hall–Petch relationship:

$$\sigma_y = \sigma_o + kd^{-1/2}$$

where σ_y is the yield strength and d is the average diameter of the grains;

k is a material constant representing the slope of the $\sigma_y - d^{-1/2}$ plot; σ_o is called the friction stress, and is the intercept on the stress axis (Fig. 4.19). σ_o is a material constant that defines the stress required to move dislocations in a single crystal without a grain boundary ($d^{-1/2} = 0$). The Hall–Petch relationship is used by engineers to calculate the grain size necessary to achieve a required level of strength. The Hall–Petch relationship is accurate for metals with a grain size between about 1 mm and 1 µm, but is not valid for materials with larger (>1 mm) or finer (<1 µm) grains.

Theoretically, metals can be made infinitely strong if their grains are infinitely small. In reality, this is impossible because the lower limit on grain size is a single unit cell of the crystal. The finest grain size in most aerospace alloys is about 1 µm. Smaller than this, the length of the dislocation approaches the size of the grain.

The grain size of aerospace metals is reduced to the size range of about one to several hundred micrometres using various techniques, including rapid cooling of the molten metal during solidification casting, addition of grain-refining elements (called inoculants), and thermomechanical processing. The control of grain size is explained in more detail in Chapters 6 and 7.

4.4.4   Dispersion strengthening

Dispersion strengthening is a hardening process where small, strong particles resist dislocation slip in metals. Dispersion-strengthened metals are alloys containing a low concentration (often under 15% volume) of tiny ceramic oxide particles (0.01 to 0.1 µm). The materials are made by mixing and fusing oxide particles with powder of the host metal using a solid-state fabrication process called mechanical alloying. The microstructure consists of a continuous phase of soft metal strengthened with a dispersion of fine particles, as shown in Fig. 4.20. An example of dispersion-strengthened metal is a nickel-based superalloy containing hard particles of yttrium oxide (Y_2O_3) which is used in jet engines.

The strengthening mechanism of dispersion strengthening operates by the hard particles blocking dislocation slip, as shown schematically in Fig. 4.21. Dislocations cannot travel through the oxide particles because their crystal structure is different from the host metal matrix. The difference in crystal structure disrupts the slip plane at the particle–matrix interface, which makes it impossible for dislocations to move through the particles. A high strain field in the lattice surrounding the particles also impedes dislocation motion. When dislocations reach a particle they are forced to bend and loop around. When adjacent loops intersect on the far side of the particle, they interact and combine in the same manner as at a Frank–Read source. The interaction leaves a dislocation loop around the particle and another dislocation which is free to move onwards. Every time a dislocation reaches a particle this process

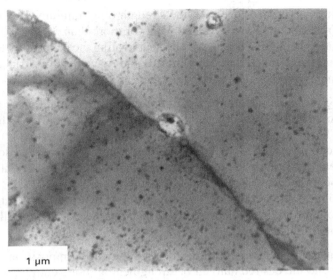

4.20 Microstructure of dispersion-strengthened metal (superalloy). The fine dark spots are dispersoid particles in the metal matrix. Photograph from M. Baloch, PhD Thesis, University of Cambridge, 1989.

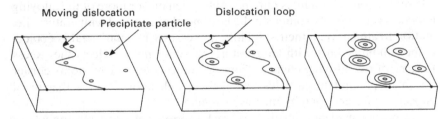

4.21 Orowan hardening mechanism for dislocation movement through a crystal structure containing particles.

is repeated, causing many disruptions to the slip process. Furthermore, the dislocation loops increase the strain field surrounding the particle, making it even more difficult for other dislocations to bend and loop around. This strengthening process is called Orowan hardening.

4.4.5 Precipitation hardening

Precipitation hardening is a strengthening process involving the formation of hard precipitate particles in the host alloy and these restrict dislocation slip. Precipitation-hardened alloys are used extensively in aircraft structures and engines, and include aluminium, magnesium, titanium and nickel alloys as well as several types of steel. Precipitation hardening is a remarkably effective strengthening process; for instance, the strength of aluminium alloy can be increased from about 150 to over 500 MPa (or more than 300%).

Precipitation hardening is achieved by heat treating the metal alloy to form a dispersion of fine precipitates that impede the movement of dislocations. Heat treatment involves the following stages:

1. Heating the metal at high temperature within the single-phase region to dissolve and disperse the alloying elements in the matrix of the host metal. This stage is known as solution treatment.
2. Rapid cooling using a cool solution (e.g. water, oil) from the solution treatment temperature to obtain a supersaturated solid solution of alloying elements in the host metal. This is known as quenching.
3. Reheating to an intermediate temperature to convert the supersaturated solid solution to finely dispersed precipitate particles. This is called the age-hardening stage.

The heat-treatment temperatures depend on the composition of the metal alloy and the desired amount of strengthening. For example, solution treatment of aluminium alloys used in aircraft is usually performed in the range 450–500 °C, quenched and then age-hardened at 160–220 °C. Titanium alloys are solution treated at about 750 °C, quenched and aged at 450–650 °C.

Precipitation hardening is only possible when one or more of the alloying elements is completely soluble in the host metal at elevated temperature (i.e. solution treatment) and their solubility decreases when the metal is cooled. As the solubility falls with temperature, the host matrix rejects the excess of atoms of the alloying element from the lattice sites. These atoms cluster into small precipitate particles that induce high lattice strains that resist dislocation slip and thereby increase strength.

Precipitation hardening involves a complex series of physical transformations in the metal alloy at the atomic and microstructural levels that induces different strengthening mechanisms over time. Figure 4.22 shows the hardening mechanism with each successive transformation in the age-hardening process. The mechanisms change with increasing age-hardening time in the order: solid-solution strengthening (SSS), Guinier–Preston (GP) strengthening, coherent precipitate strengthening, and finally incoherent precipitate strengthening.

Guinier–Preston zone strengthening

Precipitation hardening in the age-hardening stage of the heat-treatment process starts with the formation of Guinier–Preston (GP) zones. The formation of GP zones which occurs firstly by GP1 zones and then GP2 zones is illustrated in Fig. 4.23. After solution treatment and quenching, the alloying elements

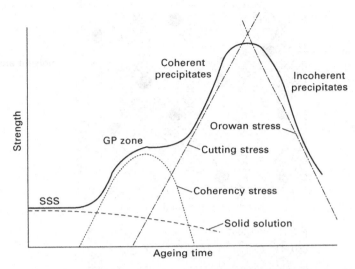

4.22 Precipitation hardening mechanisms with each successive stage in the strength–time graph.

are dissolved and dispersed as a supersaturated solid solution in the host metal. The material in this condition is relatively soft with strengthening achieved by solid solution hardening. Upon age hardening the supersaturated solid solution undergoes a transformation with the alloying atoms diffusing through the host lattice to form solute-rich clusters. The driving force for this process is the concentration of the alloying elements in solid solution being higher than their equilibrium solubility limit. Atomic diffusion and clustering of the alloying elements is assisted by vacancies in the lattice. The diffusion process is accelerated by heat treating the metal at elevated temperature; this is called thermal age hardening. The process occurs at a slower rate at room temperature with some metal alloys, and this is known as natural ageing.

The atoms of the alloying elements form a large number of tiny clusters (up to 10^{18} per cm^3) that are 10–50 nm long and only one or two atomic planes thick. These clusters are called GP zones. Figure 4.24 shows copper-rich GP zones in an aluminium alloy used in aircraft structures. The solute-rich GP zones are coherent with the crystal structure of the host metal. In other words, the arrangement of alloying atoms in the GP zone and matrix phase match at the interface plane and the two lattices are continuous across the interface. GP zones induce high strain in the surrounding lattice thus impeding dislocation slip. Dislocations are able to shear through GP zones, although the lattice strain imposes resistance against slip which strengthens the metal.

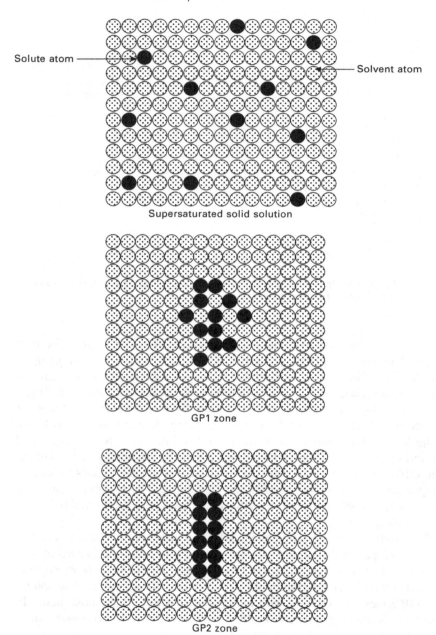

Solute atom

Solvent atom

Supersaturated solid solution

GP1 zone

GP2 zone

4.23 Schematic showing the transformation from a supersaturated solid solution to GP1 zone to GP2 zone during ageing.

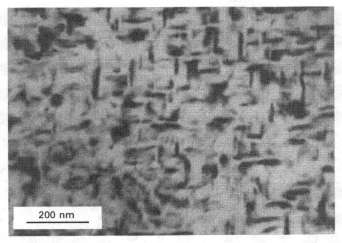

200 nm

4.24 Guinier–Preston zones (dark particles) in Al–4%Cu alloy (from G. Zlateva and Z. Martinova, *Microstructure of metals and alloys*, CRC Press, Boca Raton, 2008).

Coherent precipitate strengthening

The formation of coherent precipitate particles is the next step in the sequence of age-hardening processes. The GP zones transform into coherent precipitates in the matrix phase. Coherent precipitates also form by heterogenous nucleation at lattice defect sites such as grain boundaries and dislocations. A precipitate develops when the concentration of alloying elements clustered within a small region gives the composition of another phase. For example, copper in aluminium forms precipitate particles with a different composition (Al_2Cu) to the matrix phase. When precipitation first occurs, the particles are coherent with the host lattice (Fig. 4.25). That is, the lattice planes to the precipitate particles are continuous with the planes to the matrix, and there is no distinct interface between the particles and matrix. An abrupt change in composition occurs across the precipitate–matrix interface, but there is no change to the crystal structure. For instance, coherent Al_2Cu precipitates have an fcc crystal structure similar to the host aluminium metal.

Coherent precipitates distort the surrounding matrix which resists dislocation slip. Once the dislocation has overcome the lattice strain in the matrix and reached the precipitate, it can move through the particle because the slip planes are coherent with the matrix. However, the particles resist the slip process, thus increasing strength. The strength gained by coherent precipitates increases with time during the ageing process as their population increases and they become larger.

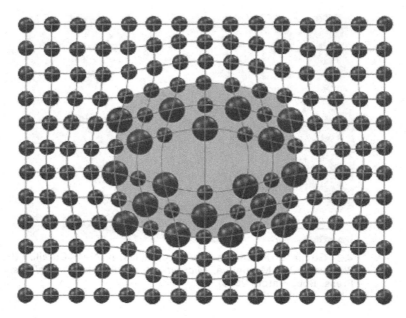

4.25 Coherent precipitate structure. The shaded region indicates region of high lattice strain.

Incoherent precipitate strengthening

The final transformation in the age-hardening process is the formation of incoherent precipitate particles. These particles are created by coherent precipitates changing their crystal structure such that the lattice planes are no longer continuous with the planes of the host metal lattice. Incoherent precipitates are truly distinct second-phase particles with their own crystal structure and separated from the surrounding matrix by a well-defined interface (Fig. 4.26).

Incoherent precipitates are effective obstacles against dislocation slip because the large distortion of the matrix surrounding the particles interacts strongly with the stress field of the dislocations. The precipitates also obstruct dislocation movement because slip cannot occur across the particle–matrix interface owing to the change in crystal structure. Strengthening by incoherent precipitates occurs by the Orowan hardening mechanism whereby dislocations must bend and loop around the particles (Fig. 4.21). The combination of high lattice strain and Orowan hardening results in incoherent precipitates being very effective in strengthening age-hardenable alloys.

The maximum improvement in strength gained by precipitation hardening occurs immediately upon the coherent precipitates transforming into incoherent particles. During the ageing process the particle size increases

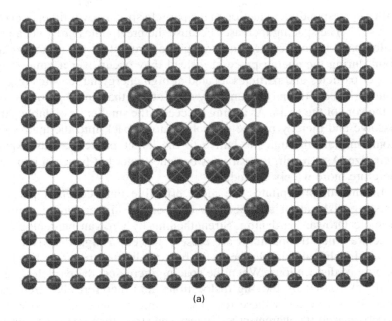

(a)

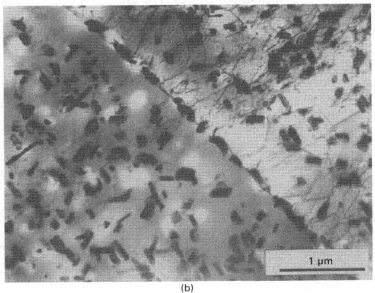

(b)

4.26 Incoherent precipitates. (a) Structure; (b) in aluminum alloy (from G. Zlateva and Z. Martinova, *Microstructure of metals and alloys*, CRC Press, Boca Raton, 2008).

with a corresponding reduction in the particle spacing over time, as shown in Fig. 4.27. The spacing increases because the precipitates grow via a process called particle coarsening. The precipitates are not all the same size at any point during the ageing process; instead they occur over a range of sizes. Larger particles and particles located at high energy lattice sites (e.g. grain boundaries) are thermodynamically more stable than small particles located in the core of the grain. As ageing proceeds, the smaller, less stable particles dissolve and thereby release the solute atoms back into the matrix. These atoms migrate through the lattice to the larger particles, thus expanding their size. As a result, the particles become larger but fewer in number and therefore more widely spaced.

Incoherent precipitates are more effective than coherent particles at resisting dislocation slip because they create higher lattice strains and promote Orowan hardening. Strengthening by Orowan hardening is most effective when the particles are closely spaced; a high density of tightly packed incoherent particles being the best condition to resist the bending and looping of dislocations. When the spacing increases, the particles are more easily by-passed by moving dislocations, and therefore the strengthening effect diminishes. For these reasons, maximum strengthening occurs with closely spaced incoherent precipitates. The strengthening effect diminishes with prolonged ageing because the incoherent particles coarsen and thereby Orowan hardening becomes less effective. The age-hardenable alloys used

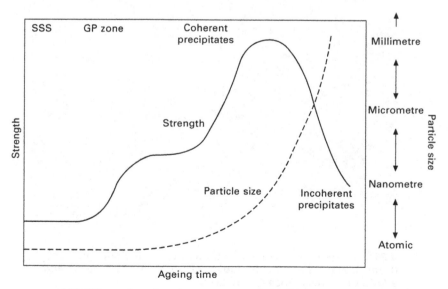

4.27 Effect of ageing time on strength (solid line) and particle size (dashed line) during the thermal ageing process.

in aircraft structures are usually heat-treated to close to maximum strength. More information on the age hardening of aerospace alloys is provided in later chapters.

4.5   Summary

The design ultimate load limit (defined by the yield strength) of metal alloys used in aircraft structures and engines is increased by numerous mechanisms that occur at the atomic, nanometre, micrometre, microstructural and millimetre scales. The main strengthening mechanisms of structural materials such as aluminium, titanium and steel are strain hardening, grain boundary hardening, solid solution strengthening and precipitation hardening. Nickel-based superalloys used in jet engines are also strengthened by dispersion hardening. The common feature of these various strengthening processes is that they operate by restricting or stopping the movement of dislocations and, therefore, the applied stress needed to induce plastic flow increases resulting in a corresponding increase in strength.

Dislocations have a strong influence on the yield strength, ultimate strength, ductility, fatigue strength, toughness and creep resistance of metals. The strength properties increase whilst the ductility decreases with increasing concentration of dislocations. The dislocation density in high-strength metals is typically in the range of 10^{12} to 10^{14} per cubic centimetre.

The strengthening mechanism of strain hardening involves plastic deformation (cold working) of the metal during fabrication of the aircraft component to create a high density of dislocations. Forging processes such as extrusion, rolling and stamp forming are used to plastically deform and shape the metal. During processing a high concentration of dislocations is created and they become entangled, thus causing a restriction of dislocation slip, which increases the strength properties (but lowers ductility and toughness).

The strength properties are increased by solid solution strengthening, which is achieved by the addition of soluble alloying elements to the base metal. Impurity elements in the metal can also cause solid-solution strengthening. The solute elements occupy interstitial or substitutional lattice sites in the crystal structure of the host metal, depending on their atomic size and electron valence number. The interstitial and substitutional atoms induce a localised elastic strain field in the surrounding lattice owing to the mismatch between their atomic size and the size of the base metal atoms. This strain field repulses the movement of dislocations, and thereby raises the strength properties. The strengthening effect increases with the atomic size difference between the solute and solvent atoms as well as the solubility limit of the solute.

The strength properties are increased by grain boundary hardening, which requires the grain size of the metal to be as small as possible. This mechanism operates by grain boundaries impeding the movement of dislocations, thereby raising the yield strength. The strength improves when the grain size is reduced because the distance that dislocations must travel to reach a grain boundary is shortened. The grain size in aerospace metals is controlled in several ways, including rapid cooling of the metal during casting, addition of grain refining elements, and thermomechanical processing during fabrication of the metal components. The average grain size in aerospace structures is typically in the range of one micrometre to several hundred micrometres. An advantage of the grain boundary hardening process is that yield strength is improved without any significant reduction to toughness or ductility, unlike the other strengthening processes. Grain boundary hardening is not normally used to strengthen jet engine materials because a high density of grain boundaries accelerates high-temperature creep deformation.

Dispersion hardening involves the inclusion of small, hard particles in the metal, thus restricting the movement of dislocations, and thereby raising the strength properties. This strengthening process is applied to nickel-based superalloys used in jet engine components. Tiny ceramic oxide particles resist dislocation slip, even at high temperatures. The particles also improve the high-temperature creep strength by pinning grain boundaries which restricts sliding.

Precipitation hardening is an important strengthening mechanism in most types of aerospace metals. Precipitates form in the metal when the concentration of the alloying element exceeds its solid solubility limit in the base metal. Precipitation hardening is achieved by thermomechanical processing (often called thermal ageing), which involves dissolving the alloying elements into solid solution at high temperature, quenching, and then thermally ageing the metal under controlled temperature and time conditions to create a fine dispersion of incoherent intermetallic precipitates in the base metal. The precipitates restrict the movement of dislocations and thereby increase strength. Maximum strengthening is achieved at the point when the precipitates transform from coherent to incoherent particles. The strength properties are reduced by 'over-ageing' caused by coarsening of the precipitate particles, and this must be avoided during thermomechanical processing.

4.6   Terminology

Age hardening: An increase in strength and hardness occurring over a period of time, owing to the formation of GP zones and precipitate particles.

Cold working: Any plastic deformation process in which the metal does not recrystallise but becomes progressively harder and less ductile up to some limit.

Dislocation: A line imperfection in the crystal structure caused by misalignment of the lattice planes.

Grain boundary: Thin band separating adjoining grains caused by the misalignment in the lattice grain structures.

Hall–Petch relationship: The relationship between grain size and yield strength of a polycrystalline metal

$$\sigma_y = \sigma_o + kd^{-1/2}$$

Interstitial defect: A point defect produced when an atom (alloying or impurity) is located at a site within the crystal that is normally not a lattice point.

Mechanical alloying: The production of a homogenous material by mixing and fusing of powders using mechanical methods (e.g. ball milling).

Metallic bonding: The bonding in metals characterised by the valence electrons forming a free cloud (or sea) in the crystal lattice rather than being attached to individual atoms.

Natural ageing: Age hardening at ambient temperature.

Point defect: Imperfection localised to a single point in the lattice structure.

Polycrystalline metal: Metal composed of many grains.

Slip: Deformation by planes of atoms in the crystal lattice sliding over each other which allows dislocations to move.

Slip planes: The planes in the crystal lattice across which slip can occur.

Solubility limit: The maximum concentration of the solute that can dissolve within the solvent without the formation of a second phase rich in solute.

Solution treatment: Heating to transform most, if not all, of the second phase into solid solution in the primary phase.

Substitutional defect: A point defect produced when a foreign atom (alloying or impurity) is located at a lattice point in the crystal structure, thereby substituting an atom of the base metal.

Supersaturated solid solution: An excess (above the equilibrium solubility limit) of the solute is retained in the solvent, usually achieved by rapid cooling.

Unit cell: The smallest region of the lattice that provides a full description of the crystal structure.

Vacancy: A lattice point in the crystal structure where an atom is missing.

4.7   Further reading and research

Abbaschia, R., Abbaschia, L. and Reed-Hill, R. E., *Physical metallurgy principles*, CenCage Learning, 2009.

Ashby, M. F. and Jones, D. R. H., *Engineering materials 1: an introduction to their properties and applications*, Butterworth–Heinemann, 1996.

Dieter, G. E., *Mechanical metallurgy*, McGraw-Hill, London, 1988.
Gordon, J. E., *The new science of strong solids or why you don't fall through the floor*,
    Penguin Science, 1991.
Hull, D. and Bacon, D. J., *Introduction to dislocations*, Elsevier, Oxford, 2001.

5

Mechanical and durability testing of aerospace materials

5.1   Introduction

The selection of materials for aircraft structures and engines is assessed according to a multitude of parameters such as cost, ease of manufacture, weight and a host of other factors. Central to the selection of materials is their mechanical properties such as stiffness, strength, fatigue resistance and creep performance. The durability properties of structural and engine materials in the aviation environment is also critical to their selection. Metals must be resistant to corrosion and oxidation whereas fibre–polymer composites must resist absorbing an excessive amount of moisture from the atmosphere. Aerospace materials should be durable enough to resist degradation and damage over the design life of the structural component, which may range from several hours for rocket engine parts to longer than thirty years for airframes.

Materials are selected by matching their properties to the design specifications and service conditions of the aircraft structure or engine component. The preliminary design of new structures or engines requires the aerospace engineer to analyse the performance requirements for the materials. Key information on the property requirements for the materials is determined early in the design process. For example, in the design of an aircraft wing, the minimum stiffness, strength and toughness properties needed by materials used in the spars, stringers, skins and other load-bearing components must be known. The environmental conditions in which the materials operate must also determined during the design process. For example, materials used in satellites and spacecraft must be built using materials that have high mechanical properties at extremely low temperatures and are not damaged by ionising radiation, micro-meteor impact or the low pressure conditions of the space environment.

The aerospace engineer must understand the mechanical and durability properties of materials to ensure they function over the design life of the aircraft component without the need for excessive maintenance and repair. Unfortunately, many of the mechanical and durability properties of materials cannot be calculated using mathematical models and therefore must be measured. For example, it is not possible to calculate the strength and

91

hardness of metals or the fracture toughness and fatigue life of fibre–polymer composite materials. Likewise, the corrosion resistance of metals or the durability of composites in hot and moist environments cannot be calculated. No theory exists to determine most of the mechanical or durability properties of metals. Calculating the properties of metals is too difficult because they are dependent on too many factors, such as their alloy content, crystal structure, microstructure, heat treatment and processing conditions. Similarly, the complexity of the microstructure, residual stress state and damage modes of composite materials make it difficult to calculate many of their mechanical and durability properties. It is possible to calculate the elastic properties of composites using theoretical models, but many other important properties, strength, fracture toughness, fatigue, creep, and so on, cannot be accurately calculated. Because many of the mechanical and durability properties of metals and composites cannot be calculated, they must be measured under standardised, controlled test conditions.

The testing procedures used to measure the properties of materials are performed under conditions specified by standards organisations, such as the American Society for Testing and Materials (ASTM) or the International Organization for Standardization (ISO). The aerospace industry uses these standards and, in some cases, uses their own specialist test procedures when a standardised method does not exist.

The aim of this chapter is to introduce the mechanical and durability properties of materials and describe the tests used to measure the properties. The main mechanical properties for aerospace materials, including the elastic modulus, yield strength, ultimate strength, ductility, hardness and fracture toughness are described as well as the test methods used to measure these properties. The methods examined include the tension, compression, flexure, fatigue, hardness, fracture toughness and creep tests. Several test methods used to measure the durability properties of materials are introduced. Tests to measure the corrosion resistance of metals and the moisture resistance of fibre–polymer composites are described. Towards the end of the chapter we examine how test data on the mechanical and durability properties of materials is used for certification of new aircraft and the structural modification of existing aircraft. The information given in this chapter provides an understanding of the engineering properties of materials; how the properties are measured; and how the property data is used in aircraft certification.

5.2   Tension test

5.2.1   Basics of the tension test

The tension test is one of the most common and important methods for measuring the mechanical properties of materials. The tension test is popular because a large number of properties can be determined in a single test: elastic modulus, strength, ductility, and other properties. The test is also popular because it is simple, quick and inexpensive. An important aspect of the tension test is that it is fully standardised. The test is performed under a controlled set of conditions using standardised equipment which ensures reliable results that are consistent and repeatable when measured anywhere in the world.

The tension test measures the resistance of a material to a slowly applied pulling force. Figure 5.1 shows the design and equipment for the test. A coupon specimen made of the test material is held at the two ends by the grips of the tensile machine. A tension (pulling) force is applied to the specimen by holding one end firm and forcing the other end away. A strain gauge or extensometer is attached to the specimen to measure the material extension under increasing force. The reaction of the material to the pulling force until it breaks is recorded and analysed. This data is used to quantify how the material responds to tensile forces applied in practical situations such as airframes and engines.

Tension specimens can be flat, round or dog-bone in shape, and are usually in the size range of 5–20 mm wide and 100–200 mm long. The flat specimens are either continuous or contain an open centre hole, as shown

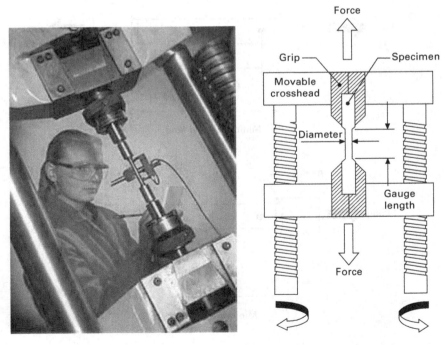

5.1 Tensile test.

in Fig. 5.2. Specimens without the hole are used to measure the tensile properties of materials. Specimens containing an open hole are used by the aerospace industry to determine the tensile properties of materials affected by geometric stress raisers, such as fastener holes, which reduce the tensile strength.

5.2.2   Tension stress–strain curve

During the tension test, the reaction of the material to an increasing applied load is measured until the sample breaks. The machine records the extension of the specimen with increasing load, and this is plotted as the applied force (or load)–extension curve. Figure 5.3 shows force–extension curves for aluminium measured using specimens with different sizes and shapes. As expected, the larger the specimen the greater the force that must be applied to elongate and deform the material. The force–extension curve has limited use in engineering design because its data is dependent on the dimensions and geometry of the test piece. This means the force–displacement curve measured for a material cannot be used to understand the tension properties

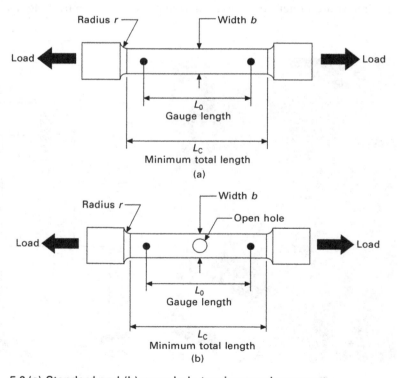

5.2 (a) Standard and (b) open hole tension specimens.

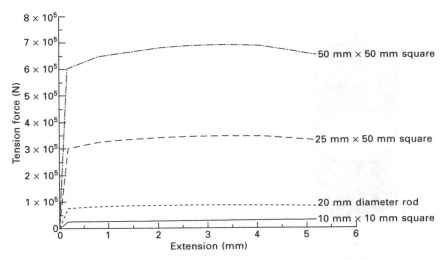

5.3 Load–extension graphs for aluminium using specimens with different sizes.

of the same material used in a smaller component (e.g. aircraft fastener) or larger component (e.g. wing or fuselage) with a different shape.

To overcome this problem, the tensile load is converted into tensile stress (force per unit cross-sectional area) and the extension is converted into strain (or percent elongation). Expressed mathematically, the tension stress (σ) is calculated using:

$$\sigma = \frac{P}{A} \qquad [5.1]$$

where P is the applied force and A is the load-bearing area of the test specimen (see Fig. 5.4).

Strain can be expressed in two ways: engineering strain and true strain. Engineering strain is the easiest and most common expression of strain, and is the ratio of the change in specimen length to the original length:

$$e = \frac{L - L_o}{L_o} = \frac{\Delta L}{L_o} \qquad [5.2]$$

where L is the specimen length under load and L_o is the original length (before loading). The true strain is calculated using the instantaneous length of the specimen as the tension test progresses to failure, and is calculated using:

$$e = \ln\left(\frac{L_i}{L_o}\right) \qquad [5.3]$$

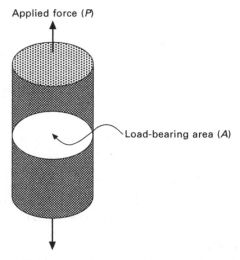

Applied force (P)

Load-bearing area (A)

5.4 Load-bearing area of the tensile specimen.

where L_i is the instantaneous specimen length, which is measured with a strain gauge or extensometer attached to the specimen.

The benefit of converting a force–displacement curve to a stress–strain curve is that direct comparisons can be made on test results from specimens of different sizes and shapes. The stress–strain curve is independent of the specimen size and geometry, and its data can be used to determine the tensile reaction of the material used in any situation. For instance, the same stress–strain curve can be used to assess the tensile properties of a tiny fastener or large wing panel made using the same material.

Tensile stress–strain curves that typify brittle materials (e.g. carbon-fibre composites and ceramics) and ductile materials (e.g. most metals and polymers) are presented in Fig. 5.5. Actual graphs for several aerospace structural materials are given in Fig. 5.6. The curve for a brittle material shows a linear (elastic) relationship between stress and strain to the failure point. The curve for a ductile material can be divided into the linear (elastic) and the nonlinear (plastic) regimes. The initial linear portion of the stress–strain curve defines the elastic regime. Within the elastic regime the material stretches when stress is applied, and then relaxes back to its original shape when the stress is removed. The material does not experience any permanent deformation or damage when loaded within the elastic regime for a short period of time. The plastic regime covers the nonlinear section of the graph between the elastic regime and point of final failure. The material in the plastic regime is permanently deformed under the applied stress, which causes a nonrecoverable change in shape when the stress is removed. When the material is deformed too much it breaks.

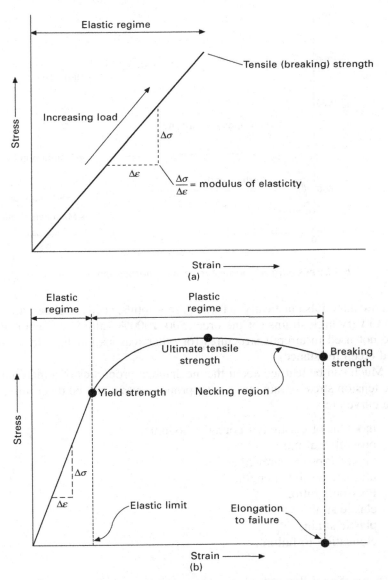

5.5 Typical tensile stress–strain graphs for (a) a brittle material (e.g. fibre–polymer composite, ceramic) and (b) a ductile material (e.g. metal, plastic).

The stress applied to materials used in aerospace structures must always remain within the elastic regime to ensure they are not permanently deformed. All brittle and ductile solids are linear elastic at low strain (typically less than 0.1%), but at higher strains brittle materials suddenly break whereas

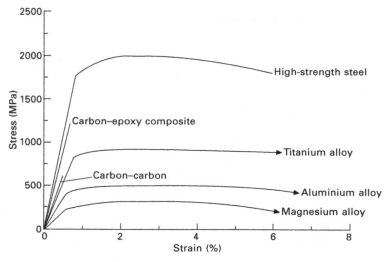

5.6 Stress–strain graphs for various aerospace materials.

ductile materials plastically deform. A few solids, such as rubber, are elastic up to very high strains (of the order 500–1000%), although such materials are not used in aircraft structures for other reasons (such as low strength and creep resistance).

Much can be learned about the mechanical properties of a material from the tension stress–strain curve. The properties that can be determined from the curve are:

- modulus of elasticity (Young's modulus),
- proportional limit,
- yield (or proof) strength,
- ultimate tensile strength,
- Poisson's ratio,
- elastic strain,
- plastic strain,
- elongation-to-failure.

5.2.3   Modulus of elasticity

The modulus of elasticity or Young's modulus is the measure of stiffness, and is one of the most important engineering properties for aircraft structural materials. The elastic modulus defines how much a material stretches under an applied tension stress. The greater the modulus, the less the material elastically deforms under the application of a given stress. For instance, materials with low elastic modulus, such as rubber or plastic, are more flexible than higher modulus materials, such as steel or titanium, under the same applied stress.

There is a requirement for aircraft structures and engine components to have high stiffness to resist excessive deformation under load. Therefore, materials with high elastic modulus are used, such as aluminium, titanium, steel and carbon fibre–epoxy composite. Occasionally, there is a requirement to use low modulus materials in aircraft, such as rubber seals for doors and other openings, but they are not required to carry high loads.

The Young's modulus (E) is calculated from the gradient of the straight line region of the stress–strain curve. In this rectilinear region, the material obeys the relationship defined as Hooke's law, where the ratio of the applied stress (σ) to strain (ε) is constant:

$$E = \frac{\sigma}{\varepsilon} \qquad [5.4]$$

The elastic modulus is calculated using this equation by simply dividing the stress by strain within the elastic regime.

The elastic modulus of metals is closely related to the binding energy between their atoms. The binding energy describes the magnitude of the attraction force between atoms. The elastic modulus of metals increases with the binding energy. A steep slope to the linear region of the stress–strain curve indicates that a high force is required to separate the atoms and cause the material to stretch, thereby resulting in high elastic modulus. The binding energy between atoms is constant and cannot be changed, and therefore the elastic modulus of a metal is constant. The elastic modulus of metals is one of very few properties not sensitive to the microstructure. This is because the modulus is determined by the strength of atomic bonds, and not by microstructural features such as dislocations or grain structure. The elastic modulus is not changed significantly by heat treatment, cold working, grain size and other microstructural features that have a large influence on other tensile properties such as strength and ductility. The elastic modulus can be changed by the addition of alloying elements, although their concentration must usually be high to have a noticeable effect. The elastic modulus of structural polymers and fibre–polymer composites is determined by other factors, which are described in chapters 13 and 15, respectively.

The elastic modulus for engineering materials is shown in Fig. 5.7; they range over six orders of magnitude from around 0.001 to 1000 GPa. The main types of aircraft structural materials have elastic modulus of 40–400 GPa: magnesium (45 GPa), aluminium (72 GPa), titanium (110 GPa), steel (210 GPa) and carbon–epoxy composite (anywhere from about 70 to over 300 GPa depending on the type, volume content and orientation of the fibres).

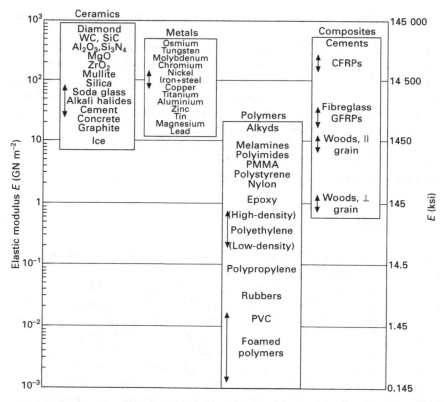

5.7 Elastic modulus values for engineering materials (reproduced with permission from M. F. Ashby and D. R. H. Jones, *Engineering materials 1*, Elsevier, 2005).

5.2.4   Poisson's ratio

Poisson's ratio defines the amount of lateral contraction versus the amount of axial elongation experienced by a material under the action of an applied elastic load. The Poisson effect is shown in Fig. 5.8. The Poisson ratio v is calculated using:

$$v = -\frac{\varepsilon_x}{\varepsilon_z}$$

[5.5]

where ε_x and ε_z are the elastic strains in the lateral and longitudinal directions at the same tensile stress. The Poisson's ratio for most metals is about 0.3 and for carbon fibre–epoxy composites is typically 0.2–0.3. Poisson's ratio is an important engineering property for aerospace materials because of the need for close tolerances in aircraft structures and engines. For example, a material having a high Poisson's ratio (i.e. $v \rightarrow 1$) used in an engine turbine

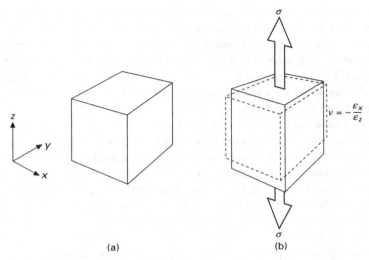

5.8 The Poisson effect in solids, (a) unloaded and (b) loaded in tension (from J. F. Shackelford, *Introduction to materials science for engineers* (7th ed.), Prentice–Hall, 2002).

blade contracts laterally by an excessive amount when under load, resulting in a large loss in propulsion efficiency.

5.2.5   Yield and proof strengths

All materials have an elastic limit beyond which something happens and its original shape can never be recovered. The stress–strain curve is rectilinear over the entire strain range up to the elastic limit, which is also the failure strain limit for brittle materials. Ductile materials behave differently beyond the elastic limit; they permanently change shape owing to plastic deformation. The point when a material changes from elastic to plastic behaviour is called the yield point (also known as the proportional limit). Before the yield point, a material returns to its original shape when the load is removed. After the yield point, the material undergoes permanent plastic deformation, and when the load is removed it cannot relax back into the original shape.

The yield point is the point on the stress–strain curve where it is no longer rectilinear. The yield strength is defined at this point as the amount of stress needed to start plastic deformation. In metallic materials, the yield strength is determined by the amount of stress needed to initiate dislocation slip. Dislocations cannot move when the stress is below the yield point, and begin to move when the applied stress exceeds the yield stress. Metals that contain dislocations that can move easily have low yield strength whereas high-strength metals contain dislocations that are highly resistant to slip.

(The process of dislocation slip was described in the previous chapter). In polymeric materials the yield strength is determined by the stress needed to begin permanent disentanglement and sliding of the polymer chains. (The behaviour of polymers under tensile loading is explained in Chapter 13).

Figure 5.5 shows the tensile stress–strain curve for a ductile material with well-defined yield strength owing to an abrupt change from elastic to plastic deformation. The occurrence of a sharp yield point depends on the sudden increase in the number of mobile dislocations. However, not all stress–strain curves show an abrupt change from elastic to plastic behaviour. The curve for many metals and polymers shows a gradual transition from the elastic to plastic regimes, as shown in Fig. 5.9, and the exact point where permanent deformation begins is hard to locate. The yield strength of materials which display this behaviour is determined from the curve using the offset method. The method involves specifying an offset percentage strain, which is usually 0.2% for metals. A line starting from the offset value is drawn parallel to the linear (elastic) portion of the curve. The stress corresponding to the intersection of this line with the stress–strain curve is the offset yield strength or (more commonly called) the proof strength. The proof strength is used to define the stress required to cause plastic deformation to materials which lack a sharp yield point on their stress–strain curve.

Figure 5.10 shows yield and proof strength for many types of materials. Similar to the Young's modulus, the yield strength ranges over nearly six orders of magnitude (from 0.1 to 10 000 MPa). Unlike the modulus, however,

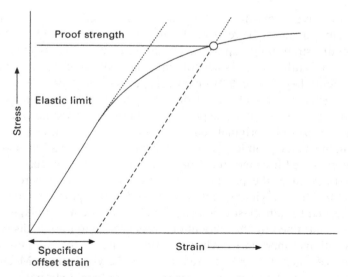

5.9 Offset method to determine the proof strength of ductile materials.

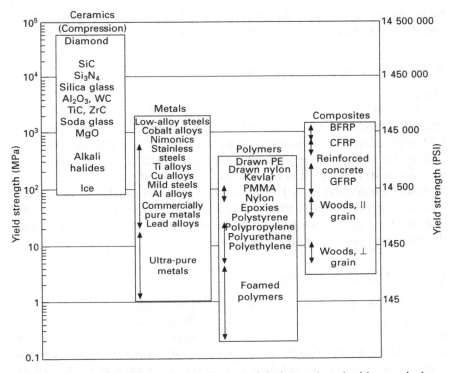

5.10 Yield strength values for materials (reproduced with permission from M. F. Ashby and D. R. H. Jones, *Engineering Materials 1*, Elsevier, 2005).

the yield strength of metals is sensitive to the alloy content, type of heat treatment, amount of cold working, grain size and other microstructural features. By appropriate alloying and processing it is possible to greatly increase the yield strength of metals. The strengthening of metals is critical to the design of weight-efficient aircraft structures (i.e. high strength-to-weight ratio). Without the ability to increase the yield strength of metals by alloying, heat-treatment, work hardening and other strengthening processes then aircraft structures would be heavy and bulky because of the need to increase their load-bearing section thickness.

Yield strength is an important property in the design of aerospace structures. It is essential that aerospace materials are not subject to stress levels exceeding their yield point, otherwise the structure permanently deforms. The yield strength is used in aerospace design to define the upper stress limit of the material (called the design limit load). However, aerospace structures are always designed to operate at stresses well below the design limit load to avoid permanent damage caused by unexpected overloading of the airframe.

Safety factors are used in the calculation of the maximum operating stress and strain limits on materials to avoid overloading and damage.

5.2.6   Ultimate tensile strength

The stress–strain curve for ductile materials increases between the yield point and the point of maximum stress. This maximum point is the ultimate tensile strength (UTS, more often called the tensile strength) of the material. The UTS is the maximum stress the specimen can sustain during the tension test. The tensile strength is the parameter most often quoted from the results of a tension test; although it is of little importance in engineering. Aerospace structures are never designed using materials that are loaded to their ultimate tensile strength; otherwise there is a high risk of failure. The yield strength is a more meaningful property than the ultimate strength for determining the maximum stress loading on an aircraft structure.

5.2.7   Necking and failure

During tensile testing, the deformation of a ductile specimen is uniform along its length up to the ultimate tensile stress. When loaded beyond this point, the deformation becomes nonuniform. At some point along the specimen, one region plastically deforms (or stretches) more rapidly than the other regions, usually at a local imperfection (e.g. void or second-phase precipitate). There is a local decrease in the cross-sectional area in this region, which develops into a neck (Fig. 5.11). When the specimen is subjected to increasing strain beyond the point of ultimate tensile stress the cross-sectional area of the necked region becomes progressively thinner. A lower force is then required to continue the deformation, and therefore the engineering stress–strain curve decreases until the failure point.

Tensile failure of ductile materials is a complex process, and often involves the formation of submicrometre-sized cavities within the neck region when stretched beyond the point of ultimate tensile strength. The number and size of the cavities increases with the amount of strain, and the cavities eventually link-up into a single crack which causes the specimen to break (Fig. 5.12).

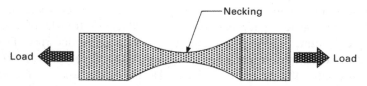

5.11 Localised deformation of ductile materials during the tensile test produces a necked region where final failure occurs.

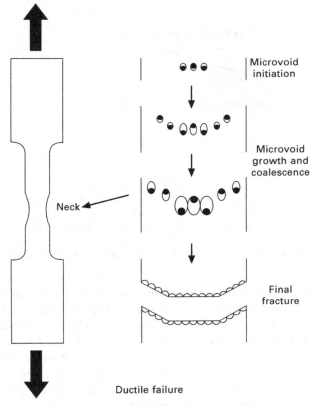

5.12 Schematic of ductile failure caused by the formation, growth and coalescence of microvoids in a ductile material under tensile loading.

The amount of extension experienced by the material up to the point of failure is called the elongation-to-failure, and this property is used to define the ductility of the material. The elongation-to-failure for the specimen is taken from the stress–strain curve at the point of maximum strain.

5.2.8   True stress–true strain curve

There are two types of tensile stress–strain curves for ductile materials: the engineering stress curve and the true stress curve. A comparison of the two curves for the same material is shown in Fig. 5.13. The engineering stress–strain graph is based entirely on the original dimensions of the specimen. It is assumed that the load-bearing area of the specimen does not change during testing, and therefore the tensile stress is calculated using equation [5.1] by dividing the applied force P by the original load-bearing area A_o.

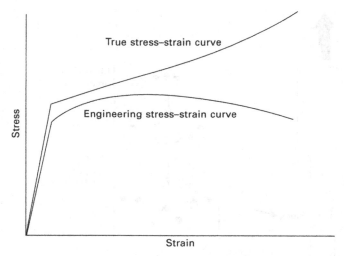

5.13 Comparison of engineering and true stress–strain graphs.

In reality, the load-bearing area reduces continuously over the course of the test. The area becomes progressively smaller as the material is stretched under increasing load to the ultimate tensile stress. The loaded area then reduces rapidly between the points of ultimate stress and final failure owing to necking. The true stress–strain curve takes into consideration the reduction in the load-bearing area. The true stress is determined (again using equation 5.1) by dividing the force P by the actual (or true) cross-sectional area of the specimen A_i. When the stress is calculated in this way, the true curve rises continuously up to final failure owing to strain hardening of the material. Both engineering and true stress–strain curves are important in determining the tensile properties of aerospace materials.

5.3   Compression test

The compression test determines the mechanical properties of materials under crushing loads. There are many aircraft structures that carry compression loads, such as the undercarriage during take-off and landing or the upper wing surface during flight, and therefore the mechanical behaviour of their materials must be determined by compression testing. It is often assumed that the tension and compression properties of materials are the same, yet this is only true for the elastic modulus. Other important mechanical properties, such as yield strength, may be different under tension and compression. It is essential that the mechanical properties of materials used in compression-loaded aerospace components are measured under compression loading, rather than assuming they are identical to the tensile properties.

The compression test involves squashing a small material specimen under increasing load until the point of mechanical instability. Compression specimens are usually short, stubby rods with a short length L to wide diameter D ratio (usually $L/D < 2$) to prevent buckling and shearing modes of deformation. The reaction of the specimen against an increasing compression load is measured during the test and, from this, the compression stress–strain curve is determined. The compression curve is simply the reverse of the tension curve at small strains within the elastic regime because the elastic modulus is the same. As the strain increases the difference between the compression and tension curves often becomes significant. As the specimen is squashed, it becomes shorter and fatter, and the load needed to keep it deforming rises. The specimen does not experience necking and so the stress increases until an instability occurs, such as cracking. The main mechanical properties which are determined from the compression test are the compression modulus and compression yield strength.

5.4   Flexure test

The flexure test measures the mechanical properties of materials when subjected to bending load. A flat rectangular specimen is loaded at three or four points, as shown in Fig. 5.14. The load causes the specimen to flex, thus inducing a compressive strain on the concave side, tensile strain on the convex side, and shear along the mid-plane. The separation distance between the support

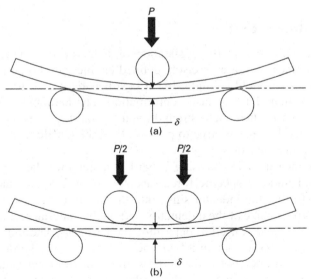

5.14 Flexure test performed under (a) three-point and (b) four-point bending.

points must be sufficiently large to avoid the generation of high shear stress. This is achieved by setting the ratio of the length of the outer span L to the specimen height h to 16:1 or greater. This ensures failure occurs by tension or compression failure at the surface, rather than by shear.

During the flexure test, the reaction of the specimen to increasing force is measured to calculate the bending stress-strain curve. The flexure strain is related to the vertical displacement of the specimen (δ) by:

$$\varepsilon_f = \frac{6\delta h}{L^2} \qquad\qquad [5.6]$$

where L is the support span between the two outer points, and w and h are the width and height of the specimen, respectively. The flexure stress is determined by:

$$\sigma_f = \frac{3PL}{2wh^2} \text{ (3-point bending)} \qquad\qquad [5.7]$$

$$\sigma_f = \frac{3PL}{4wh^2} \text{ (4-point bending)} \qquad\qquad [5.8]$$

where P is the applied force.

Flexure testing can determine several important mechanical properties, including the flexural modulus, yield stress and breaking stress. Flexure tests are often done on brittle materials and fibre–polymer composites, but only rarely on ductile materials such as metals.

5.5   Hardness test

Simply stated, hardness is the resistance of a material to permanent indentation. Hardness is not a precisely defined engineering property, such as elastic modulus or yield strength, but it is still widely used to describe the resistance of materials to plastic deformation. The hardness of ductile materials is related to their yield strength, and the hardness test (which is simpler, faster and less destructive to perform than the tensile test) is often used to obtain a measure of strength.

Hardness is measured by pressing a hard indenter into the surface of the test material under a specific force (usually about 2 N), as shown in Fig. 5.15. The further the indenter sinks into the material, the softer is the material. The depth or size of the indentation left on the material surface after the indenter is removed is used to determine the hardness. The indentation should be large enough to obtain a bulk measurement of the hardness but small enough that it does not damage the surface finish or act as a stress raiser. The indentations produced by the hardness test are usually 0.1–1 mm wide and less than 0.5 mm deep.

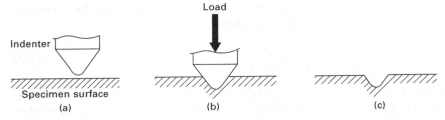

5.15 Hardness test. The indentation left by the indenter is used to determine the hardness value.

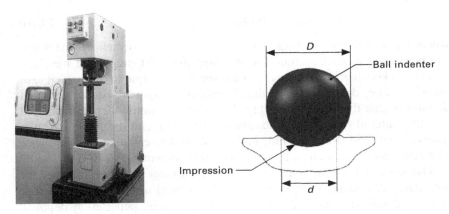

5.16 Brinell hardness test (photograph courtesy of directindustry. com).

Several methods are used to measure the hardness of metals, and the most common are the Brinell, Vickers and Rockwell tests. These tests are popular because they are quick, easy, inexpensive, nondestructive and can be performed on components of virtually any shape and size. However, hardness is not an intrinsic material property and cannot be used to describe the structural behaviour of a material under load. Instead, hardness numbers are used mainly as a quantitative basis for comparison between materials and to determine the consistency of the same material when produced in batches.

Figure 5.16 shows a Brinell test machine and the hardness indentation. The test involves pressing a hardened steel ball into the test material. The ball is 1, 5 or 10 mm in diameter and is pressed into the surface with an applied load of 30, 300 or 750 kgf. The ball size and load used increase with the hardness of the material. The ball is pressed into the material for a fixed period of time, usually 30 s, and then removed. The diameter of the indentation is measured using a microscope and the Brinell hardness number (HB) is calculated from the expression:

$$\text{HB} = \frac{2F}{\pi D(D - \sqrt{D^2 - d^2})} \qquad [5.9]$$

where F is the applied force (in kg), D is the diameter of the indenter (mm), and d is the diameter of the indentation (mm). The Brinell hardness number has the units of stress (Pa). Although hardness testing should only be used for quantitative comparisons between materials, the Brinell hardness number is closely related to the tensile yield strength of many metals by the simple relationship:

$$\text{Yield strength (psi)} = 500 \text{ HB} \qquad [5.10]$$

where HB has the units of kg mm^{-2}. This means the Brinell hardness test can be used as a quick method for assessing the yield strength of metals.

The Vickers hardness test works on a similar principle to the Brinell test, but with the key difference being that the indenter is a square-based pyramidal diamond rather than a hard ball (Fig. 5.17). The pyramid indenter is pressed into the material for a fixed time (usually 10–15 s), and the size of the two diagonals (d_1 and d_2) in the indentation are used to calculate the hardness. The hardness is reported as the Vickers hardness number (VHN).

The Rockwell hardness test uses a small diameter steel ball indenter for soft materials and a diamond cone indenter for hard materials. The indenter is pressed into the material for a fixed time and its penetration depth is measured automatically by the Rockwell test machine and converted into the Rockwell hardness number (HR).

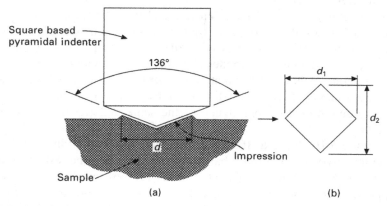

5.17 Vickers hardness test: (a) Vickers indentation; (b) measurement of impression diagonals.

5.6   Fracture test

Fracture toughness is an engineering property that defines the resistance of a material against cracking. Tough materials require large amounts of energy to crack whereas low toughness materials have little resistance against cracking. For the materials used in aircraft structures, fracture toughness is just as important as other mechanical properties such as elastic modulus and strength. Aerospace materials need high toughness to resist the growth of cracks initiating at damage sites (e.g. corrosion pits, impacted regions) or sites of high stress concentration (e.g. fastener holes, windows, doors and other access points in the aircraft).

There are several test methods to measure fracture toughness, and the most common is the single-edge notch bend (SENB) test. The SENB specimen is a rectangular block of the test material containing a machined notch with a V-shaped tip (Fig. 5.18). The notch extends about 50% through the specimen. The SENB test involves applying a three-point load to the specimen to generate a tensile stress at the notch tip. The applied load required to grow a crack from the notch tip through the specimen is used to calculate the fracture toughness. Tough materials require a high load to cause complete fracture of the specimen. Another popular fracture test is the compact tension (CT) method, which involves tension loading a block-shaped specimen containing a sharp notch (Fig. 5.19). The load required to break the specimen is used to calculate the fracture toughness.

The toughness values for materials vary over a wide range (five orders of magnitude), from the very tough to extremely brittle. Tough metals have a fracture energy of 100 kJ m^{-2} or more, whereas those of weak brittle

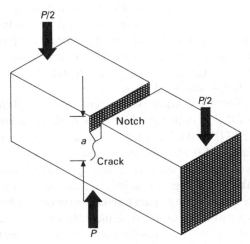

5.18 Single-edge notch bend specimen for fracture toughness testing.

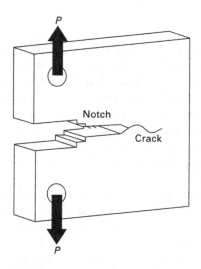

5.19 Compact tension specimen for fracture toughness testing.

materials are under 0.01 kJ m^{-2}. Most high-strength alloys, including those used in aircraft structures, have moderately high toughness (20–100 kJ m^{-2}). Fibre–polymer composites have anisotropic toughness properties because of their microstructure, and the highest toughness (10–30 kJ m^{-2}) is when the direction of crack growth is perpendicular to the fibre orientation. The fracture toughness properties of aerospace materials are fully explained in chapters 18 and 19.

The toughness properties of materials are also measured using impact tests. These tests involve measuring the energy required to fracture the material specimen when impacted at high velocity by a heavy object. The impact energy absorbed by the specimen is related to its toughness; tough materials require higher impact energies than brittle materials. The most popular methods are the Charpy and Izod impact tests. Both tests involve striking the specimen containing a V-notch with a pendulum travelling at a set speed. Impact loading of the specimen produces much higher strain rates than those generated in the SENB and CT fracture toughness tests. Therefore, the Charpy and Izod tests are useful for determining the dynamic toughness of materials under high loading rates, such as those experienced during an impact event (e.g. bird strike). However, the tests are qualitative; the result can only be used to compare the relative impact toughness properties between materials. Impact test results cannot be used to calculate the intrinsic fracture toughness of the material, although the test is used as a simple, accurate and inexpensive method for screening materials for toughness.

5.7   Drop-weight impact test

Fibre–polymer composites are susceptible to damage from impact events such as bird strike, dropped tools during aircraft maintenance, tarmac debris kicked-up by the wheels during take-off or landing, and large hail stones. Impact testing is performed at different impact energy levels to screen composite materials for damage resistance and damage tolerance. The most common and simplest test for measuring impact resistance is the falling-weight test, as shown in Fig. 5.20. A flat specimen panel of the composite material is impacted perpendicular to the surface by a hard object, usually a hemispherical steel tub. The weight is dropped from a height of 1–2 m with a mass of 4.5–9 kg to replicate a low-velocity impact event on an aircraft. The amount of damage caused by the impact event is measured (often by a nondestructive inspection method such as ultrasonics or radiography) to determine the damage resistance. The residual mechanical properties, such as compression strength or fatigue life, may also be measured to determine the damage tolerance. Civil aviation authorities specify no reduction to the mechanical properties following an impact at an energy level in the range of 35–50 J whereas the USAF specifies a minimum energy level of 135 J.

Other impact tests are often included at higher energy levels to fully characterise the impact resistance of aircraft composite components. These may include ballistic impact for critical military structures (e.g. main rotor blades), ice-hail simulation and bird strike simulation.

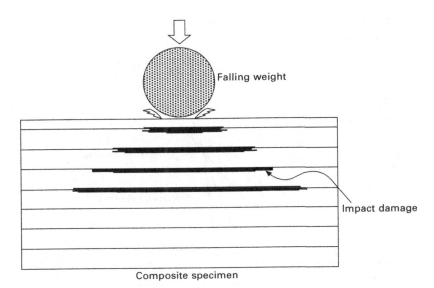

5.20 Impact testing of composites.

5.8   Fatigue test

Fatigue tests measure the resistance of materials to damage, strength loss and failure under the repeated application of load. Aerospace materials must withstand repeated loading for long periods of time, which is in the order of 15 000–20 000 flight hours for modern jet engine materials and anywhere from 80 000 to 120 000 h for airframe materials. Materials can be damaged by repeated loading, which causes a loss in strength and eventually leads to complete failure. Fatigue tests are performed to measure the reduction in stiffness and strength of materials under repeated loading and to determine the total number of load cycles to failure.

Fatigue tests are performed by repeated tension–tension, compression–compression, tension–compression or other combinations of cyclic loading. The fatigue stress is applied repeatedly to the specimen using a variety of load waveforms, as shown in Fig. 5.21. The shape of the loading wave is usually sinusoidal, although triangular and block loads are also used. The frequency of the load cycles is generally low, typically 1–20 Hz (i.e. load cycles per second), to avoid heating of the specimen, which can affect the fatigue results.

The basic method of determining the fatigue resistance of materials is the fatigue life (S–N) curve. This curve is a plot of the maximum fatigue stress S against the number of load cycles-to-failure of the material N. The curve is plotted with the fatigue stress as a linear scale and load cycles-to-failure as a log scale. Data for the curve is generated by fatigue testing specimens

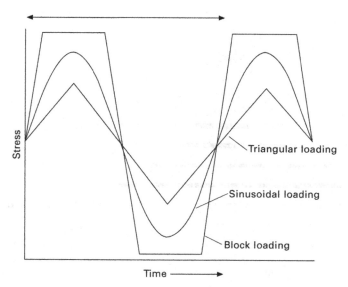

5.21 Examples of fatigue-loading waveforms.

at different fatigue stress levels to measure the number of load cycles-to-failure. Examples of fatigue life curves for aluminium alloy and carbon–epoxy composite are presented in Fig. 5.22. A fatigue test can be stopped after a specific number of load cycles and then the specimen is loaded to fail to measure the residual fatigue strength. Chapter 20 provides a full description of the fatigue properties of aerospace materials.

5.9   Creep test

Creep is a plastic deformation process that occurs when materials are subjected to elastic loading for a long period of time, often at high temperature. Engineering materials do not plastically deform when loaded within the elastic regime for short times. However, when the elastic load is applied for a sufficient period the material eventually deforms plastically by creep. The amount of creep experienced by structural and engine materials used in aircraft is negligible at room temperature. However, the creep rate increases rapidly with temperature (usually above 40% of the absolute melting temperature for metals) and therefore creep is an important engineering property for materials required to operate at high temperature, such as jet and rocket engine materials.

The creep properties of materials are determined using the creep test. The test involves measuring the elongation of a material specimen while under a constant applied tensile stress at high temperature (Fig. 5.23). The specimen is heated inside a thermostatically controlled furnace attached to a creep testing machine that applies the tensile load. The level of applied stress is below the

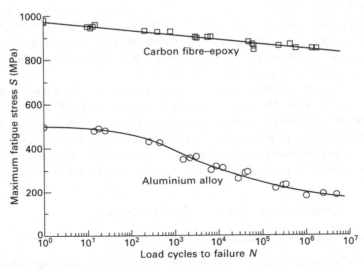

5.22 Fatigue life (*S–N*) curves.

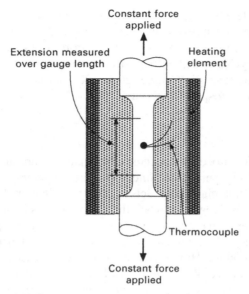

5.23 Creep test.

yield strength of the material. The elongation of the specimen over time is measured using an extensometer or other device capable of measuring the strain. Creep tests are usually performed over long periods, with the loading times typically ranging from 1000 to 10 000 h. The extension of the specimen must be measured using a very sensitive device because the actual amount of deformation before failure may be less than a few percent. Ideally, the test should be performed under conditions that replicate, as close as practical, the temperature, stress and time-scale of the aircraft material when used in service. For example, creep tests performed on materials for engine turbine blades should be performed close to the operating stress and temperature of the blades, which is about 180 MPa and 1200 °C.

The test results are plotted as strain against time to give a creep graph, as shown in Fig. 5.24. The curve for most materials can be divided into three regions: primary creep, secondary creep, and tertiary creep. A series of creep tests performed at different stress levels and temperatures are performed to obtain a complete assessment of the creep properties for a material. The creep properties of aerospace metals and composites are described in chapter 22.

5.10   Environmental durability testing

5.10.1 Corrosion testing of metals

One of the most damaging environmental effects of aerospace metals is corrosion. Corrosion of the metal alloys used in aircraft structures and engines

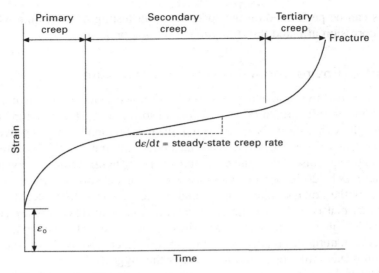

5.24 Creep graph.

occurs in many forms, including general corrosion, stress corrosion, pitting corrosion, crevice corrosion and exfoliation corrosion. Chapter 21 describes the corrosion properties of metals. There is no single, universal corrosion test; instead there are a large number of tests used to measure the resistance of metals to different types of corrosion. The tests are often performed under extreme corrosion conditions to accelerate the corrosion rate and thereby determine the long-term corrosion performance of metals.

Most tests have been designed to measure (either quantitatively or qualitatively) the resistance of metals to a specific type of corrosion or corrosive environment. For example, tests are available to determine the resistance of metals to pitting corrosion whereas other tests are used to measure the resistance of metals to crevice corrosion. Aerospace companies may sometimes use their own corrosion test methods, such as exposing structural metals to aviation fuels, lubricants or paint strippers that may contain potentially corrosive chemicals. Jet engine manufacturers perform specialist corrosion tests of the materials used in turbine blades and discs that involve hot combustion gases produced by aviation fuel.

The selection of corrosion tests for metals used in aircraft should be based on the types of corrosion and corrosive environments most likely to be experienced in service. An example of a test used to measure the corrosion rate of aircraft metals is the salt spray test. This test involves exposing metal specimens to a dense saline fog within a closed chamber. The fog is produced by producing a fine mist of salted solution (usually water containing sodium chloride to replicate seawater). The appearance of corrosion products and damage on the specimen is evaluated after a period of time. Mechanical tests can be performed on the specimen after testing to determine whether the corrosion affected the strength and fatigue properties.

5.10.2 Environmental testing of composites

The greatest environmental problem with fibre–polymer composite materials used in aircraft structures is moisture absorption. Water moisture in the atmosphere (humidity) is absorbed into the polymer matrix causing swelling, damage and softening of the material under extreme conditions. Moisture absorption can also reduce the maximum operating temperature of a composite, sometimes by 20 °C or more. Composites typically absorb anywhere from 0.5 to 2% of their own weight in water, and this resides as water molecules in the polymer matrix and along the fibre–polymer interface. The moisture content is often higher in sandwich composites because the water can condense in the core material. Composites can also absorb other types of fluids, including aviation fuel, although airborne water is the biggest problem.

Material specimens are tested under conditions representative of extreme cases of environment: cold temperature and dry; room temperature and dry; hot and dry; and hot and wet. Of these conditions, the worse case for polymer composites is hot and wet representative of a tropical environment. Specimens are placed inside a sealed chamber which simulates a severe tropical environment with a temperature of 71 °C (160 °F) and high humidity (85–95% water). At regular intervals the specimens are removed from the chamber to be weighed. The change in specimen weight indicates the amount of moisture absorbed by the composite. Figure 5.25 shows the weight gain for several types of carbon-fibre composite materials during exposure to hot–wet conditions. The mechanical properties of the specimens are measured at increasing intervals of exposure time, particularly the matrix-dominated properties such as compression strength. This data is used to assess likely changes to the mechanical behaviour of structural composites over the life of an aircraft. Other environmental tests may also be performed on composites, including exposure to ultraviolet radiation.

5.11   Certification of aerospace materials

5.11.1 Pyramid approach to aircraft certification

The certification of structural and engine materials is one of the most important issues with the testing and evaluation of new aircraft. Certification is also performed when new materials are used in major structural refits of existing aircraft, usually for life extension. Certification is essential to ensure the materials are safe, reliable, durable and functional in their structural application. New and improved materials cannot be introduced into aircraft

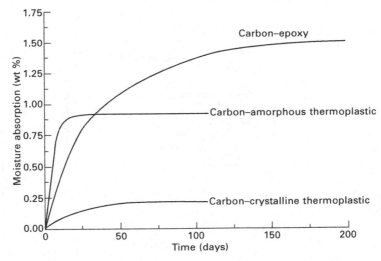

5.25 Moisture uptake–time curves for several aerospace composite materials (adapted from F. C. Campbell, *Manufacturing technology for aerospace structural materials*, Elsevier, 2006).

without thorough analysis, testing and evaluation. A rigorous engineering assessment of the structural materials must be undertaken and passed before they are certified to use on aircraft and helicopters. The certification assessment involves mechanical and environmental (durability) testing of the material together with computation analysis using finite element modelling and other analytical methods. Aircraft certification is a complex, expensive and time-consuming process.

Certification regulations for structural and engine materials are specified by aviation regulatory agencies such as the Federal Aviation Authority (FAA) in the United States. Aerospace companies must ensure the materials pass the certification compliance standards specified by the regulation agency before being introduced to civil aircraft. Military organisations often use their own certification specifications and, although generally not as demanding as the civil specifications they still require extensive physical testing of the materials.

The certification procedure is often represented by the testing pyramid shown in Fig. 5.26. Often called the 'building-block' approach, this method is widely used by the aerospace industry to establish mechanical property data, property knock-down factors, and validation of critical design features for structures. Certification begins at the bottom of the pyramid. Mechanical properties of the material are determined by a series of tests at the 'coupon level', which means sample sizes about 100–200 mm long and 10–50 mm wide. Coupon tests are performed to determine basic property data, such

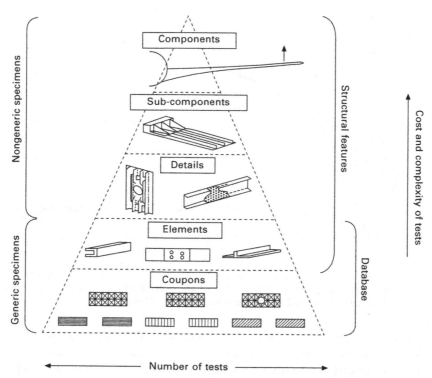

5.26 Testing pyramid for certification of aerospace structural and engine materials.

as Young's modulus, strength, fracture toughness, fatigue life and so on. The coupon tests are carried out under a standardised set of conditions which specify test parameters such as the sample size and loading rate. The test conditions are specified by standardisation organisations, such as the American Society for Testing and Materials. Large aerospace companies also have their own specifications for certain mechanical or durability tests not covered by the standards organisations.

Coupon tests are divided into quantitative or qualitative. Quantitative tests provide data that can be used for design purposes as well as certification. Examples of quantitative tests are the tension, compression and creep tests. Qualitative tests give results that can only be used for comparison purposes. For example, the hardness test and Charpy impact test provide a simple 'go/no go' assessment of materials. Certification testing at the coupon level involves measuring the material properties under different load conditions (e.g. tension, compression, bending) and operating environments (e.g. corrosive fluids, humidity, temperature). Simple drop weight impact tests are also performed, although other impact tests are often included at higher levels of the building block approach. These may include ballistic impact for critical military structures (e.g. main rotor blades), ice-hail simulation, bird strike simulation, and other program-specific impact tests.

5.11.2 Material allowables

The number of coupon tests performed on a new material can range from several hundred to many thousand, depending on the material itself and its intended aircraft structural application. For example, Table 5.1 shows the types of tests and numbers of specimens for a composite material intended for a new aircraft design. A large number of coupon tests are necessary to provide a statistical database on the mechanical properties. This is because many properties, including yield strength, fracture toughness and fatigue life, are sensitive to small variations in the material. Minor differences in the alloy content, microstructure and defects (e.g. gas holes) between metal samples can result in significant differences in the properties. When a large number of coupon tests are performed on the same material then the amount of scatter in the mechanical properties is quantified. For example, Fig. 5.27 shows the variation in the mechanical property of a material which has a Gaussian-type distribution. The mean property value is 100, although the measured properties vary from about 72 to 128.

Large databases which quantify the amount of scatter in the mechanical properties are essential for safe aerospace design. The databases are used to determine the so-called material allowables that are used in design analysis. The material allowables are often given under the headings of A or B. The A basis allowable defines the mechanical property values for materials used in safety critical aircraft structures such as the fuselage, wings and undercarriage. The confidence required in the property values for these materials is critical to aircraft safety. The mechanical property value using the A basis allowable

Table 5.1 Example of the number and types of coupon and element tests used to evaluate a new composite material for an aircraft structure

Test type	Number of tests
Bolted joints (stiffness, strength)	3025
Laminate strength (tension, compression, etc.)	2334
Durability and long-term exposure	585
Interlaminar shear strength	574
Defect effects on mechanical properties	494
Crimping	271
Bonded repairs (stiffness, strength, etc.)	239
Ply properties (Young's modulus, Poisson's ratio, etc.)	235
Stress concentrations (open-hole tension, etc.)	118
Radius details	184
Total	8059

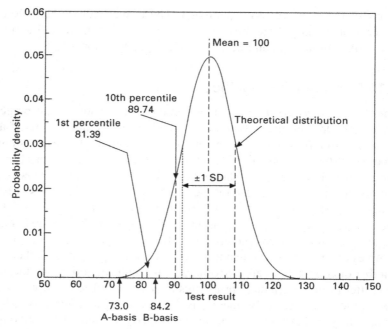

5.27 Representation of statistical variations in the mechanical properties of materials.

is defined as the value above which 99% of the population of values fall within the distribution with a confidence of 95%. The B basis allowable is less stringent, and is applied to components where material failure does not result in the loss or excessive damage to the aircraft. The property value using the B allowable is the value above which 90% of the population of values is expected to fall with a confidence of 95%.

5.11.3 Structural certification

After the mechanical and environmental properties of the materials have been determined by an exhaustive series of coupon tests, the structural and environmental properties of aircraft components built using the materials are then measured by more testing on a larger scale. As shown in the certification pyramid, structural elements, details and subcomponents that represent increasingly complex and more complete sections of the final aircraft are tested. The final component (e.g. wing or fuselage) is constructed of different subcomponents which, in turn, are assembled from many structural details which contain a large number of elements. The elements, details and subcomponents contain structural design features not present in the coupon specimens, such as cut-outs, stringers, rib attachments, changes in section thickness, bolted or bonded connections. Tests that replicate the actual loading on the final component are performed on the elements, details and subcomponents to ensure they comply with the design specifications (Fig. 5.28).

5.11.4 Full-scale structural testing

Testing of the entire aircraft is the final stage of the certification process. The full-scale test is one of the most important ways of proving how well the aircraft meets its performance requirements. The test is extremely important because it tests all the components and materials of the aircraft in the most realistic manner by simulating actual flight conditions. A full-scale structural test is usually performed on one of the first four aircraft built. Full-scale tests are also performed on in-service aircraft that have undergone a major structural design change. Full-scale tests are used to ensure the aircraft is structurally sound after all the materials, elements, details and subcomponents have been fully integrated. The full-scale test is also important to determine the effect of secondary loading caused by complex out-of-plane loads, which may not be determined in earlier testing. Such loads arise from eccentricities, stiffness changes and local buckling which may not be fully predicted or eliminated in design nor represented by the structural detail specimen. Another important aspect of the full-scale test is the confidence that the aircraft is safe.

The full-scale component is tested under conditions given in the certification specifications to ensure it is fully compliant. For example, wing testing by Boeing of their B787 aircraft involved bending a complete wing upwards beyond the ultimate load (or 1.5 times the load limit) to the point of destruction (Fig. 5.29). The point was reached when the wing tips deflected by about 7.6 m above their normal position while enduring loads equal to about 500 000 lbf. In addition to this full-scale test, the aircraft was tested for 120 000 simulated flights, or double the expected life cycle, for fatigue performance. The simulated flights lasted about 4 min each and replicated taxiing, climb, cabin pressurisation and depressurisation, descent and landing. To add realism, the artificial flight conditions varied from completely smooth and level to extremely turbulent. It is only after such an exhaustive series of passed tests from the coupon to final component levels that the aircraft is certified by the aviation safety authority.

5.12 Summary

The mechanical properties of aerospace materials should be measured under the stress conditions experienced in service. As examples, the properties of materials used in tension loaded structures, such as the fuselage, are

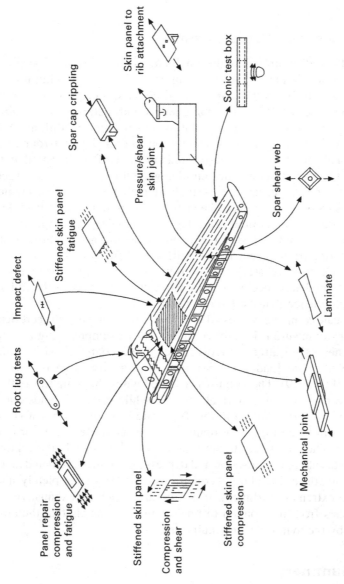

Panel repair
compression
and fatigue

Root lug tests

Impact defect

Stiffened skin panel
fatigue

Spar cap crippling

Skin panel to
rib attachment

Pressure/shear
skin joint

Sonic test box

Spar shear web

Stiffened skin panel

Compression
and shear

Stiffened skin panel
compression

Mechanical joint

Laminate

5.28 Types of element and detail tests performed on a subcomponent of an aircraft wing.

5.29 Structural certification testing of the B787 wings.

determined by tensile testing whereas materials used in compression loaded structures are measured by compression testing.

Mechanical property tests are classified as quantitative and qualitative. Examples of quantitative tests include tension and compression tests which give absolute measurements for properties such as elastic modulus and yield strength. Qualitative tests, such as Charpy impact and hardness tests, measure property values which should only be used for ranking materials.

Durability tests are performed on aerospace materials to ensure they do not lose functionality and degrade when used in the aviation environment. The main tests for metals determine the corrosion and oxidation (for engine applications) properties whereas the main tests for fibre–polymer composites determine the moisture absorption behaviour.

The properties of engineering materials are rarely constant, and there is usually scatter in the measured values. The certification testing of aerospace materials requires a large number of mechanical property tests to be performed to quantify the variability in properties, which is then expressed as an A- or B-basis allowable in the design of safety critical structures.

The mechanical and durability testing of materials forms an important part of the certification of aircraft. Coupon tests performed using small samples of the material are used to obtain basic data on the mechanical and durability properties. Further tests are then performed at the element, structural detail, subcomponent and final component (which may be the entire aircraft) levels as part of the complete structural certification process.

5.13   Terminology

A-basis allowable: The minimum property value for an aerospace material above which 99% of the population of values exceed the minimum value with a confidence limit of 95%.

B-basis allowable: The minimum property value for an aerospace material above which 90% of the population of values exceed the minimum value with a confidence limit of 95%.

Ductility: The amount a material can yield and plastically deform before failing. Determined from the elongation-to-failure value of the material.

Elastic limit: The maximum strain applied to the specimen without causing plastic deformation. The elastic limit is the transition point between elastic and plastic deformation. Also called the proportional limit.

Elastic modulus (also called Young's modulus): The ratio of the stress to strain when the material is below the proportional limit.

Elastic strain: The change in shape of a material that disappears instantly when the stress is removed.

Elongation-to-failure: The percentage strain at final failure. Also used to describe the ductility of a material.

Engineering strain: The average linear strain determined by dividing the elongation of the specimen by the original gauge length.

Engineering stress: Applied force divided by original cross-sectional area of the specimen.

Fracture toughness: Mechanical property that describes the resistance of a material to cracking.

Neck: Region of localised thinning in a tensile specimen when strained beyond the ultimate tensile stress.

Offset method: Method used to calculate the proof strength of a material that does not have a sharp yield point. A straight line is drawn parallel to the linear elastic part of the stress–strain curve at a specific strain offset value. The intercept point of the offset line and the stress–strain curve defines the proof strength of the material.

Plastic strain: The permanent change in shape of a material when the stress is removed. Determined by subtracting the elastic strain from the total strain of the specimen.

Poisson's ratio: The ratio of the transverse (lateral) strain to the corresponding axial (longitudinal) strain resulting from uniformly distributed axial stress below the proportional limit of the material.

Proof strength (also called offset yield strength): The stress at which a material undergoes a specified amount of deformation (usually 0.2% strain). The stress used to define the onset of plastic deformation in ductile materials that do not have a clearly defined yield point on the stress–strain curve.

Proportional limit: The highest stress at which stress is directly proportional to strain. It is the highest stress at which the stress–strain curve is rectilinear, and marks the point between elastic and plastic deformation.

Tensile toughness: Property that describes the amount of elastic work to the material before the onset of plastic deformation. Determined from the area under the stress–strain curve in the elastic regime.

True strain: The instantaneous strain in the specimen when under load.

True stress: The applied load divided by the actual cross-sectional area of the specimen.

Ultimate tensile strength (also called tensile strength): The ultimate or final (highest) stress sustained by a specimen in a tension test.

Work of fracture: Property that describes the total amount of work to cause failure. Determined from the total area under the stress–strain curve.

Yield strength: The lowest stress at which a material undergoes plastic deformation.

5.14   Further reading and research

Askeland, D. R. and Phulé, P. P., *The science and engineering of materials*, Thomson, 2006.

Callister, W., *Materials science and engineering, an introduction*. John Wiley & Sons, NY, 1985

Dieter, G. E., *Mechanical metallurgy*, McGraw-Hill, London, 1988.

Kuhn, H. and Medlin, D. (eds.), *ASM handbook volume 08: mechanical testing and evaluation*, ASM International, 2000.

Production and casting of aerospace metals

6.1   Introduction

There are many stages in the manufacture of aircraft structures and engine components using metals, and it begins with the production and casting of the metal alloy. The mechanical properties, corrosion resistance and many other properties that influence the selection and performance of metals in aircraft are determined by their method of production and casting. The production process involves the melting of the base metal, addition of selected alloying elements in carefully controlled concentrations to the molten metal, and then casting of the molten metal into a solid product for subsequent forming and machining into the final aircraft component. The structure and properties of the metal alloys are highly sensitive to the processes used in their production and casting.

In this chapter, we study the processes used to produce and cast metal alloys for aircraft structural components. The making of metal alloys by the addition of alloying elements is described, including the selection and solubility of the elements into the base metal. The key aspects of metal casting are examined, including the process of solidification, the structure and properties of cast metal, the formation of defects and damage which can affect structural performance, and the process methods used to ensure high quality casting. The main methods used to cast metal alloys for aircraft structures and jet engine components are described, including permanent mould casting, pressure die casting, sand casting, investment casting, directional solidification casting, and single crystal casting.

6.2   Production of metal alloys

6.2.1   Addition of alloying elements to metal

Metals are not used in aerospace applications in their pure (unalloyed) condition. Pure metals are too soft to use in aircraft structures and engines, and must be alloyed with other elements to produce high-strength materials. For instance, the addition of a few percent of copper to aluminium increases the strength by 500–600%: about 80 MPa for pure Al to 450–500 MPa for the Al–Cu alloy. As another example, adding just 0.8% carbon to iron makes high-strength steel more than 1000% stronger than the pure metal.

There are many important reasons for alloying. The most common reason is to increase the mechanical properties of the base metal, such as higher strength, hardness, impact toughness, creep resistance and fatigue life. Alloying can also affect the other properties of metals: magnetic properties, electrical conductivity and corrosion resistance. In a few cases, alloying has the benefit of lowering the density of the metal, such as the addition of lithium to reduce the weight of aluminium alloys. In the majority of cases, however, alloying has little or no major effect on density unless a high concentration is added to the base metal. Another reason for alloying is to increase the maximum working temperature of metals or improve their toughness at very low temperatures. Certain metals are alloyed to improve their corrosion resistance and durability in harsh environments. Alloying may also reduce the material cost when a cheap alloying element is added to an expensive base metal, although this should be considered a side benefit rather than the main reason for alloying.

The process of adding alloy elements to a base metal is relatively simple, although aerospace alloys are subject to stringent quality control that requires their production under carefully regulated conditions. The base metal is melted inside a large crucible within a temperature-controlled furnace. The metal is usually in a pure form (>99% purity), although it often contains trace amounts of impurity elements that were not fully removed during the smelting and refining of the ore. All the base metals used in aircraft structures and engines, aluminium, magnesium, titanium, iron and nickel, contain impurities. For example, aluminium alloys used in aircraft structures contain low concentrations of silicon and iron impurities from the bauxite ore. In general, the effect of impurities is deleterious whereas the effect of alloying elements on properties is beneficial. Provided the concentration of impurity elements is kept low then they do not pose a problem and, in some materials, may be beneficial to the processing or mechanical properties. For example, small amounts of silicon increase the fluidity of molten aluminium, which makes it easier to cast without the formation of pores and cavities. As another example, low concentrations of iron and atomic oxygen impurities in titanium increases the yield stress by solid solution strengthening.

The base metal is heated inside the furnace to a temperature sufficient to melt and dissolve the alloying elements into the melt. The furnace environment is controlled during the melting of reactive base metals, such as titanium, nickel or magnesium, to stop oxidation and contamination from the air. The melting of reactive metals is performed under vacuum or inert gas (e.g. argon). Vacuum induction melting (VIM) is one of the most common methods, whereby melting of the base metal and alloying elements is performed under high vacuum to minimise oxidation and remove dissolved gases (such as hydrogen and nitrogen) from the melt. Alloying elements are added to the molten base metal in measured amounts to produce the metal alloy melt. The elements are usually added as powder or small pellets, which are dissolved into the liquid metal and distributed through the melt by stirring and mixing. Once the alloying elements have dissolved in the molten metal it is ready for casting.

6.2.2   Solubility of alloying elements

Alloying is based on the principle that the alloying elements are soluble in the base metal. That is, the alloying element can dissolve as individual atoms when added to the molten base metal, and may remain dissolved when the material solidifies and cools to room temperature. There are two types of solubility: unlimited solubility and limited solubility.

Unlimited solubility means that the alloying element completely dissolves in the base metal, regardless of its concentration. For example, nickel has unlimited solubility in copper in concentrations less than 30% by weight. After solidification, the copper and nickel atoms do not separate but instead are dispersed throughout the material (Fig. 6.1). The structure, properties and composition are uniform throughout the metal alloy, and there is no interface between the copper and nickel atoms. When this occurs the alloy is called a single-phase material.

Unlimited solubility occurs when several conditions, known as the Hume–Rothery rules, are met:

- Size factor: The atomic sizes of the base metal and alloying element must be similar, with no more than a 15% difference to minimise the lattice strain.
- Crystal structure: The base metal and alloy element must have the same crystal structure.
- Electronegativity: The atoms of the base metal and alloying element must have approximately the same electronegativity.

The Hume–Rothery conditions must be met for two elements to have unlimited solid solubility. Even when these conditions are met, however, there can be instances when unlimited solubility still does not occur.

Limited solubility means that an alloying element can dissolve into the base metal up to a concentration limit, but beyond this limit it forms another phase. The phase which is formed has a different composition, properties and (in some cases) crystal structure to the base metal. Most of the alloying elements used in aerospace base metals have low solubility. For example, when copper is present in aluminium it is soluble at room temperature up to about 0.2% by weight. At higher concentrations the excess of copper reacts with the aluminium to form another phase ($CuAl_2$). As another common example, when carbon is added to iron in the production of steel it has a solubility limit of under 0.005% by weight at 20 °C. In higher concentrations,

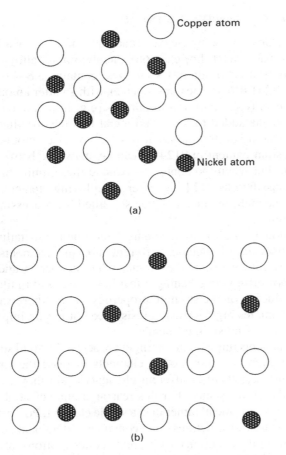

6.1 Unlimited solubility of nickel in copper: (a) molten alloy; (b) solid alloy.

the excess of carbon atoms form carbide particles (e.g. Fe_3C) which are physically, mechanically and chemically different to the iron.

The limited solubility of alloying elements is an important means of increasing the strength of aerospace metals by precipitation hardening. The concentration of alloying elements used in the main types of base metals for aircraft are often well above their solubility limit to promote strengthening by precipitation and other hardening mechanisms. However, there is a practical limit to the amount of nonsoluble alloying elements that can be added to a base metal. When there is too much of the alloying element then the metal can become brittle and not useful for structural applications.

6.2.3   Selection of alloying elements

The alloy produced is determined by the types and concentrations of alloying elements used in the base metal. For example, the aluminium alloy known as 2124 Al, commonly used in aircraft structures, contains 3.8–4.0% (by weight) copper and 1.2–1.8% magnesium together with smaller amounts of other alloying elements (e.g. zinc) and impurities (e.g. iron, silicon). The amount of alloy elements added to the liquid metal is often specified over a narrow range because it is often impractical to add precise amounts. For example, the magnesium content in 2124 Al can be anywhere between 1.2 and 1.8% by weight, but when the content is outside these limits then the alloy is no longer classified as 2124 Al. After the alloying elements have fully dissolved into the melt, the metal is then solidified for processing into an aircraft component.

The types and amounts of alloying elements determine the metallurgical and mechanical properties of metals. Different alloying elements alter the properties in different ways. For example, when copper is added to aluminium it has a powerful strengthening effect but when used in titanium it has virtually no influence on the strength properties. As another example, chromium in steel promotes high corrosion resistance but has no impact on the corrosion properties of most other metals.

Certain alloying elements improve the strength properties by solid solution hardening or precipitation hardening, other elements increase the strength by refining the grain size, whereas different elements again may enhance the corrosion or oxidation resistance. For this reason, a number of alloying elements are used in the same metal rather than a single element. For instance, titanium alloys used in aircraft structures often contain two dominant alloying elements (aluminium and vanadium) with small concentrations of other elements (e.g. tin, zirconium, molybdenum). Another example is nickel superalloys that contain large amounts of many different alloying elements to increase the maximum working temperature, including iron, chromium, molybdenum, tungsten, cobalt, aluminium and niobium. Some alloying elements have several functions, such as chromium in stainless steel that increases both strength and corrosion resistance.

The concentration of alloying elements is also critical to controlling the mechanical and durability properties. The properties of metals do not necessarily increase steadily with greater additions of alloying elements; instead there is an optimum concentration. For example, the optimum copper content in aluminium for maximum strengthening is 3–5%; below this range the metal is too soft and above this range too brittle. The types, approximate concentrations and main functions of the principal alloying elements used in aerospace structural metals are given in Table 6.1.

Table 6.1 Alloying of the main types of base aerospace metals (SSS = solid solution strengthening; PH = precipitation hardening)

Base metal	Alloy element	Alloy content (wt%)	Main functions of alloying element
Aluminium	Copper	1–5	Increased strength by SSS and PH ($CuAl_2$)
	Zinc	5–8	Increased strength by SSS and PH ($AlZn$)
	Magnesium	0.4–1.8	Increased strength by SSS and PH (Al_2CuMg)
	Manganese	0.2–1.0	Increased strength by SSS and PH ($Al_{20}Cu_2Mn_3$)
	Lithium	2–3	Reduced density and increased strength by SSS and PH
Magnesium	Zinc	4–6	Increased strength by SSS and PH ($MgZn_2$)
	Aluminium	5–6	Increased strength by SSS and PH ($Mg_{17}Al_{12}$)
	Rare Earth elements	1–2	Increased strength by SSS and PH (e.g. $Mg_{12}NbY$, $MgCe$)
Titanium	Aluminium	5–6	Stabilise α-phase, increase strength by SSS and PH
	Vanadium	4–10	Stabilise β-phase, increase strength
	Tin	2–11	Increase strength by SSS
	Niobium	0.7–1.0	Increase strength by grain size refinement
Steel	Carbon	<0.8	Increased strength by SSS and carbide hardening (Fe_3C)
	Manganese	<1	Increased strength by SSS
	Nickel	2–18	Increased strength by SSS and PH
	Molybdenum	0.1–0.5	Increased strength by SSS
	Chromium (in stainless steel	12–18	Increased corrosion resistance and strength
Nickel	Iron	2–20%	Increased strength by SSS
	Carbon	<0.2%	Increased creep resistance and strength by carbide hardening
	Chromium	10–20%	Increased corrosion resistance and strength by PH
	Molybdenum	2–9%	Increased creep resistance and strength by SSS
	Cobalt	1.5–15%	Increased strength by SSS
	Aluminium	0.2–5%	Increased creep resistance and strength by PH
	Tungsten	0.5–12.5%	Increased creep resistance and strength by SSS
	Niobium	1.5–5%	Increased creep resistance and strength by PH

6.3   Casting of metal alloys

6.3.1   Shape and ingot castings

Casting is the operation of pouring molten metal into a mould and allowing it to solidify. The pouring temperature is usually 50–180 °C above the melting point of the metal alloy. There are two broad classes of casting operations known as shape casting and ingot casting. Shape casting involves pouring the liquid metal into a mould having a shape close to the geometry of the final component. The shape casting is removed from the mould after solidification, and then heat treated and machined into the finished component. Ingot casting, on the other hand, involves pouring the metal into a mould having a simple shape, such as a bar or rod. After cooling, the ingot is strengthened and shaped by working processes such as forging, extrusion or rolling. Both shape and ingot castings are used in the production of aerospace components. However, primary aircraft structures are most often fabricated from ingot castings.

6.3.2   Solidification of castings

Solidification of the metal within shape or ingot casting moulds has a strong influence on the microstructure and mechanical properties. The solidification process is complex and does not simply involve the metal changing immediately from liquid to solid when cooled to the freezing temperature. When molten metal is cooled below the equilibrium freezing temperature there is a driving force for solidification and it might be expected that the liquid would spontaneously solidify. However, this does not always occur – pure metals can be cooled well below their equilibrium freezing temperature without solidifying. This phenomenon is known as undercooling or supercooling. For example, liquid aluminium can be undercooled to 130 °C below the equilibrium freezing temperature (660 °C) and held there indefinitely without solidifying. Similarly, pure nickel can be supercooled as much as 480 °C below the freezing temperature (1453 °C) before becoming solid. The reason for this behaviour is that the transformation from liquid to solid begins with the formation of tiny solid particles or nuclei within the melt. The nuclei develop spontaneously in the liquid as nano-sized particles composed of several dozen atoms arranged in a crystalline structure. The nuclei are freely suspended in the melt and are surrounded by liquid metal. The creation of solid nuclei in this way is called homogenous nucleation. The nuclei often dissolve back into liquid before they grow to a critical size that is thermodynamically stable, and it is this instability that allows the large supercooling of pure metals. It is only when nuclei grow beyond a critical size, which requires cooling of the liquid well below the equilibrium freezing temperature, that the solidification process stabilises and the metal completely solidifies.

In practice, large undercooling does not occur with metal alloys because solidification occurs by a process called heterogeneous nucleation. Most alloys freeze within about 1 °C of their equilibrium melting temperature because the nucleation of ultra-fine solid nuclei occurs at free surfaces, such as the mould walls or solid impurity particles in the melt. The heterogeneous nucleation of solid nuclei at pre-existing surfaces gives them greater stability than the nuclei which develop by homogenous nucleation. Once the nuclei have formed by heterogeneous nucleation they grow by atoms within the liquid attaching to the solid surface. At the equilibrium freezing temperature, a metal contains an extremely large number of nuclei. Each nucleus forms the embryo to a single grain, and the grains grow until no liquid remains and the metal has completely transformed into a polycrystalline solid. The sequence of events in the solidification of a metal under the conditions of heterogeneous nucleation is presented in Fig. 6.2.

6.3.3   Structure of castings

Chill, columnar and central zones

The solidification of shape and ingot castings often occurs in three separate phases, with each phase developing a characteristic arrangement of grain sizes and shapes. The grain structures through the section of an ingot from the mould wall to the centre are shown in Fig. 6.3. From the surface to core these regions are called the chill zone, columnar zone and central zone, and each has a distinctive grain structure. The solidification process that leads to these different grain structures is shown schematically in Fig. 6.4.

The chill zone is a thin band of randomly orientated grains at the surface of the casting. This is the first phase to develop during solidification because the mould surface rapidly cools the metal below the melting temperature. Many solid nuclei particles are created by heterogeneous nucleation at the mould wall and grow into the liquid to form the chill zone. Each nucleation event produces an individual crystal, or grain, in the chill zone, which then grows. The grains grow until they impinge on another grain, when the growth process stops.

When the pouring temperature of the molten metal into the mould is too low, the entire casting rapidly cools below the melting temperature. The nuclei particles that develop at the mould walls break away and are swept throughout the melt under the turbulence created by pouring. When this occurs the chill zone, which consists of equiaxed grains, extends through the entire casting and no other zones develop. However, most casting is performed with a high pouring temperature that keeps the metal at the centre

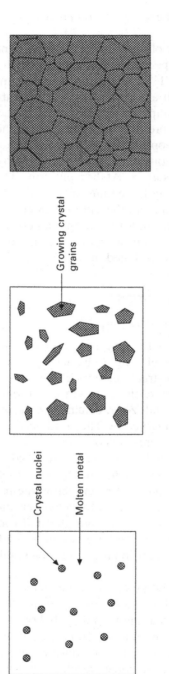

Crystal nuclei

Molten metal

Growing crystal grains

Decreasing temperature

6.2 Solidification sequence for a metal.

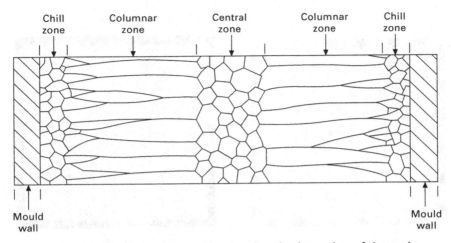

6.3 Section through a casting showing the formation of the grain structures during solidification (reproduced with permission from R. E. Reed-Hill, *Physical metallurgy principles* (2nd edition), Van Nostrand Company, New York, 1973).

above the melting temperature for a long time. Under these conditions, the columnar zone develops from the chill zone.

The columnar zone forms by the solidification of liquid metal into elongated grains. The columnar grains grow perpendicular to the mould wall by a solidification process involving dendrites. When the temperature falls in the liquid just ahead of the solid/liquid interface then it becomes unstable and crystalline stalks known as dendrites grow out from the interface. Secondary and tertiary dendrite arms grow out from the primary stalk, and the branched dendrite structure often appears like a miniature pine tree. (Dendrite comes from the Greek word dendron that means tree). Dendrites have a primary arm that extends from the solid–liquid interface, secondary arms that branch from it, and tertiary arms that branch from the secondary arms. During solidification a number of dendrites of almost equal spacing are formed and they grow parallel to each other as depicted in Fig. 6.5. The growth direction of dendrites depends on the crystal structure of the metal; fastest growth occurs along the most energetically favourable direction. Dendrites develop into columnar grains which grow in the direction opposite the heat flow, or from the coldest towards the hottest regions of the casting. Each crystal that is formed consists of a single dendrite. During solidification the turbulent motion of molten metal ahead of the solid–liquid interface may break off dendrites and carry them into the melt. These dendrites act as nuclei for more grains and multiply the number of grains that nucleate in the liquid.

The metal sometimes continues to solidify in a columnar manner until all the liquid has solidified. More often, however, a central zone (also called

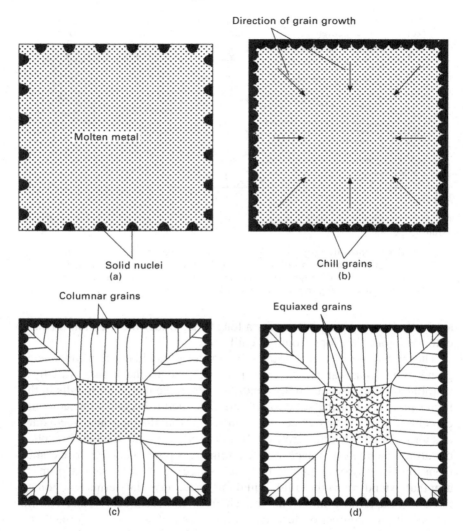

6.4 Solidification process comprising: (a) formation of solid nuclei at the mould walls by heterogeneous nucleation; (b) growth of the chill zone from the mould walls into the melt; (c) growth of columnar grains; and (d) final solidification of casting by formation of equiaxed grains in the central zone.

the equiaxed zone) develops at the core of the casting. The equiaxed zone contains relatively round grains with a random orientation, and they often stop the growth of columnar grains. The amount of the final cast structure that is columnar or equiaxed depends on the cooling rate and alloy composition. A sharp thermal gradient across the solid–liquid interface owing to rapid

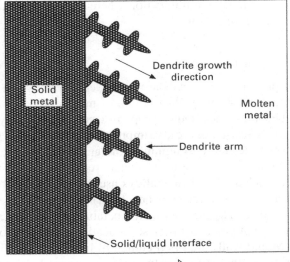

6.5 Dendritic growth into the liquid from the solid/liquid interface: (a) schematic, (b) photograph.

cooling encourages columnar solidification whereas a low thermal gradient promotes equiaxed solidification.

Grain refinement of castings

The mechanical properties of metals are dependent on the grain structure. As described in chapter 4, the yield strength of metals increases when the grain structure becomes finer. Other properties such as ductility, fracture toughness and fatigue resistance are also improved by reducing the grain size. Therefore, it is often desirable to minimise the grain size in cast structural alloys. (The exception is the casting of metals requiring creep strength at high temperature, such as nickel superalloys used in gas turbine blades and discs, when a coarse grain structure is beneficial).

Grain size is controlled by the pouring temperature and solidification rate; lower pouring temperatures and faster solidification rates promote a finer grain structure. Mould walls are often chilled to increase the solidification rate and thereby refine the grain structure. Grain size is also controlled by the use of inoculates or grain refiners, which are chemicals added to the melt to promote many sites for heterogeneous nucleation during solidification. Inoculates act as the nuclei to which the atoms of the molten metal attach to transform from the liquid to solid states. For example, small amounts of titanium (0.03% by weight) and boron (0.01%) are added to liquid aluminium alloys to promote grain refinement. The elements react to create tiny inoculating particles of aluminium titanium (Al_3Ti) and titanium diboride (TiB_2) which cause a large reduction in grain size. Grain refinement is also an important strengthening process for magnesium alloys used in aerospace components. The Mg–Al and Mg–Al–Zn alloys are grain refined with carbon-containing compounds which form inoculating particles of aluminium carbide (Al_4C_3 or $AlN.Al_4C_3$).

6.3.4   Casting defects

Stringent regulations apply to the quality of metal castings used as the stock material for primary aircraft structures. Castings must not contain defects that compromise structural integrity and could cause unexpected failure of components. Casting defects that can cause structural failures include porosity, shrinkage and intermetallic inclusion particles. The nonuniform distribution of alloying elements in the casting, which is called segregation, is another problem because it leads to nonuniformity in the mechanical properties. In other words, discrete regions of soft and hard metal can exist within the same component owing to alloy segregation.

Porosity and shrinkage

Porosity occurs in the form of irregularly shaped cavities known as blowholes or long tubular cavities called wormholes (Fig. 6.6). Molten metals often contain dissolved gases such as hydrogen and nitrogen which develop into gas bubbles during cooling of the liquid and then become trapped by the solid growing around them. Blowhole porosity consists of small and irregularly shaped cavities filled with trapped gas. Blowholes can also form by the shrinkage of the metal during solidification. The amount of shrinkage experienced by aerospace alloys is given in Table 6.2. Shrinkage is a major problem with aluminium castings because the material contracts by anywhere from 3.5 to 8.5% during solidification, resulting in a significant number of

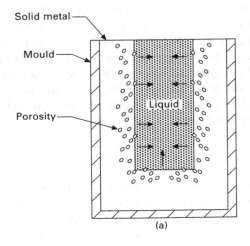

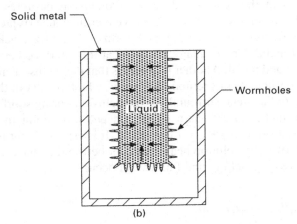

6.6 Two forms of gas porosity: (a) blowholes and (b) wormholes formed during solidification of metal castings.

Table 6.2 Average shrinkage values during
solidification of aerospace metals

Material	Shrinkage (%)
Aluminium	7.0
Magnesium	4.0
Titanium	1.8
Low-carbon steel	2.7
High-carbon steel	4.0

blowholes and cavities. To avoid shrinkage, a reservoir of molten metal (called a feeder or riser) must be placed on the casting to feed liquid metal into the mould cavity.

Wormhole porosity occurs when gas bubbles grow in length at the same rate as the liquid–solid interface moves. The gas bubbles form worm-shaped holes that grow in the heat flow direction from the casting. Both blowholes and wormholes are detrimental to the structural integrity of the casting, but can be removed under the extreme pressure of hot working (described in chapter 7) which squashes the holes and welds their sides together. When the welds are successful, the holes are eliminated and any gas inside the cavities is absorbed into the solid casting.

Inclusion particles

Large inclusion particles are another defect often found in castings. These particles may form during the casting of metals that are strengthened by precipitation hardening, which includes most of the structural aerospace alloys. During the casting and solidification process, alloying and impurity elements react with the base metal to form inclusion particles that are extremely brittle. When the particles are above a critical size (typically 1–2 mm) they easily fracture under load, which forms the site for crack growth through the cast metal. Cracks may also occur at the interface between the inclusion particle and metal. Another problem is that the inclusion and metal have different coefficients of thermal expansion, which results in thermally induced stresses in the metal surrounding the inclusion during solidification. As a result, the inclusion acts as a stress concentration point in the cast metal which may lead to cracking at low stress. It is essential that inclusion particles are removed by solution heat treatment before the metal is used in an aircraft, otherwise cracking and failure can occur.

Segregation of alloying elements

Segregation of the alloying elements is another problem with castings. There are two types of segregation, macrosegregation and microsegregation, which occur at different levels in the casting. Macrosegregation is where the average alloy composition varies over a large distance through the casting. Macrosegregation often occurs between the surface and core of the casting, with the alloy composition of the surface (which freezes first) being different from the centre owing to diffusion of alloy elements ahead of the solid/liquid interface. Alloying elements can either diffuse from the liquid into the solid, which raises the alloy content near the surface, or the elements can migrate from the solid into the liquid which enriches the central region of the casting. The mechanical properties vary from the surface to centre of the cast metal owing to the change in alloy content. The effect of macrosegregation on the properties can be minimised by hot working the cast metal; this involves plastically deforming the casting at high temperature and redistributing the alloying elements. The hot working of cast metal is described in chapter 7.

Microsegregation is the local variation in alloy composition on a scale smaller than the grain size. Microsegregation occurs over short distances, often between the dendrite arms which are typically spaced several micrometres apart. The dendrite core, which is the first solid to freeze, is richer in the alloying elements with higher melting temperatures.

Segregation can be removed by a heat treatment process known as homogenising anneal, which involves heating the solid metal to just below the melting temperature for an extended period of time to allow the alloying elements to disperse throughout the casting. Homogenisation treatments are generally effective in producing a levelling of micro-scale concentration differences in alloying elements, although some residual minor differences may remain.

Section 6.8 presents a case study of casting defects causing engine disc failure on United Airlines flight 232.

6.4   Casting processes

There are many commercial processes to produce castings for processing into aircraft components. Casting processes are usually classified into two broad groups distinguished by the mould type: reuseable casting moulds and single-use casting moulds. This section briefly describes the processes commonly used for casting metals for aircraft structures and engines. The processes are permanent mould casting, pressure-die casting and single-crystal casting, which are reusable moulding methods; and sand casting and investment casting, which are single-use moulding methods. These processes account for the majority of castings for aircraft structural components.

6.4.1   Permanent mould casting

Permanent mould casting is a process for producing a large number of castings using a single reusable mould. The casting process simply involves pouring molten metal into a mould where it cools and solidifies. The mould is then opened, the casting removed, and the mould is reused. The mould is made from a high-temperature metallic material, such as cast iron or hot work die steel, which can withstand the repeated heating and cooling involved with large volume production.

Permanent mould casting produces metal with better dimensional tolerance, superior surface finish, and higher and more uniform mechanical properties when compared with metal solidified using sand casting (which is described in section 6.4.3), which is another popular casting process. Permanent mould castings have relatively high strength, toughness and ductility owing to the mould walls rapidly removing heat from the liquid metal. This generates a fast solidification rate which produces a fine grain structure in the cast metal. Disadvantages of permanent mould casting is the high cost of the reusable mould, and that the casting process is usually viable only when high-volume production can offset the cost or it is critical that high-quality castings are produced. The most common application of permanent mould casting in the aerospace industry is the casting of aluminium, titanium and steel ingots, which are then processed by working operations (e.g. forging, rolling) into aerospace structural components.

6.4.2   Pressure die casting

Pressure die casting involves squeezing molten metal into a mould cavity under high pressure and then holding the metal under pressure during solidification, as shown in Fig. 6.7. Pressure is applied by the action of a hydraulic piston ram that injects molten metal through a steel die and into the mould. The metal casting is removed from the mould following solidification, and the process repeated using the same permanent mould. The process is suitable for very high rate production because of the fast solidification rate of the casting. Many aluminium and magnesium alloys are cast using this method.

6.4.3   Sand casting

Sand casting involves the pouring of molten metal into a cavity-shaped sand mould where it solidifies (Fig. 6.8). The mould is made of sand particles held together with an inorganic binding agent. After the metal has cooled to room temperature, the sand mould is broken open to remove the casting. The main advantage of sand casting is the low cost of the mould, which is a large expense with permanent mould casting methods. The process is

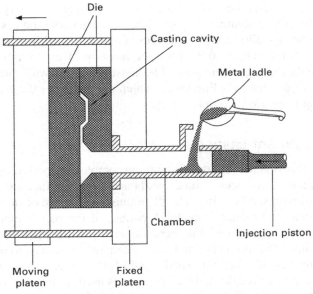

6.7 Pressure die casting (image from The Open University).

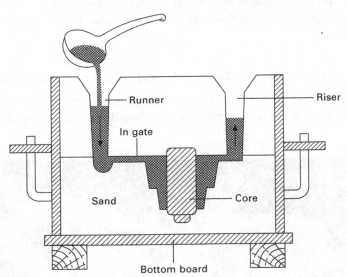

6.8 Sand mould casting (image from The Open University).

suitable for low-volume production of castings with intricate shapes, although it does not permit close tolerances and the mechanical properties of the casting are relatively low owing to the coarse grain structure as a result of slow cooling rate.

Sand casting is the process of choice for the aerospace industry for the economical production of small lot sizes. The industry uses sand casting for producing magnesium alloy and certain types of aluminium alloy components with complex shapes. Figure 6.9 shows the housing of an aircraft engine made using sand cast magnesium alloy. The housing is a thin wall component containing small diameter cooling holes, which demonstrates the capability of sand casting to produce complex shapes.

6.4.4   Investment casting

Investment casting is generally used for making complex-shaped components that require tighter tolerances, thinner walls and better surface finish than can be obtained with sand casting. The distinguishing feature of investment casting is the way the mould is made. A pattern of the part is made with wax, which is then dipped into fine ceramic slurry that contains colloidal silica and alumina. The mould is dried and heated inside an oven to melt out the wax leaving behind a ceramic shell mould for casting. The investment casting method, also called the lost wax process, is used for precision casting of aerospace components such as gas turbine blades. The investment casting of aluminium alloys results in lower strength than sand castings, but offers tighter tolerances, better surface finish and the capability to produce thin-walled sections.

6.4.5   Directional solidification and single-crystal casting

Many mechanical properties are improved by reducing the grain size. A metal with a fine grain structure has higher strength, ductility, toughness and

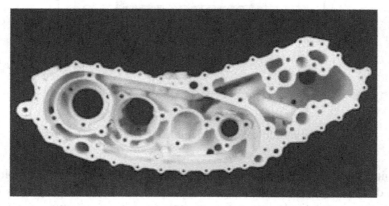

6.9 Sand cast magnesium gearbox housing (from www.tagnite.com/applications/).

fatigue resistance than the same metal containing coarse grains (as explained in chapter 4). An important mechanical property which opposes this trend is creep, which is the plastic deformation of a material when sustaining an applied elastic load for a long period of time. The resistance of metals against creep deformation increases with their grain size. This occurs because an important process driving creep is sliding of the grain boundaries at high temperature. Increasing the grain size reduces the number of grain boundaries per unit volume of material, and this reduces grain boundary sliding and thereby improves creep resistance. (A complete description of the creep properties of aerospace alloys is given in chapter 22). Therefore, materials that require high creep strength, such as the nickel superalloys used in the hottest sections of jet engines, must have a coarse grain structure. Specialist casting methods are used to produce very large grains during the solidification of superalloys and other metals for high-temperature applications. Two methods used to improve the creep properties of high-temperature metal alloys are directional solidification and single-crystal casting (Fig. 6.10).

Directional solidification involves slowly withdrawing the cast metal from the furnace to produce a coarse, columnar grain structure. Molten metal is rapidly cooled directly onto a chill plate immediately upon leaving the furnace. The chill plate generates a steep temperature gradient along the casting, with the temperature dropping several hundred degrees within a short distance from the exit of the furnace. The chill plate moves away from the furnace at a speed of only a few inches per hour. The sharp temperature gradient and slow solidification rate generate the conditions that allow the solid/liquid interface to advance slowly in the heat flow direction, thus forcing the grains to grow as very long columns. Directionally solidified metal consists of long, thin grains in the solidification direction with very few or no transverse grain boundaries. When the cast metal is externally loaded in the grain direction it is the transverse grain boundaries which cause creep deformation, and minimising their presence by directional solidification improves the creep resistance.

An advance on the directional solidification process is single-crystal casting, which is used to produce metals with exceptional creep resistance. The development of single-crystal casting has led to a 100 °C improvement in temperature capability over conventionally cast alloys with a polycrystalline grain structure. More than 1.5 million high-pressure gas turbine blades made by single-crystal casting are currently in service on civil and military aircraft. Single-crystal casting is increasingly being used for producing nozzle guide vanes. Single-crystal metals are produced by allowing just one grain to grow into the main body of the casting. The mould end has a constriction in the shape of a corkscrew through which only one crystal can pass. A single crystal grows through the constriction and into the main section of the casting without the formation of grain boundaries. The solidified casting consists of

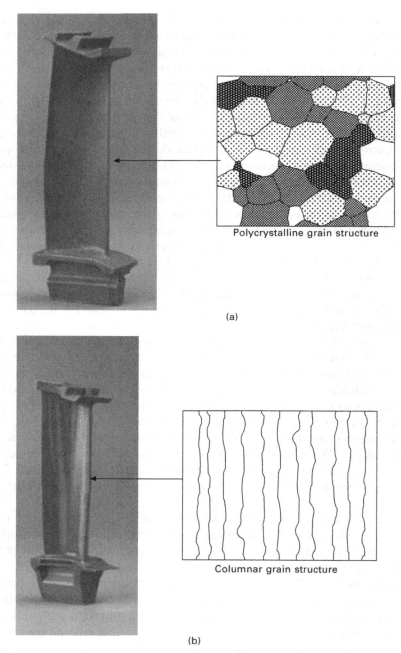

(a)

Polycrystalline grain structure

(b)

Columnar grain structure

6.10 Turbine blade made with nickel-based superalloy using: (a) investment casting resulting in a polycrystalline grain structure, (b) directional solidification resulting in a columnar grain structure and (c) single crystal solidification resulting in no grain boundaries.

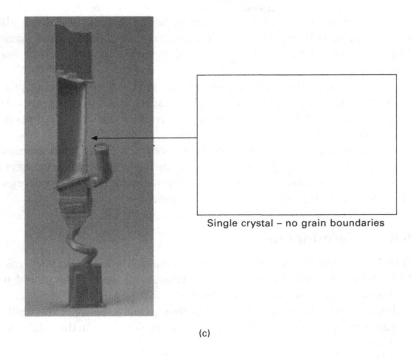

Single crystal – no grain boundaries

(c)

6.10 Continued

a single-crystal metal that has exceptional resistance to creep softening and thermal shock owing to the absence of grain boundaries.

6.5   Summary

The mechanical properties of cast metals are not uniform because of the development of chill, columnar and equiaxed zones during solidification inside a mould. Primary aircraft structures must have uniform properties and, therefore, metal alloys must be homogenised by heat treatment and working operations after casting.

The method used for casting is dependent on several factors, with the most important being the mechanical properties and volume of production. Permanent-mould casting methods generally produce higher strength castings than nonpermanent moulds because of their finer grain structure. Permanent mould methods are suited for a large number of castings whereas nonpermanent moulds are more economical for a small volume production.

A fine grain structure is often desirable in aircraft structural alloys because it results in high strength, ductility, toughness and fatigue resistance. The grain size of cast metals is reduced by rapid cooling and the use of inoculators.

A coarse grain structure is needed in jet engine components to provide high creep resistance. Directional solidification and single-crystal casting methods are often used for producing superalloys for engine components with exceptional creep strength.

Defects that reduce the structural integrity of cast metals must be eliminated. Porosity and macrosegregation are removed by hot working; large intermetallic precipitates are eliminated by solution heat treatment; and microsegregation is overcome by homogenising anneal treatment.

6.6   Terminology

Blowhole: Irregular shaped cavity in cast metal formed by trapped gas.

Chill zone: Region within cast metals (usually closest to the mould wall) where the cooling rate is the fastest.

Columnar zone: Region within cast metals (usually between the chill and equiaxed zones) where the grain structure is elongated in the solidification direction.

Dendrites: Fine branches of solid material that project into the melt zone during solidification.

Equiaxed zone: Region within cast metals (usually the core of the casting) where the grains are approximately the same size in every direction.

Equilibrium freezing temperature: Temperature when solidification of molten material occurs under standard thermodynamic conditions.

Heterogeneous nucleation: Formation of nuclei of solid particles within molten metal that occurs at a pre-existing nucleation site (e.g. mould wall, inoculate particle).

Homogenising anneal: Heating process used to eliminate clustering of alloying elements caused by microsegration and to disperse the elements through the metal. The process is performed on the solid alloy at high temperature, but does not involve melting.

Homogenous nucleation: Formation of nuclei of solid particles within molten metal that occurs without a pre-existing nucleation site.

Hume–Rothery rules: Conditions that determine whether an alloy or impurity element occupies an interstitial or substitutional lattice site. The rules are based on differences in the atomic size, crystal structure and electron valence state of the solute and solvent. Substitutional lattice sites are usually occupied when the atomic size of the solute is within 15% of the solvent and the electron valence number of the solute and solvent is within two. When these conditions are not met the solute is usually interstitial.

Ingot casting: Casting process that uses a mould with a simple shape (usually flat bar) to produce an ingot of the cast metal.

Inoculates (also called grain refiners): Elements or compounds added to molten metal which serve as the nucleation sites of grain growth during solidification. Inoculates are used to refine the grain structure of cast metal.

Limited solubility: Solute can dissolve in solid solvent to a limiting concentration, and at higher amounts the excess of solute forms a second phase within the solvent.

Macrosegregation: Nonuniform distribution of alloying elements in metals above the microstructural level.

Microsegregation: Nonuniform distribution of alloying elements in metals at the microstructural level.

Shape casting: Casting process that uses a mould shaped close to the geometry of the final metal component.

Solution heat treatment: Heating process used to dissolve second-phase particles and disperse the alloying elements evenly throughout the cast metal. Process performed on the solid alloy at high temperature, but does not involve melting.

Undercooling (also called supercooling): Process by which metals remain liquid when cooled below the equilibrium freezing temperature.

Unlimited solubility: Solute can dissolve into the solid solvent to any concentration without the formation of a second phase.

Wormhole: Elongated cavity in cast metal formed by trapped gas.

6.7   Further reading and research

Ammen, C. W., *Metal casting*, McGraw-Hill, 2000.
Cantor, B., and O'Reilly, K., *Solidification and casting*, Institute of Physics, 2003.
Campbell, F. C., *Manufacturing technology for aerospace structural materials*, Elsevier, Amsterdam, 2006.
Lampman, S. (ed.), *Casting design and performance*, ASM International, 2009.
Stefanescu, D. M., *Science and engineering of casting solidification*, Springer, 2009.

6.8   Case study: casting defects causing engine disc failure in United Airlines flight 232

The United Airlines flight 232 accident tragically highlights what happens when defects are present in safety-critical titanium components. Flight 232 was a scheduled flight of a Douglas DC-10 from Denver to Philadelphia on July 19, 1989. The aircraft suffered an uncontained failure of the number two engine (mounted in the tail) while cruising at 38 000 feet. The titanium (Ti–6Al–4V) fan disc, which is the size of a large truck tyre and weighs 150 kg, broke suddenly while spinning at several thousand revolutions per minute. Engine fragments pierced the tail section in many places, including both horizontal stabilizers. Pieces of shrapnel punctured the right horizontal stabilizer and then severed all three hydraulic lines, allowing fluid to quickly drain away causing a loss in the flight control system. The pilots lost hydraulic control of the ailerons, rudder, flaps and other control surfaces. The flight crew performed an amazing feat by flying the aircraft by adjusting power to the two wing-mounted engines and managed an emergency landing at Sioux City Airport, Iowa. The lack of flight control caused the aircraft to crash and break-up on landing, killing 110 of the 285 passengers and 1 of the 11 crew members (Fig. 6.11).

The Federal Aviation Administration launched an immediate investigation into the cause of the engine fan disc failure. The investigation discovered that the disc failed because of an intermetallic particle, called a hard alpha inclusion, in the titanium alloy (Fig. 6.12). An excess of nitrogen, present in the titanium ingot during processing, created the hard alpha inclusion. The inclusion cracked and then fell out during final machining, leaving a small cavity (about the size of a sand grain). A fatigue crack grew from the cavity during normal operation of the engine until eventually, after around 18 years of operation, it was large enough for the titanium disc to catastrophically fail. Titanium is a safe, reliable material for fan discs, and it only failed because the cavity was not found during manufacture and the fatigue crack not detected during routine engine checks. This accident shows the need for stringent quality control procedures when manufacturing aircraft components and the requirement for careful, thorough examination over the aircraft life.

6.11 Wreckage from flight 232.

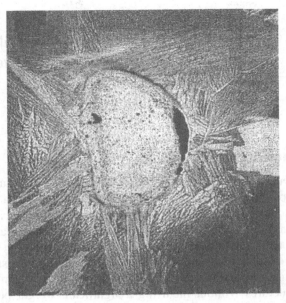

6.12 Hard alpha inclusion in titanium alloy (from F. C. Campbell, *Manufacturing technology for aerospace structural materials*, Elsevier, 2006).

7

Processing and machining of aerospace metals

7.1   Introduction

Metals are not used in their as-cast condition for safety-critical aircraft structures or load-bearing jet engine components. The use of cast metal in the primary structures of commercial airliners is virtually nonexistent, and it is used sparingly in military aircraft. Cast metals are used in a small number of nonstructural components on aircraft and helicopters, but not in heavily loaded structures which require high strength and toughness. The mechanical properties of cast metals are not good enough for them to be used in aircraft structures without incurring a significant weight penalty. Instead, metals are processed after casting by plastic-forming operations which increase the strength properties. Materials processed in this way are called wrought metals.

Wrought metals used in aircraft structures are produced using various processes, with the most common being forging, extrusion, rolling and sheet forming. Other processes used by the aerospace industry are superplastic forming and hot isostatic pressing. Common to all wrought processes is the plastic deformation of the metal which increases strength while simultaneously changing the shape. Most forming processes are designed to simultaneously harden and deform the metal into a simple shape, such as a sheet, plate or bar, which is then heat-treated and machined into the final component. The plastic forming of wrought metal increases the mechanical properties (except creep resistance) by the strengthening mechanisms of work hardening, grain refinement, elimination of casting defects (such as porosity), and the breakdown of alloy segregation in the casting. The forming processes used to fabricate wrought metals are a major operation in aircraft manufacturing.

The superior properties of wrought metals over as-cast metals are shown in Table 7.1 and Fig. 7.1. The yield strength of a wrought metal is typically 50–200% higher than the cast metal with the same alloy composition. Other material properties important for aircraft structures, such as fatigue life and fracture toughness, are also higher for wrought metals. The mechanical properties are not only better for wrought metals, but their properties also tend to be more consistent through a component than as-cast metal. The properties in cast metal components are more variable because of regional

154

Table 7.1 Comparison of typical properties for cast and forged aluminium and magnesium alloys

Material	Condition	Yield strength (MPa)	Ultimate strength (MPa)	Elongation-to-failure (%)
2XXX Aluminium (T6)	Cast	200–230	300–400	1–4
	Forged	300–530	400–580	10–18
7XXX Aluminium (T1)	Cast	130–180	230–250	3–4
	Forged	350–490	500–680	9–14
Magnesium (AZ80)	Cast	80	140	3
	Forged	200	290	6
Magnesium (HK31)	Cast	90	180	4
	Forged	180	255	4

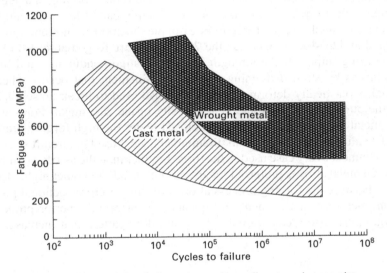

7.1 Comparison of the fatigue properties of cast and wrought titanium alloy (Ti–6Al–4V).

differences in grain structure and segregation of alloying elements. Also, porosity is more likely to be higher in cast metals because casting pores are eliminated in the forming operations used to produce wrought metals. Wrought metal is the material of choice over cast metal in aircraft structures and engines owing to higher and more consistent mechanical properties and fewer casting defects.

In this chapter, we study the main forming processes used by the aerospace industry to produce wrought metals for aircraft structures and jet engine components. The changes to the metallurgical and mechanical properties of metals during the forming operations, including cold and hot working, are examined. Following this, the processes used for machining and drilling the wrought metal into the finished aerospace component are described.

7.2   Metal-forming processes

7.2.1   Forging

Forging is a forming process used to strengthen and shape thick sections of cast metal. The process involves plastically deforming the metal under high compressive forces inside a die cavity, as shown in Fig. 7.2. Deformation usually occurs by repeated strokes or blows applied using hammers, mechanical presses or hydraulic presses. The metal is plastically deformed into the shape of the cavity, and is then removed for machining into the finished part. Forging is performed at room temperature, called cold working, or at elevated temperature to make the metal soft and ductile, called hot working. Most aerospace metals are hot forged; for example the 2000 aluminium alloys are forged at 425–460 °C whereas the 7000 alloys are forged at 400–440 °C.

Forging improves the strength, ductility, fatigue endurance and impact toughness by strain hardening and grain refinement that occurs when the metal is plastically deformed. Forging is also used to break up segregation of the alloying elements and heal porosity in the cast metal. Pores within the metal are squashed and welded closed under the high forging pressure.

The aluminium, magnesium, titanium and steel used in aircraft structures are in almost every case used in the forged (rather than the as-cast) condition. Forged metal alloy is preferred for aircraft bulkheads and other highly loaded parts because the forging process allows for thinner cross-sectional product forms before heat treatment and quenching, enabling superior properties. Forging can also create a favourable grain flow pattern that increases both

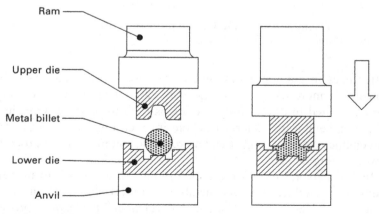

7.2 Schematic of forging.

fatigue life and fracture toughness of the various metal alloys used in aircraft. The only metal type that is not always forged are the nickel-based superalloys used in jet engines because brittle intermetallic inclusions in the metal fracture under the high forming loads and these fractures act as failure initiation sites so that alternative forming processes must be used.

7.2.2   Extrusion

The forming process of extrusion involves pushing metal through a die under high compressive force, as shown in Fig. 7.3. Extrusion is somewhat analogous to squeezing paste out of a toothpaste tube. The cast metal (called a billet), which is hot or cold, is placed inside the die cavity and then squeezed through an annular opening in the die under force applied by a hydraulic ram. The metal effectively squirts out as a continuous bar having the same cross-sectional shape as the die opening. The extrusion is then fed onto a run-out table where it is straightened by stretching and cut to length. Extrusion is used to produce components of uniform cross-section such as beams and rods. Extruded 'T', 'H' 'J' and 'L' shaped sections are used for wing stringers and fuselage frames. High strength aerospace alloys, including 2000 and 7000 aluminium alloys, are referred to as 'hard' alloys because they are difficult to extrude. These extrusions are often performed at high temperature to soften the metal and thereby aid plastic flow through the die opening.

7.2.3   Roll forming

Rolling is a process that plastically deforms thick metal under heavy rollers into thin sheet. Rolling is used to produce high-strength sheet metal for skin panels, stringers, frames and other thin-walled structures in aircraft. Figure 7.4 shows the rolling process, which involves feeding metal stock between

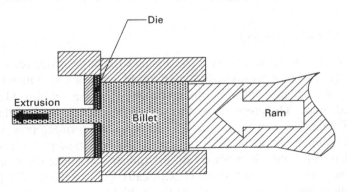

7.3 Extrusion process.

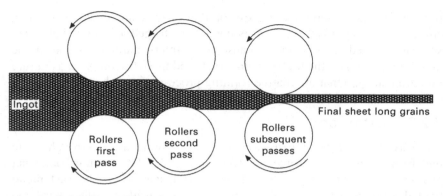

7.4 Rolling of metal plate.

counter-rotating rollers which squash it into a thin, long strip. The rollers are spaced so that the distance between them is slightly less than the reduced thickness of the metal. The rolling process usually begins with hot rolling of the ingot to close any pores and break down the coarse grain structure of the casting. The thickness of the ingot is reduced progressively by a series of rolling operations. The reduction in thickness with each pass between the rollers is determined by the type of material and other factors. A large reduction in thickness with one pass is possible with soft metals whereas multiple passes are required for high-strength materials to achieve the same decrease in thickness. Rolled products are classified as either sheet (with a thickness between 0.15 and 6.3 mm) or plate (thickness greater than 6.3 mm). Widths can range up to about 2500 mm. Rolled aluminium sheet and plate in the range of about 1 to 10 mm is used in the fuselage skin and stringers. Thicker plate (25 to 50 mm) is used for wing skins, and the thickest plate, up to 200 mm, is used for bulkheads, wing spars and supporting structures.

7.2.4   Sheet forming

Sheet forming is used to produce curved panels for large structures such as the fuselage. The process involves clamping the ends of rolled metal sheet (usually thinner than 6 mm) and then stretching over a forming block to the desired shape. The pressure used to stretch the sheet is applied through male or female dies or both. Figure 7.5 shows a sheet forming process that combines a punch (or metal male die) and a rubber female pad to produce curved plate. The sheet is pressed into the rubber pad under force applied through the punch. The flat sheet is plastically deformed into the punch shape without any major change in the sheet thickness.

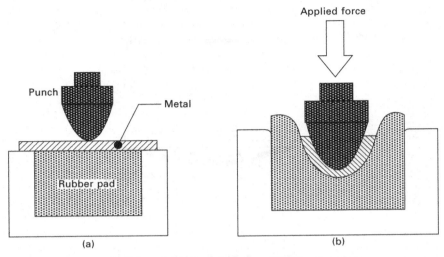

7.5 Sheet forming: (a) before forming; (b) after forming.

7.2.5   Superplastic forming

Superplastic forming is a specialist process used for deforming metal sheet to extremely large plastic strains to produce thin-walled components to the near-net shape. Stretching of the sheet during superplastic forming is much higher than with rolling and sheet forming. Superplastic forming involves stretching the material at least 200% beyond its original size, although the deformation can exceed 1000% with some metals. Figure 7.6 shows the basic steps involved in superplastic forming of metal sheet. The sheet is placed within a die cavity and then heated while high gas pressure is evenly applied to plastically deform the metal at very large strains into a complex shaped, single-piece component. Figure 7.6 illustrates the simplest type of superplastic forming; however, there are many variations to the process depending on the shape and thickness of the part. During superplastic deformation the metal becomes uniformly thinner over the entire part, and does not experience necking (or local thinning) which often leads to tensile fracture. Also, during the forming process, the material does not develop internal cavities, which is another common cause of tensile failure.

Metals must have very high plasticity to be processed by superplastic forming. Metals with superplastic properties have a fine and equiaxed grain structure, typically less than 10–20 μm. The grain structure must be stable at the high temperatures used in the superplastic forming process, and this is often achieved using a dispersion of thermally stable particles in the metal. These particles pin the grain boundaries and thereby retain the fine grain

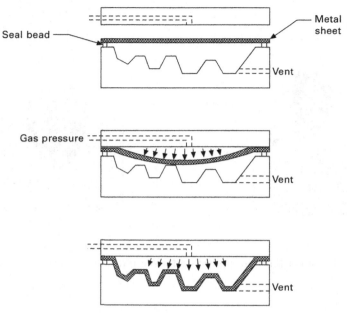

7.6 Superplastic forming process.

structure at high temperature while also providing high resistance to cavitation (void formation). Metals with these properties become superplastic at high temperature, typically one-half of their absolute melting temperature. For example, titanium is superplastic at about 900 °C and aluminium at 450–520 °C. The superplastic metal is soft and ductile at a temperature high enough to allow it to plastically stretch over large strains without tearing or breaking. The mechanism of superplasticity relies on the sliding of grains past each other in a creep-type deformation process together with atomic diffusion processes within the crystal structure that aid plastic flow.

Superplastic forming is used by the aerospace industry as an economical process to fabricate complex components to the near-net shape as a single piece. This reduces the number of individual parts and the assembly cost often associated with making complicated components. For example, a single sheet can be formed into a complex arrangement of ribs, stiffeners and skin as a single-piece part. Superplastic forming is credited with weight and cost savings on virtually every modern aircraft. Figure 7.7 shows an inner door skin for an aircraft made by superplastic forming of aluminium alloy. This component replaced the conventional door component that was built up from 15 to 25 parts. A weight reduction of 20% and cost saving of 40% was achieved using superplastic forming rather than conventional manufacturing of the door panel. However, superplastic forming is not always the most

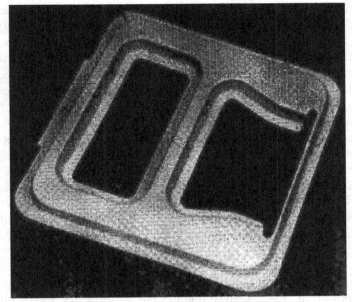

7.7 Superplastic formed access-door panel for an aircraft (reproduced with permission from F. C. Campbell, *Manufacturing technology for aerospace structural materials*, Elsevier, 2006).

economical process. The superplastic forming process is slow and material costs are higher than for conventional alloys. Superplastic forming usually offers an economic advantage when a small to medium number of complex parts, which are normally manufactured with expensive materials having low formability, are required.

7.3   Hot and cold working of metal products

7.3.1   Hot working

Plastic forming of wrought metals by process operations such as forging, extrusion and rolling is performed under hot working or cold working conditions. Hot working is the plastic deformation and shaping of a metal above its recrystallisation temperature. A large number of dislocations are generated by strain hardening during the plastic forming of cast metal. The dislocation density may rise from 10^6–10^7 dislocations cm^{-3} in the cast metal to 10^{11}–10^{12} dislocations cm^{-3} or more after heavy working. A high density of dislocations reduces the ductility and toughness and, in the worst case, can transform the wrought metal into a brittle-like solid. The loss in ductility limits the amount of shape forming before the metal begins to tear and crack.

Hot working is used to retain high ductility in the wrought metal by eliminating dislocations as soon as they form during plastic forming. Hot working can also produce a finer grain structure in the deformed metal owing to recrystallisation, which is a heat-activated process that involves the formation of new grains that are strain-free. The temperature to activate recrystallisation depends on many variables and is not a fixed temperature. Recrystallisation occurs above about 150 °C for aluminium, 200 °C for magnesium and 450 °C for most steels.

Hot working above the recrystallisation temperature is the initial step in the shape forming of most forged, extruded and rolled metals used in aerospace structures. The annihilation of dislocations and formation of strain-free grains provides the hot worked metal with high ductility. This allows thick ingots to be reduced to thin-walled components without tearing or cracking. Wrought metal components produced by hot working have a very fine equiaxed grain structure that provides high strength, fatigue resistance and ductility.

7.3.2   Cold working

Cold working is the plastic shape forming of metals below their recrystallisation temperature, and the process is usually performed at room temperature. During cold working the microstructure is permanently deformed with the grains being elongated in the direction of the applied forming stress, as shown in Fig. 7.8. The grains elongate along certain crystallographic directions in the forming direction which gives a so-called textured or fibrous grain structure. Under high amounts of cold working the grains are stretched into long, thin crystals. An important difference between cold- and hot-working operations is that the texture is retained in the wrought metal after cold working but is largely removed by recrystallisation in hot-worked metal.

The change in grain structure during cold working alters the mechanical properties of the metal. The strength properties increase with the amount of cold working because of the higher density of dislocations formed by strain hardening. However, the strength properties do not increase by the same amount in each direction of the cold worked metal. The strength properties become increasingly more anisotropic (i.e. not equal in all three dimensions) as the grain structure becomes more distorted and elongated in the forming direction.

The three directions of grain orientation in cold worked metal are called:

- longitudinal (L), which is parallel to the direction of working or the main direction of grain flow;
- long transverse (T), which is perpendicular to the direction of working; and

- short transverse (S), which is the shortest dimension of the wrought metal perpendicular to the direction of working.

Figure 7.9 shows these three directions in a cold-worked metal. Hot-worked metal can also have a textured grain structure, although not as severe as cold-worked metal subject to the same amount of plastic deformation.

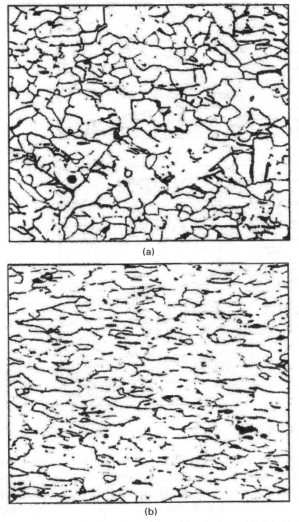

(a)

(b)

7.8 Progressive change from an equiaxed to fibrous (textured) grain structure of steel with increasing amount of cold working: (a) 10% cold work, (b) 30% cold work, (c) 60% cold work and (d) 90% cold work (source: *ASM handbook Vol. 9, metallography and microstructure*, ASM International, Materials Park, OH, 1985).

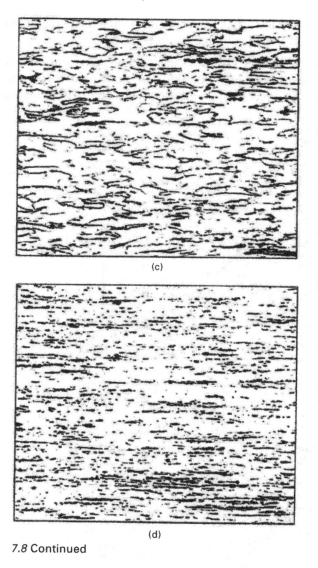

(c)

(d)

7.8 Continued

Figure 7.10 shows the effect of grain texture on the mechanical properties of several aerospace alloys. In general, properties such as yield strength, fatigue life and fracture toughness are highest in the longitudinal and long transverse directions and lowest in the short transverse (or through-thickness) direction. Highly textured grain structures also have an impact on the corrosion resistance of high-strength aluminium alloys. The longitudinal direction is the most resistant to stress corrosion cracking, followed by the long transverse direction, with the short transverse being the most susceptible to corrosion.

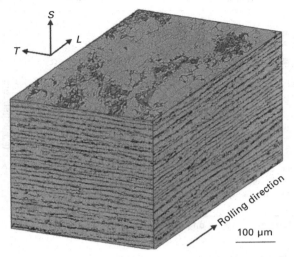

7.9 Textured grain structure showing the longitudinal (*L*), long transverse longitudinal (*T*) and short transverse (*S*) directions (from A. A. Korda, Y. Mutoh, T. Sadasue and S. L. Mannan, 'In situ observation of fatigue crack retardation in banded ferrite-pearlite microstructure due to crack branching', *Scripta Materialia*, **54**, 2006, 1835–1840).

The anisotropic properties of worked metals have important practical significance to their use in aerospace structures. Metals should be orientated in aircraft structures to ensure that the direction of highest mechanical performance (which is usually the longitudinal and long transverse directions) is parallel to the principal loading direction. For example, aircraft designers take advantage of the ~70 MPa higher strength of extruded aluminium products used in wing stringers and spars relative to rolled plate used for wing covers.

During cold-working operations, a small amount of the applied forming stress, perhaps about 10%, is stored as elastic strain energy within the tangled network of dislocations. This stored energy is called residual stress, and the magnitude of this stress increases with the amount of cold working. Residual stress within metal aircraft components may be beneficial or undesirable depending on the type of stress. The residual stress is not evenly distributed throughout the cold worked metal; compressive stresses occur at some locations and balancing tensile stresses occur elsewhere. Compressive residual stresses are advantageous, particularly when present at the surface or sites of stress concentration (e.g. fastener holes, cut-outs), because they counteract applied tension loads which cause fatigue cracking problems in metal structures. When the residual stress is tensile, however, it can facilitate the initiation and growth of fatigue damage. Residual stress should be removed from the metal

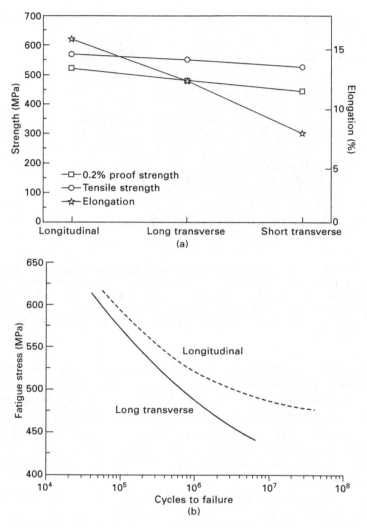

7.10 Examples of anisotropic mechanical properties in cold-worked alloys used in aircraft structures: (a) tensile properties of 7075 Al (from P. J. E. Forsyth and A. Stubbington, *Metals Technology*, **2** (1975), 158); (b) fatigue life of titanium alloy (Ti–6Al–4V) (from Bowen, A. W., in *Titanium science and technology*, eds R. J. Jaffee and H. M. Burte, Plenum Press, New York, Vol 2, 1973).

component before it is built into the aircraft. Residual stress is relieved by a heat treatment process called stress-relief anneal, which involves heating the cold-worked metal to moderate temperature to allow the highly strained crystal lattice to relax.

7.4   Powder metallurgy for production of aerospace superalloys

Forging, extrusion, rolling and other working operations are used in the forming and shaping of aluminium, magnesium, titanium and steel structural components for aerospace applications. However, it is difficult to shape nickel-based superalloys for jet engine components using these same working operations. Superalloys are so-named because of their very high alloy content; typically 40–60% of alloying elements (such as iron, chromium and molybdenum) in a nickel matrix. The high concentration of alloying elements is needed to create a large volume fraction of hard intermetallic precipitate particles. These particles provide the superalloy with high mechanical performance at the operating temperatures of jet engines. Owing to these particles, however, superalloys cannot be shaped using conventional working operations without tearing and cracking.

Since the mid-1970s, the aerospace industry has manufactured superalloy components using a powder metallurgy process called hot isostatic pressing (HIP). Powder metallurgy is a fabrication technique that involves three major processing stages: (i) production of metal powder, (ii) compaction and shaping of the powder, and (iii) consolidation and fusing of the powder into a solid metal component under high temperature and pressure. The first process step involves the production of fine spherical superalloy powder using a gas atomisation process. Powder is produced by pouring molten superalloy through a narrow hole to produce a thin liquid stream. High-pressure argon gas is blown into the metal to break up the stream into tiny droplets which rapidly solidify at $\sim 10^6$ °C s^{-1}. The fine spherical superalloy powder is then injected into a high pressure container to form a weak compact with a shape close to the final component. Air within the container is removed to avoid any contamination of the powder during the final processing stage of hot isostatic pressing. The powder is consolidated in the HIP process under high temperature (1100–1300 °C) and high pressure (15 000–45 000 psi). Pressure is applied to the compacted powder from all directions, hence the term 'isostatic pressing'. The high temperature and pressure fuse the superalloy particles into a dense solid by plastic flow and diffusion bonding. The finished component has a fine grain structure that is virtually free of porosity.

One of the main advantages of the HIP process is the ability to fabricate components to the near-net shape that requires little machining. Conventional working operations do not usually form products to the near-net shape, and for many aircraft components anywhere from 70–90% of the material must be removed by machining to achieve the final shape. A component made by HIP typically requires the removal of only 10–20% of the material and, as a result, the machining time and cost are reduced.

7.5   Machining of metals

The final stage in the fabrication of aircraft metal structures involves the machining operations of milling, routing, trimming and drilling. Metal components are rarely in their final shape at the end of processing, and it is necessary to remove excess material using a variety of machining processes. Increasingly, these processes are being carried out automatically using numerically controlled machines to increase production rates and accuracy. Some components are fabricated to the near-net shape and require only a small amount of material (under 5–10% of the weight) to be removed. Components with intricate shapes can require a large amount of machining (up to 90%) and the milling, routing, trimming and drilling processes which are used are critical to the production rate and quality. Material removal generally becomes more difficult with increasing hardness, and the machining of hard metals such as titanium is performed using specialist methods such as laser assisted milling. Table 7.2 ranks materials in order of increasing difficulty and cost of machining.

Milling is a common machining process that involves the use of a multi-tooth cutter to remove metal from the workpiece surface to create flat and angular surfaces and grooves. Routing refers to the shaping of apertures or edge trimming of components in flat and shaped panels, and is performed using a cylindrical, flat-nose cutting tool. Trimming involves removing excess material around the edges of a fabricated component, and is performed using various methods from simple techniques such as cutting blades and routers to precision methods such as high-powered laser cutting and water-jet cutting.

Drilling is a major operation in the pre-assembly of aircraft because of the large number of fasteners, bolts, screws and rivets required to connect

Table 7.2 Machining energy requirements and machining costs for various metals and composites used in aircraft

Material	Yield strength (MPa)	Approximate machining energy (W s)/mm^3	Machining cost index*
Magnesium alloy	180–280	0.7	0.4
Aluminium alloy	450–550	0.9	0.4
1020 steel (finished)	300–400	1.5	1.0
Carbon–epoxy composite	800–1000	2.7	2.2
4140 steel (hot rolled)	500–600	3.0	2.3
Stainless steel (PH)	1300–1500	3.0	5.1
Maraging steel	1700–2300	3.5	3.4
Titanium alloy (Ti-6Al-4V)	900–1000	3.5	5.1
Nickel superalloy	550–800	4.0	3.4

*Machining cost index relative to high-strength aluminium alloy. If it costs $1 to machine aluminium then it costs $3.40 to machine the maraging steel.

the many airframe components. Drilling of panels and subsequent fastener installation is one of the more time-consuming operations in aerospace manufacture (for both metals and composites). Drilled holes are also a common source of damage in aircraft structures, particularly cracks growing from fastener holes, and therefore correct drilling and fastener installation is essential for structural integrity. A commercial airliner can have as many as 1.5 to 3 million fasteners whereas a typical fighter aircraft might have up to 300 000. Drilling is broadly classified as either hand, power feed, automated drilling or automated riveting. The majority of fastener holes in aircraft structures are produced using automated drilling equipment, which ensures high accuracy in hole tolerance and rapid production. Automated drilling units are custom designed and built for specific aircraft structures. The drilling is automatically controlled with the capability to change speed and feed rates when boring through different materials.

Hole drilling of most metals is performed using standard twist drill bits. High hardness steel bits are used for drilling aluminium and magnesium components. Hard-coated or cobalt grades of high-strength steel are used in drills for hard metals such as titanium alloys to ensure long drill life. Drilling is performed at a carefully controlled speed to avoid overheating and other damage to the hole. As a general rule, the rotational speed of the drill bit should be reduced as the hole size gets larger, or when the metal gets thicker or harder. Typical maximum drill speeds for aluminium range from 1000 rpm for 12 mm wide holes to 5000 rpm for 4 mm holes. Titanium alloys are drilled at slower speeds because of their higher hardness: 150 rpm for 12 mm holes up to 700 rpm for 4 mm holes.

Automated riveting machines are widely used throughout the aerospace industry, and have the capability to drill the hole, inspect the hole for quality and tolerance, grip the rivet, apply sealant over the rivet (if required), and then install the rivet by squeezing. Drilling and fastener installation can be achieved in just a few seconds for each hole using automated processes involving high-speed robotic machines.

After the hole is drilled it is deburred (if necessary) then cleaned of any swarf and other contamination before the fastener is installed. Many different fasteners are used in aircraft assembly, with the most prevalent being solid rivets, pins with collars, bolts with nuts, and blind fasteners. The selection of the fastener type depends on its ability to transfer load across the joint connection, compatibility with the metal structure (to avoid corrosion damage), and ease of installation. Aluminium rivets are by far the most widely used fasteners for joining metal structures. Rivets should only be used in joints that are primarily loaded in shear; tension loads must be avoided to minimise the risk of rivet pull-out. The most common alloy used in rivets is 2117-T4 aluminium (called an AD rivet). Other types of aluminium rivets, such as 2024Al-T4 and 7075-T73, are used instead of the AD rivet when higher strength is required. Steel fasteners are used in undercarriage and other highly loaded components.

Fatigue cracking from fastener holes is a major problem with old aircraft. The aerospace industry uses both cold working of the fastener hole and interference fit fasteners to resist fatigue damage by forming a residual compressive stress in the metal surrounding the hole. Cold work is introduced by pulling a mandrel through the hole that expands sufficiently to plastically deform by compressing the material surrounding the hole. An interference fit fastener is then installed into the cold-worked hole. This type of fastener expands during installation and thereby exerts a compressive stress onto the material bordering the hole. The ability of cold working and interference fit fasteners to improve the fatigue life of metals is shown in Fig. 7.11 for an aircraft-grade aluminium alloy. These methods can increase the fatigue life above that of the parent metal structure (without a hole) because of the residual compressive stress resisting fatigue cracks that grow under tension loads.

7.6   Summary

Metal alloys for primary aircraft structures and engine components are used in the wrought condition, and not the as-cast condition. Wrought metals generally have superior mechanical properties to cast metals, including higher yield strength, fatigue life and toughness. Furthermore, the properties

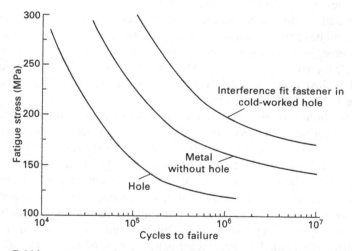

7.11 Improvement in the fatigue life of 2024-T851 aluminium owing to cold working and interference fit fasteners (adapted from F. C. Campbell, *Manufacturing technology for aerospace structural materials*, Elsevier, 2006).

of wrought metals are more consistent owing to the elimination of casting defects such as porosity and segregation of alloying elements.

Wrought aluminium, magnesium, titanium and steel used in aircraft structures are shaped using the processes of forging, extrusion, rolling and sheet forming. Nickel superalloys used in jet engines are susceptible to tearing when shaped using these processes. Instead, superalloys are processed using powder metallurgy methods that involve hot isostatic pressing.

High-strength metals tear and crack when plastically deformed at room temperature by cold-working operations. High-strength aerospace metals are often hot worked to maintain high ductility. Hot working eliminates dislocations created by the working operation and produces strain-free grains.

The mechanical properties of cold-worked metals are anisotropic owing to the texture of the grain structure. The grains are stretched and elongated in the forming direction, causing the properties to be different in the three directions of the worked metal. Textured metals used in aircraft structures should be orientated to ensure the direction with the highest mechanical properties is parallel with the main loading.

7.7   Terminology

Hot working: Plastic deformation performed above the recrystallisation temperature of the material.

Longitudinal grain direction. The orientation along which a material, usually in the form of a forged sheet or plate, is principally extended during a working operation.

Long transverse grain direction: The orientation along the plane of a material which is normal to the principal working direction.

Recrystallisation: The process of forming a new grain structure in a solid material, usually during heat treatment of a work-hardened material.

Recrystallisation temperature: The minimum temperature at which recrystallisation (new grain growth) commences.

Short transverse: The through-thickness direction of a material during a working operation.

Wrought metal: Metal that has been shaped by a plastic forming process (e.g. rolling, extrusion), as opposed to casting.

Stress-relief anneal: Heating process used to remove residual stress in a component.

Residual stress: The stress remaining in a component after all external influences (e.g. working, temperature) have been removed. Residual stresses exist as an internal strain system with compressive stress at one location balancing tensile stress at another location. Residual stress can be induced in a component by treatments such as solidification, plastic deformation, machining, and heating or cooling.

7.8   **Further reading and research**

Black, J. T., DeGarmo, E. P. and Kohser, R. A., *DeGarmo's materials and processes in manufacturing,* Wiley, 2007.

Campbell, F. C., *Manufacturing technology for aerospace structural materials*, Elsevier, Amsterdam, 2006.

Dieter, G. E., *Mechanical metallurgy*, McGraw–Hill, London, 1988.

Groover, M. P., *Fundamentals of modern manufacturing*, John Wiley & Sons, Inc., 2010.

Lascoe, O. D., *Handbook of fabrication processes*, ASM International, 1988.

8

Aluminium alloys for aircraft structures

8.1   Introduction

Aluminium has been an important aerospace structural material in the development of weight-efficient airframes for aircraft since the 1930s. The development of aircraft capable of flying at high speeds and high altitudes would have been difficult without the use of high-strength aluminium alloys in major airframe components such as the fuselage and wings. Compared with other major aerospace materials, such as magnesium, titanium, steel and fibre-reinforced polymer composite, aluminium is used in greater quantities in the majority of aircraft. Aluminium accounts for 60–80% of the airframe weight of most modern aircraft, helicopters and space vehicles. Aluminium is likely to remain an important structural material despite the growing use of composites in large passenger airliners such as the Airbus 380 and 350XWB and the Boeing 787. Around 400 000 tonne of aluminium is used each year in building military and civil aircraft. Many types of airliners continue to be constructed mostly of aluminium, including aircraft built in large numbers such as the Boeing 737, 747 and 757 and the Airbus A320 and A340. Competition between aluminium and composite as the dominant structural material is likely to intensify over the coming years, although aluminium remains central to weight-efficient airframe construction.

Figure 8.1 shows examples of the present day use of aluminium alloys in passenger and military aircraft: the Boeing 747 and *Hornet* F/A-18. The weight percentages of structural materials used in the Boeing 747 is typical of passenger aircraft built between 1960 and 2000. Most of the airframe is constructed using high-strength aluminium alloys, and only small percentages of other metals and composites are used. Aluminium is used in the main structural sections of the B747, including the wings, fuselage and empennage. The only major components not built almost entirely from aluminium are the landing gear (which is made using high-strength steel and titanium) and turbine engines (which are made using various heat-resistant materials including nickel-based superalloys and titanium).

The amount of aluminium alloy used in modern fighters varies considerably between aircraft types, although many are built almost entirely using aluminium. In general, the percentage of the airframe made using aluminium is lower for military attack aircraft than for civil aircraft. The airframe of most modern military aircraft consists of 40–60% aluminium, which is less

173

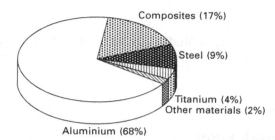

(a)

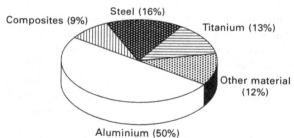

(b)

8.1 Use of aluminium alloys and other structural materials in (a) the Boeing 747 and (b) *Hornet* F/A-18. (a) Photograph supplied courtesy of M. Tian. (b) Photograph supplied courtesy of M. Nowicki.

than the 60–80% used in commercial airliners. The *Hornet* is typical of military aircraft built over the past thirty years in that aluminium is the most common structural material.

Aluminium is a popular aerospace structural material for many important reasons, including:

- moderate cost;
- ease of fabrication, including casting, forging and heat-treatment;
- light weight (density of only 2.7 g cm^{-3});
- high specific stiffness and specific strength;
- ductility, fracture toughness and fatigue resistance; and
- good control of properties by mechanical and thermal treatments.

As with any other aerospace material, there are several disadvantages of using aluminium alloys in aircraft structures including:

- low mechanical properties at elevated temperature (softening occurs above ~150 °C);
- susceptibility to stress corrosion cracking;
- corrosion when in contact with carbon-fibre composites; and
- age-hardenable alloys cannot be easily welded.

This chapter describes aluminium alloys, with special attention given to those alloys that are used in aircraft, helicopters, spacecraft and other aerial vehicles. The following aspects of aluminium alloys are covered: the benefits and problems of using aluminium alloys in aircraft; the different grades of aluminium alloys and the types used in aircraft; and engineering properties of aerospace aluminium alloys and their heat treatment.

8.2   Aluminium alloy types

8.2.1   Casting and wrought alloys

Aluminium alloys are classified as casting alloys, wrought non-heat-treatable alloys or wrought heat-treatable alloys. Casting alloys are used in their as-cast condition without any mechanical or heat treatment after being cast. The mechanical properties of casting alloys are generally inferior to wrought alloys, and are not used in aircraft structures. Casting alloys are sometimes used in small, non-load-bearing components on aircraft, such as parts for control systems. However, the use of casting alloys in aircraft is rare and, therefore, these materials are not discussed.

Nearly all the aluminium used in aircraft structures is in the form of wrought heat-treatable alloys. The strength properties of wrought alloys can be improved by plastic forming (e.g. extrusion, drawing, rolling) and heat treatment. Heat treatment, in its broadest sense, refers to any heating and cooling operation used to alter the metallurgical structure (e.g. crystal structure, grain size, dislocation density, precipitates), mechanical properties (e.g. yield strength, fatigue resistance, fracture toughness), environmental durability (e.g. corrosion resistance, oxidation resistance) or the internal residual stress state. However, when the term 'heat treatment' is applied to wrought aluminium alloys it usually implies that heating and cooling operations are used to increase the strength via the process called age (or precipitation) hardening.

There are two major groups of wrought aluminium alloys: non-age-hardenable and age-hardenable alloys. The distinguishing characteristic of non-age-hardenable alloys is that when heat treated they cannot be strengthened by precipitation hardening. These alloys derive their strength from solution solid strengthening, work hardening and refinement of the grain structure. The yield strength of most non-age-hardenable alloys is below about 300 MPa, which is inadequate for aircraft structures. Age-hardenable alloys are characterised by their ability to be strengthened by precipitation hardening when heat treated. These alloys achieve high strength from the combined strengthening mechanisms of solid solution hardening, strain hardening, grain size control and, most importantly, precipitation hardening. The yield strength of age-hardenable alloys is typically in the range of 450 to 600 MPa. The combination of low cost, light weight, ductility, high strength and toughness makes age-hardenable alloys suitable for use in a wide variety of structural and semistructural parts on aircraft.

8.2.2   International alloy designation system

There are over 500 different aluminium alloys, and for convenience these are separated into categories called alloy series. The International Alloy Designation System (IADS) is a classification scheme that is used in most countries to categorise aluminium alloys according to their chemical composition. This system is used by the aerospace industry to classify the alloys used in aircraft. All aluminium alloys are allocated into one of eight series that are given in Table 8.1. The main alloying element(s) is used to determine into which one of the eight series an alloy is allocated. The main alloying element(s) for the different series are given in Table 8.1. The 8000 series is used for those alloys that cannot be allocated to the other series, although the principal alloying element is usually lithium.

Each alloy within a series has a four-digit number: XXXX. The first digit indicates the series number. For example, 1XXX indicates it is in the 1000 series, 2XXX is a 2000 series alloy, and so forth. The second digit indicates the number of modifications to the alloy type. For example, with the alloy 5352 Al the second digit (3) indicates that the alloy has been modified three times, but has a similar composition to earlier versions 5052 Al, 5152 Al and

Table 8.1 Wrought aluminium alloy series

Alloy series	Main alloying element(s)	
1000	Commercially pure Al (>99% Al)	Not age-hardenable
2000	Copper	Age-hardenable
3000	Manganese	Not age-hardenable
4000	Silicon	Age-hardenable (if magnesium present)
5000	Magnesium	Not age-hardenable
6000	Magnesium and silicon	Age-hardenable
7000	Zinc	Age-hardenable
8000	Other (including lithium)	Mostly age-hardenable

5252 Al. The last two numbers in the four-digit system only have meaning for the 1000 series alloys. In this series, the last two digits specify the minimum purity level of the aluminium. As examples, 1200 Al has a minimum purity of 99.00% and 1145 Al is at least 99.45% pure. The last two digits in the 2000 to 8000 series has no meaningful relationship to the alloy content and serves no purpose other than to identify the different alloys in a series.

When an alloy is being developed it is prefixed with an X to signify it has not yet been fully evaluated and classified by the IADS. For example, the alloy X6785 indicates it is a new 6000 series alloy that is being tested and evaluated. When the evaluation process is complete, the prefix is dropped and the alloy is known as 6785 Al.

Most countries use the IADS to classify aluminium alloys. However, some nations use a different classification system or use the IADS together with their own system. For example, in the UK the IADS is used by the aerospace industry, although sometimes the British Standards (BS) system is also used to classify aluminium alloys. There are three principal types of specifications used in the UK: (i) BS specifications for general engineering use, (ii) BS specifications for aeronautical use (designated as the L series), and (iii) DTD (Directorate of Technical Development) specifications for specialist aeronautical applications.

8.2.3   Temper designation system

A system of letters and numbers known as the temper designation system is used to indicate the type of temper performed on an aluminium alloy. Temper is defined as the forging treatment (e.g. cold working, hot working) and thermal treatment (e.g. annealing, age-hardening) performed on an aluminium product to achieve the desired level of metallurgical properties. The temper designation system has been approved by the American Standards Association, and is used in the USA and most other countries. The system is applied for all wrought and cast forms of aluminium (except ingots).

Basic temper designations consist of individual capital letters, such as 'F' for as-fabricated and 'T' for age-hardened. Major subdivisions of basic tempers, when required, are indicated by one or more numbers. The letters and numbers commonly used to describe the temper of aluminium alloys are given in Table 8.2. The temper designation follows the alloy designation, and is separated from it by a hyphen. As examples, 1100 Al-O means the aluminium has been heat treated by annealing and 2024 Al-T6, shown in Table 8.3, means that the alloy has been tempered to the T6 condition, which involves solution treatment followed by artificial ageing as given in Table 8.2.

Table 8.2 Temper designations for aluminium alloys

F	As-fabricated (eg. hot-worked, forged, cast, etc.)
O	Annealed (wrought products only)
H	Cold-worked (strain hardened) H1x – cold-worked only (x refers to amount of cold working and strengthening) H2x – cold-worked and partially annealed H3x – cold-worked and stabilised at a low temperature to prevent age-hardening
W	Solution-treated
T	Age-hardened T1 – cooled from fabrication temperature and naturally aged T2 – cooled from fabrication temperature, cold-worked and naturally aged T3 – solution treated, cold-worked and naturally aged T4 – solution treated and naturally aged T5 – cooled from fabrication temperature and artificially aged T6 – solution-treated and artificially aged T7 – solution-treated and stabilised by over-ageing T8 – solution-treated, cold-worked and artificially aged T9 – solution-treated, artificially aged and cold-worked T10 – cooled from fabrication temperature, cold-worked and artificially aged

Table 8.3 Composition of 2000 alloys used in aircraft

Alloy	Cu	Mg	Zn	Mn	Cr (max)	Si (max)	Fe (max)
2017	3.5–4.5	0.4–0.8	0.25	0.4–1.0	0.1	0.8	0.7
2018	3.5–4.5	0.45–0.9	0.25	0.2	0.1	0.9	1.0
2024	3.8–4.9	1.2–1.8	0.3	0.3–0.9	0.1	0.5	0.5
2025	3.9–5.0	0.05	0.3	0.4–1.2	0.1	1.0	1.0
2048	2.8–3.8	1.2–1.8	0.25	0.2–0.6		0.15	0.2
2117	2.2–3.0	0.2–0.5	0.25	0.2	0.1	0.8	0.7
2124	3.8–4.0	1.2–1.8	0.3	0.3–0.9	0.1	0.2	0.3

8.3   Non-age-hardenable aluminium alloys

The use of non-age-hardenable wrought alloys in aircraft is limited because they lack the strength, fatigue resistance and ductility needed for structural components such as skin panels, stiffeners, ribs and spars. The proof strength of most tempered non-age-hardenable alloys is below 225 MPa, which is inadequate for highly stressed aircraft structures. However, these alloys are used in some nonstructural aircraft parts and therefore it is worthwhile to briefly examine these materials.

The 1000, 3000, 5000 and most of the 4000 alloys cannot be age-hardened via heat-treatment processing. Solid solution strengthening, strain hardening and grain size control determine the strength of non-age-hardenable alloys. The improvement in strength achieved by solid solution strengthening is modest because most of the alloying elements have low solubility limits in aluminium at room temperature. The inability to dissolve large amounts of alloying elements into either interstitial or substitutional sites within the aluminium crystal structure means that very little strengthening can be achieved. Figure 8.2 shows the increase to the yield strength of high-purity aluminium from solid solution strengthening with several important alloying elements. Only a small improvement in strength is achieved. Strain hardening and grain size control are more effective mechanisms for strengthening non-age-hardenable alloys. For example, Fig. 8.3 shows the improvement in yield and ultimate tensile strengths of pure aluminium with percentage cold-working.

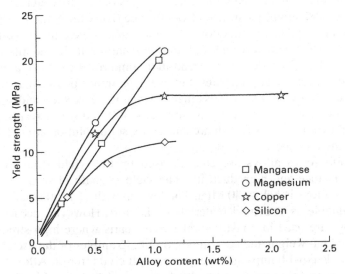

8.2 Effect of different alloying elements on the solid solution strengthening of pure aluminium in the annealed condition.

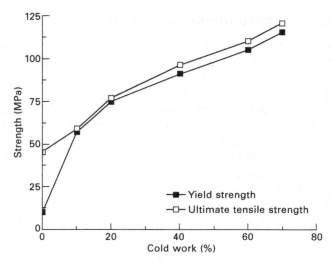

8.3 Effect of amount of cold working (strain hardening) on the yield and ultimate tensile strengths of pure aluminium.

8.3.1   1000 series aluminium alloys

The 1000 series is the highest purity form of aluminium alloys. An alloy is classified in the 1000 series when its aluminium content is greater than 99%. Only rarely are very small amounts of alloying elements deliberately added to 1000 metals. Instead, the trace amounts of impurity elements (e.g. Cu, Fe) are inadvertently extracted from the ore (bauxite) in the production of aluminium metal after extraction, refinement and processing. It is difficult and expensive to completely remove impurities, and so small amounts are left in the metal. Despite their low concentration, impurities can have a marked effect on the mechanical properties. For example, impurities such as copper, silicon and iron in amounts of less than 1% can increase the yield strength and ultimate tensile strength by as much as 300 and 40%, respectively. These improvements in strength are caused by solid solution strengthening and refinement of the grain structure.

The 1000 series alloys are characterised by low yield strength, poor fatigue resistance and high ductility. The yield strength of most annealed 1000 series alloys is below 40 MPa. The low strength of 1000 alloys makes them unsuitable as structural materials on aircraft. However, occasionally these alloys are used in nonstructural aircraft parts where high strength is not required but weight and cost are important. Examples of the use of 1000 alloys include cowl bumps and scoops on small civil aircraft. Alloy 1100 is sometimes used for aircraft fuel tanks, fairings, oil tanks, and in the repair of wing tips and tanks.

8.3.2   3000 series aluminium alloys (Al–Mn)

The main alloying element in this series is manganese, which provides some improvement in strength by solid solution hardening. It is not possible to age-harden 3000 alloys because manganese does not form precipitates in aluminium upon heat treatment. The yield strength of most 3000 alloys is below 200 MPa and, for this reason, they are rarely used in aircraft. The alloy 3003 Al, which contains 1.2% Mn, is used as a nonstructural material in fairing, fillets, tanks, wheel pants, nose bowls and cowlings in some aircraft. 3000 alloys are mostly used in non-aerospace components, such as automotive components (e.g. radiators, interior panels and trim).

8.3.3   4000 series aluminium alloys (Al–Si)

The 4000 alloys contain significant amounts of silicon. These alloys cannot be strengthened by heat treatment, unless magnesium is present to form high-strength precipitates (Mg_2Si). The use of 4000 alloys in aircraft is limited because a brittle silicon phase can form in the aluminium matrix which reduces the ductility and fracture toughness. 4000 alloys are used in non-aerospace applications, in particular as brazing and welding filler materials.

8.3.4   5000 series aluminium alloys (Al–Mg)

The main alloying element in these alloys is magnesium, which is usually present in concentrations of a few percent. Magnesium forms hard intermetallic precipitates in aluminium (Mg_2Al_3) that increase the strength of the alloy. However, the formation and growth of these precipitates cannot be controlled by heat treatment, and therefore the 5000 series alloys are not age-hardenable. As with the other non-age-hardenable alloys, 5000 alloys are used occasionally in nonstructural aircraft parts. For example, the alloy 5052 Al, which contains 2.5% Mg and 0.25% Cr, is one of the highest strength non-age-hardenable alloys that is available, and is used in wing ribs, wing tips, stiffeners, tanks, ducting and framework.

8.4   Age-hardenable aluminium alloys

The 2000, 6000, 7000 and many 8000 alloys can be strengthened by age-hardening. It is only by age-hardening that aluminium alloys obtain the strength needed for use in highly loaded structures and, therefore, this process is critical in the construction of aircraft. We first examine the composition and uses of the age-hardenable alloys used in aircraft, and then examine the age-hardening process.

8.4.1   2000 series aluminium alloys (Al–Cu)

The 2000 alloys are used in many structural and semistructural components in aircraft. The main alloying element is copper, which readily forms high-strength precipitates when aluminium is age-hardened by heat treatment. 2000 alloys are characterised by high strength, fatigue resistance and toughness. These properties make the alloys well suited for fuselage skins, lower wing panels and control surfaces.

There are many types of 2000 alloys, but only a few are used in aircraft structures. One of the most common is 2024 Al (Al–4.4Cu–1.5Mg), which has been used for many years in aircraft structures such as stringers, longerons, spars, bulkheads, carry-throughs, stressed skins and trusses. 2000 alloys are used in damage-tolerant applications, such as lower wing skins and the fuselage structure of commercial aircraft which require high fatigue resistance. The alloy is also used in nonstructural parts such as fairings, cowlings, wheel pants and wing tips. Newer alloys are being introduced with superior properties to 2024 Al. For instance, 2054 Al is 15–20% higher in fracture toughness and twice the fatigue resistance of 2024 Al. Other 2000 series used in aircraft include 2018 Al, 2025 Al, 2048 Al, 2117 Al and 2124 Al. Reducing impurities, in particular iron and silicon, has resulted in higher fracture toughness and better resistance to fatigue crack initiation and crack growth. The composition and mechanical properties of 2000 alloys used in aircraft are given in Tables 8.3 and 8.4.

The alloying elements provide important properties that aid in the processing or strengthening of aluminium. Cu, Mg and Zn provide high strength through solid solution strengthening and precipitation hardening. These elements react with aluminium during heat treatment to create intermetallic precipitates (e.g. $CuAl_2$, Al_2CuMg, $ZnAl$) that increase the strength and fatigue resistance. Mn and Cr are present in small amounts to produce dispersoid particles (e.g. $Al_{20}Cu_2Mn_3$, $Al_{18}Mg_3Cr_2$) that restrict grain growth and thereby increase the

Table 8.4 Tensile properties of 2000 alloys used in aircraft

Alloy	Temper	Yield strength (MPa)	Tensile strength (MPa)	Elongation (%)
2017	T4	275	425	22
2018	T61	320	420	12
2024	T4	325	470	20
2024	T6	385	475	10
2024	T8	450	480	6
2025	T6	255	400	19
2048	T85	440	480	10
2117	T4	165	300	27
2124	T8	440	480	6

yield strength by grain boundary hardening. The addition of trace amounts (0.1–0.2%) of Ti also reduces the grain size. Si is added to reduce the viscosity of molten aluminium, thus making it easier to cast into thick and complex shapes that are free of voids. Fe is used to reduce hot cracking in the casting. However, Si and Fe form coarse intermetallic particles (Al_7Cu_2Fe, Mg_2Si) which lower the fracture toughness, and, therefore, the amount of these elements is kept to a low concentration.

8.4.2   6000 series aluminium alloys (Al–Mg–Si)

The principal alloying elements in the 6000 series are magnesium and silicon. 6000 alloys can be age-hardened with the formation of Mg_2Al_3 and Mg_2Si precipitates. 6000 alloys are used in a wide range of non-aerospace components, such as buildings, rail cars, boat hulls, ship superstructures and, increasingly, in automotive components. However, these alloys are rarely used in aircraft because of their low fracture toughness. 6061 Al (Al–1%Mg–0.6Si) is used occasionally in wing ribs, ducting, tanks, fairing and framework, although this alloy is one of very few used in aircraft.

8.4.3   7000 series aluminium alloys (Al–Cu–Zn)

The 7000 alloys together with the 2000 alloys represent by far the most common aluminium alloys used in aircraft. The main alloying elements in 7000 alloys are copper and zinc, with the zinc content being three to four times higher than the copper. Magnesium is also an important alloying element. These elements form high-strength precipitates [$CuAl_2$, Mg_2Al_3, $Al_{32}(Mg, Zn)_{49}$] when aluminium is age-hardened.

7000 alloys generally have higher strength than 2000 alloys. The yield strength of the 7000 alloys used in aircraft is typically in the range 470 to 600 MPa as opposed to the 2000 alloys, which are between about 300 and 450 MPa. 7000 alloys are therefore used in aircraft structures required to carry higher stresses than 2000 alloy components, such as upper wing surfaces, spars, stringers, framework, pressure bulkheads and carry-throughs. The 7000 alloy most often used in aircraft is 7075 Al. Other 7000 alloys used in aircraft structures include 7049 Al, 7050 Al, 7079 Al, 7090 Al, 7091 Al, 7178 Al and 7475 Al. The composition and properties of these aluminium alloys are given in Tables 8.5 and 8.6. Figure 8.4 shows the new types of aluminium alloys and tempers used in the fuselage and wings of the Boeing 777. New high-toughness aluminium alloys for fuselage skins have enabled significant weight reductions through removal of some circumferential frames. The B777 is typical of most modern aircraft in that it uses both conventional and new aluminium alloys. The new alloy is usually superior in one or two properties over the conventional alloy. For example, the fuselage skin material is an

Table 8.5 Composition of 7000 alloys used in aircraft

Alloy	Cu	Zn	Mg	Mn	Cr (max)	Si (max)	Fe (max)
7049	1.2–1.9	7.2–8.2	2.0–2.9	0.2	0.22	0.25	0.35
7050	2.0–2.6	5.7–6.7	1.9–2.6	0.1	0.04	0.12	0.15
7075	1.2–2.0	5.1–6.1	2.1–2.9	0.3	0.28	0.4	0.5
7079	0.4–0.8	3.4–4.8	2.9–3.7	0.3	0.25	0.3	0.4
7090	0.6–1.3	7.3–8.7	2.0–3.0			0.12	0.15
7091	1.1–1.8	5.8–7.1	2.0–3.0			0.12	0.15
7178	1.6–2.4	6.3–7.3	2.4–3.1	0.3	0.35	0.4	0.5
7475	1.2–1.9	5.2–6.2	1.9–2.6	0.6	0.25	0.1	0.12

Table 8.6 Tensile properties of 7000 alloys used in aircraft

Alloy	Temper	Yield strength (MPa)	Tensile strength (MPa)	Elongation (%)
7049	T73	470	530	11
7050	T736	510	550	11
7075	T6	500	570	11
7075	T73	430	550	13
7075	T76	470	540	12
7079	T6	470	540	14
7090	T7E71	580	620	9
7091	T7E69	545	590	11
7178	T6	540	610	10
7475	T651	560	590	12

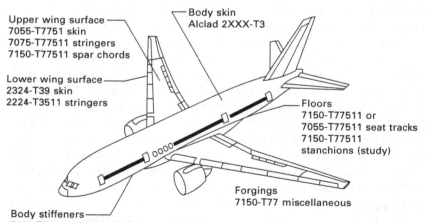

Upper wing surface
7055-T7751 skin
7075-T77511 stringers
7150-T77511 spar chords

Body skin
Alclad 2XXX-T3

Lower wing surface
2324-T39 skin
2224-T3511 stringers

Floors
7150-T77511 or
7055-T77511 seat tracks
7150-T77511
stanchions (study)

Forgings
7150-T77 miscellaneous

Body stiffeners
7150-T77511 or 7055-T77511 keel beam
7150-T77511 body stringers, upper and lower lobe

8.4 New aluminium alloys and tempers used on the Boeing 777 (adapted from E. A. Starke and J. T. Staley, Application of modern aluminium alloys to aircraft, *Progress in Aerospace Science*, Vol 32, pp. 131–172, 1996).

alclad 2XXX-T3 alloy which has higher toughness and resistance to fatigue crack growth than 2024-T3.

8.4.4   8000 series aluminium alloys (Al–Li)

Aluminium alloys that cannot be classified according to their chemical composition into any one of the 1000 to 7000 series are allocated to the 8000 series. Several 8000 alloys contain lithium, which is unique amongst the alloying elements used in aluminium because it reduces density while simultaneously increasing elastic modulus and tensile strength. (Lithium is also an important, but not the principal, alloying element in a number of 2000 alloys, such as 2020 Al, 2090 Al and 2091 Al, which are used in some aircraft). Figure 8.5 shows the effect of lithium content on the density and Young's modulus of aluminium. The density decreases by 3% whereas the modulus increases by 5% for every 1% addition of lithium. This shows that lithium in low concentrations can provide significant weight savings to large aluminium structures. In addition, Al–Li alloys generally have better fatigue properties than 2000 and 7000 alloys.

The three Al–Li alloys most often used in aircraft structures are 8090 Al (2.4%Li–1.3%Cu–0.9%Mg), 8091 Al (2.6%Li–1.9%Cu–0.9%Mg), and 8092 Al (2.4%Li–0.65%Cu–1.2%Mg). Despite the higher specific stiffness and strength gained by increasing the Li content, the alloys used in aircraft have a relatively low content (less than 3%Li). This is because Al–Li alloys can only be processed using conventional casting technology when the Li content

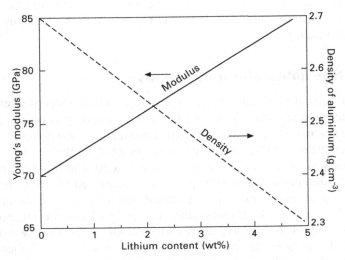

8.5 Effect of lithium content on the Young's modulus and density of aluminium.

is under 3%. Alloys containing higher amounts of Li must be processed using rapid solidification technology, whereby the molten alloy is solidified rapidly as tiny drops. The solid droplets are then compressed with a binding agent into a metal block using powder metallurgy methods. Rapid solidification processing of alloys for aircraft structures is very expensive, and therefore Li contents below 3% are used to avoid this processing route.

The aerospace industry has invested heavily in the development of Al–Li alloys since the 1980s to produce lighter, stiffer and stronger aircraft structures. However, these alloys have not lived up to their initial promise of widespread use in airframes, and have largely failed to replace conventional aluminium alloys (e.g. 2024 Al, 7075 Al) in most aerospace applications. The limited use of Al–Li alloys is the result of several problems, including the high cost of lithium metal and the high processing cost of Al–Li alloys making them prohibitively expensive for many aircraft structures. Al–Li alloys also have low ductility and toughness in the short transverse direction, which can lead to cracking.

Al–Li alloys are mainly used in military fighter aircraft where cost is secondary to structural performance. For example, Al–Li–Cu alloys are used in the fuselage frames of the F16 (*Flying Falcon*) as a replacement for 2024 Al, resulting in a three-fold increase in fatigue life, a 5% reduction in weight and higher stiffness. 8090 Al is used in the fuselage and lift frame of the EH 101 helicopter, again for improved fatigue performance and lower weight (by 180 kg), as shown in Fig. 8.6.

Al–Li alloys are also used in the super lightweight tanks for the space shuttle, which provided a weight saving of over 3 tonne, which translates directly into a similar increase in shuttle payload. The improved hydrogen tank is 5% lighter and 30% stronger than the original tank made using an Al–Cu (2119) alloy.

8.5   Speciality aluminium alloys

Occasionally, existing aluminium alloys do not have all the properties required for an aerospace application, and so the aircraft industry develops a new alloy. The practice of developing new aluminium alloys was common in the era between the mid-1930s and 1970 to meet the needs of the rapid advances in the aerospace industry. For example, a complex Al–Cu–Mg–Ni–Fe alloy known as Hiduminium RR58 was specifically developed for the Concorde. This alloy was created to maintain high tensile strength, creep resistance and fatigue endurance at the high temperatures caused by frictional heating during supersonic flight. Conventional 2000 and 7000 alloys used in the external structures of supersonic aircraft soften above about Mach 2 and, therefore, aluminium alloys with superior heat-resistant properties are required.

Aluminium–beryllium (Al–Be) alloys were developed for use in

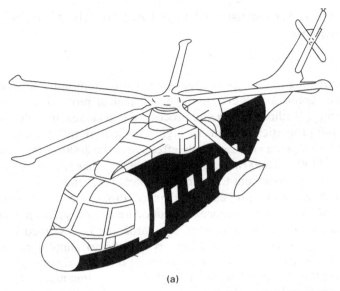

(a)

(b)

8.6 Al–Li alloy used in the EH101 helicopter. The shaded region in the illustration shows the external structures made using Al–Li alloys.

communications satellites and in load-bearing rings and brackets for spacecraft. Al–Be alloys are better than conventional aluminium alloys in terms of mechanical stability over a wide temperature range, vibrational dampening, thermal management and reduced weight, which are desirable properties for spacecraft materials.

8.6   Heat treatment of age-hardenable aluminium alloys

8.6.1   Background

The heat-treatment process of age-hardenable aluminium alloys called ageing is essential to achieve the high mechanical properties required for aerostructures. Without the ageing process, the heat-treatable alloys would not have the properties needed for highly loaded aircraft components. As mentioned, the ageing process is only effective in the 2000, 4000 (containing Mg), 6000, 7000 and 8000 alloys; the ageing of the other alloy series provides no significant improvement to their mechanical properties.

The heat-treatment process consists of three operations which are performed in the following sequence: solution treatment, quenching, artificial (or thermal) ageing. The process of age-hardening is described in chapter 4, and the explanation provided here is specific to aluminium. Solution treatment involves heating the aluminium to dissolve casting precipitates and disperse the alloying elements through the aluminium matrix. Quenching involves rapid cooling of the hot aluminium to a low temperature (usually room temperature) to avoid the formation of large, brittle precipitates. After quenching the aluminium matrix is supersaturated with solute (alloying) elements. The final operation of ageing involves reheating the alloy to a moderately high temperature (usually 150–200 °C) to allow the alloying elements to precipitate small particles that strengthen the alloy. The heat-treatment process can improve virtually every mechanical property that is important to an aircraft structure (except Young's modulus that remains unchanged). Properties that are improved include yield strength, ultimate strength, fracture toughness, fatigue endurance and hardness. The heat-treatment process is performed in a foundry using specialist furnaces and ovens, and then the alloy is delivered in the final condition to the aircraft manufacturer. Some large aerospace companies have their own foundry because of the large amount of aluminium alloys used in the production of their aircraft.

In this section, we examine the heat treatment of age-hardenable aluminium alloys. Special attention is given to the age-hardening of 2000, 7000 and 8000 alloys because of their use in aircraft. We focus on the effect of the heat treatment process on the chemical and microstructural changes to aluminium, and the affect of these changes on the mechanical properties. It is worth noting that changes caused by heat treatment are subtlely different between alloy types, and the description provided here is a general overview of the ageing process.

8.6.2   Solution treatment of aluminium

Solution treatment is the first stage in the heat-treatment process, and is performed to dissolve any large precipitates present in the metal after casting. These precipitates can seriously reduce the strength, fracture toughness and fatigue life of aluminium, and therefore it is essential they are removed before the metal is processed into an aircraft structure. The precipitates are formed during the casting process. As the metal cools inside the casting mould, the alloying elements react with the aluminium to form intermetallic precipitates. Depending on the cooling rate and alloy content, the precipitates may develop into coarse brittle particles. The particles can crack at a small plastic strain that effectively lowers the fracture toughness. The purpose of the solution treatment process is to dissolve the large precipitates, and thereby minimise the risk of fracture.

The solution treatment process involves heating the aluminium to a sufficiently high temperature to dissolve the precipitates without melting the metal. The rate at which the precipitates dissolve and the solubility of the alloying elements in solid aluminium both increase with temperature, and therefore it is desirable to solution treat the metal at the highest possible temperature that does not cause melting. The solution treatment temperature is determined by the alloy composition, and allowances are made for unintended temperature variations of the furnace. Control of the temperature during solution treatment is essential to ensure good mechanical properties. When the temperature is too low, the precipitates do not completely dissolve, and this may cause a loss in ductility and toughness. When the temperature is too high, local (or eutectic) melting can occur that also lowers ductility and other mechanical properties. The treatment temperature for most aluminium alloys is within the range of 450–600 °C. The alloy is held at the treatment temperature for a sufficient period, known as the 'soak time', to completely dissolve the precipitates and allow the alloying elements to disperse evenly through the aluminium matrix. The soak time may vary from a few minutes to one day, depending on the size and chemical composition of the part. After the alloy has been solution treated it is ready to be quenched.

8.6.3   Quenching of solution-treated aluminium

Quenching involves rapid cooling from the solution-treatment temperature to room temperature to suppress the reformation of coarse intermetallic precipitates and to freeze-in the alloying elements as a supersaturated solid solution in the aluminium matrix. Quenching is performed by immersing the hot aluminium in cold water or spraying the metal with water, and this cools thin sections in less than a few seconds. However, with aluminium components with a complex shape it is often necessary to quench at a slower rate to avoid distortion and internal (residual) stress. Slow quenching is done using hot water or some other fluid (e.g. oil, brine). Ideally, the aluminium alloy should be in a supersaturated solid solution condition with the alloying elements uniformly spread through the aluminium matrix after quenching. However, when slow cooling rates are used some precipitation can occur, and this reduces the ability to strengthen the alloy by thermal ageing. Figure 8.7 shows the effect of quenching rate on the final yield strength of 2024 Al and 7075 Al, which are alloys used in aircraft structures. The final yield strength is determined after the alloys have been quenched and thermally aged. It is seen that increasing the quenching rate results in greater final yield strength. Therefore, it is important to quench at an optimum cooling rate that maximises the concentration of alloying elements dissolved into solid solution whilst minimising distortion and residual stress.

After quenching, the aluminium is soft and ductile, and this is the best condition to press, draw and shape the metal into the final product form. For example, when manufacturing aircraft parts using age-hardenable alloys it is easiest to plastically form the components when in the quenched condition. After forming, the aluminium is ready for ageing.

8.6.4   Thermal ageing of aluminium

Ageing is the process that transforms the supersaturated solid solution to precipitate particles that can greatly enhance the strength properties. It is the formation of precipitates that provide aluminium alloys with the mechanical

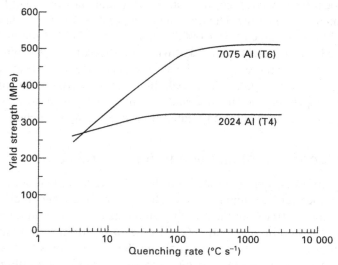

8.7 Effect of average quenching rate on the final yield strength of aerospace alloys 2024 Al and 7075 Al.

properties required for aerospace structures. Ageing can occur at room temperature, which is known as natural ageing, or at elevated temperature, which is called artificial ageing. Natural ageing is a slow process in most types of age-hardenable alloys, and the effects of the ageing process may only become significant after many months or years. Figure 8.8 shows the increase in yield strength of 2024 Al and 7075 Al alloys when naturally aged at room temperature for more than one year. The strength of 2024 Al alloy rises rapidly during the first few days following quenching, and then reaches a relatively stable condition. The strength of 7075 Al alloy, on the other hand, continues to rise over the entire period. Natural ageing can occur, albeit very slowly, at temperatures as low as –20 °C. For this reason, it is sometimes necessary to chill aluminium below this temperature immediately after quenching to suppress or delay the ageing process. It is sometimes necessary to postpone ageing when manufacturing aircraft components and, therefore, the metal must be refrigerated immediately after quenching. For example, it is common practice to refrigerate 2024 Al rivets until they are ready to be driven into aircraft panels to maintain their softness which allows them to deform more easily in the rivet hole. More often, however, the alloy is artificially aged immediately or shortly after quenching.

The artificial ageing process is performed at one or more elevated temperatures, which are usually in the range of 150 to 200 °C. The alloy is heated for times between several minutes and many hours, depending on the part size and the desired amount of hardening. During ageing, the alloy undergoes a series of chemical and microstructural transformations that have

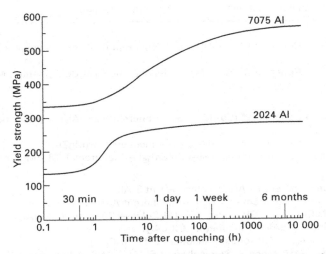

8.8 Effect of natural ageing time on the yield strength of 2024 Al and 7075 Al.

a profound impact on the mechanical and corrosion properties. The order of occurence of the transformations is:

- supersaturated solid solution (α_{ss});
- solute atom clusters (GP1 and GP2 zones);
- intermediate (coherent) precipitates; and
- equilibrium (incoherent) precipitates.

A summary of the transformations that occur to 2000, 7000 and 8000 aerospace alloys are provided in Table 8.7. It is seen that all the alloys undergo the transformation sequence: supersaturated solid solution → GP zones → intermediate precipitates → equilibrium precipitates. However, the changes that occur depend on the types and concentration of the alloying elements.

Figure 8.9 shows the different stages of the ageing process with time and the growth in the average size of the solute cluster and precipitate with time. The sizes shown are approximate and depend on the alloy composition and ageing temperature, although the trend shown is similar for most aged metals. We now examine each of the transformations in greater detail.

When aluminium in the supersaturated solid solution condition is aged, the first significant change is the formation of solute atom clusters, known as Guinier–Preston (GP) zones. GP zones develop by the solute (alloying) atoms moving over relatively short distances to cluster into solute-rich regions. When the zones first develop, the atoms of the alloying elements are randomly arranged relative to the lattice structure of the aluminium

Table 8.7 Ageing transformations of 2000, 7000 and 8000 alloys

---

2000 Alloys

α_{ss} → GP zones → Coherent θ'' ($CuAl_2$) → Semicoherent θ' ($CuAl_2$) → Incoherent θ ($CuAl_2$)

α_{ss} → GP zones → Coherent S′ (Al_2CuMg) → Incoherent S (Al_2CuMg) – high Mg content

7000 Alloys

α_{ss} → GP zones → Coherent θ'' ($CuAl_2$) → Semicoherent θ' ($CuAl_2$) → Incoherent θ ($CuAl_2$)

α_{ss} → GP zones → Semicoherent η' ($MgZn_2$) → Incoherent η ($MgZn_2$)

α_{ss} → GP zones → Semicoherent T′ [$Al_{32}(Mg,Zn)_{49}$] → Coherent T[$Al_{32}(Mg,Zn)_{49}$]

8000 Alloys

Al–Li: α_{ss} → Semicoherent δ' (Al_3Li) → Incoherent δ (AlLi)

Al–Li–Mg: α_{ss} → Semicoherent δ' (Al_3Li) → Incoherent Al_2MgLi

Al–Li–Cu (low Li:Cu): α_{ss} → GP zones → T_1 (Al_2CuLi) → Coherent θ'' ($CuAl_2$) → Semicoherent θ' ($CuAl_2$) → Incoherent θ ($CuAl_2$)

Al–Li (high Li:Cu): α_{ss} → GP zones → Incoherent T_1 (Al_2CuLi)

Al–Li–Cu–Mg: α_{ss} → GP zones → Semicoherent S′ (Al_2CuMg) → Incoherent S (Al_2CuMg)

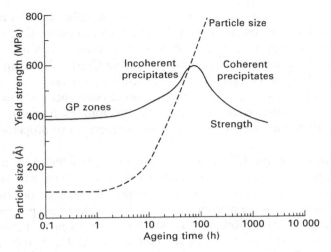

8.9 Effect of time on the ageing of an aluminium alloy.

matrix, and these are called GP1 zones. The composition of the GP zone is dependent on the alloy content. For example, GP zones formed in 2024 Al alloy are rich in copper and in 7075 Al are rich in copper and zinc. Minor alloying elements (e.g. Mn) can also be present in the zones.

With further ageing the solute atoms become arranged into an ordered pattern that is coherent with the aluminium lattice matrix, and these are known as GP2 zones. The number of GP zones is dependent on the temperature, time and alloy content, and their density can reach 10^{23} to 10^{24} m^{-3}. However, GP zones are very small, typically one or two atom planes in thickness and several tens of atom planes in length. Despite their small size, GP zones generate elastic strains in the surrounding matrix that raise the yield strength and hardness. The GP zones grow in size and decrease in number with ageing time until eventually they transform into intermediate precipitates.

During ageing the GP2 zones transform into metastable intermediate precipitates. These precipitates are coherent or semicoherent with the lattice structure of the aluminium matrix. The precipitates are often plate- or needle-shaped and grow along the crystal planes of the matrix. The precipitates nucleate at the sites of GP2 zones and this is known as homogeneous nucleation. Precipitates also grow in regions rich in solute atoms, such as dislocations and grain boundaries, and this is called heterogenous nucleation. Both homogeneous and heterogenous processes are important in the nucleation of precipitates. Following nucleation, the precipitates grow with ageing time as they scavenge solute atoms from the surrounding matrix, and they can reach 0.1 μm or more in size.

In many aluminium alloys, the intermediate precipitates undergo a number of transformations before developing into the final stable condition. For example, in aluminium containing copper, a number of intermediate $CuAl_2$ precipitates (θ', θ'') having different degrees of coherency with the matrix lattice develop before the final formation of stable $CuAl_2$ (θ) precipitates. During the nucleation and growth of intermediate particles many of the mechanical properties, such as yield strength, fatigue endurance and hardness, are improved. Eventually, the intermediate precipitates transform into stable, equilibrium particles and, at this point, the mechanical properties are maximised.

Equilibrium of the precipitates occurs when the particles reach a final chemical composition and crystal structure that does not change with further ageing. The type of equilibrium precipitates produced by ageing is determined by the composition of the aluminium alloy. The main equilibrium precipitates found in aerospace 2000, 7000 and 8000 alloys are given in Table 8.7. The single most important precipitate in 2000 alloys (Al–Cu) is θ ($CuAl_2$). Various precipitates occur in 7000 alloys (Al–Cu–Zn) including θ, η ($MgZn_2$) and T [$Al_{32}(Mg,Zn)_{49}$]. Many types of precipitates are also found in 8000 alloys (Al–Li). In Al–Li–Mg alloys, the other main precipitate is Al_2MgLi; in Al–Li–Cu alloys, the other precipitates are θ ($CuAl_2$) and T_1 (Al_2CuLi); and in Al–Li–Cu–Mg alloys the other precipitate is S (Al_2CuMg). Examples of precipitates are shown in Fig. 8.10.

The mechanical properties reach their highest value at the stage when the precipitates transform from coherent to incoherent particles. Continued ageing at too high a temperature for too long a time degrades properties such as strength and hardness as the equilibrium particles grow in size. The largest precipitates continue to grow whereas the smaller particles disappear, resulting in an increase in the average particle size and a reduction in the number of particles. The softening of an alloy as a result of particle coarsening is called over-ageing, and it must be avoided if optimum properties are required.

The optimum ageing condition is achieved by heat treating the aluminium alloy in a foundry at the correct temperature and time. The optimum heat-treatment condition is governed by the composition of the alloy and geometry of the part. However, it is possible that natural ageing of the alloy occurs after the part has been put into service on an aircraft, which may cause over-ageing. Although this is not a significant issue for subsonic aircraft, it may be problem with supersonic aircraft when frictional heating of the aluminium skins at high flight speeds may cause over-ageing. Surface temperatures in excess of 150 °C occur at the leading edges of aircraft during supersonic flight, and this has the potential to weaken the skins. However, structural failures of aluminium alloys on supersonic aircraft caused by over-ageing do not occur because of the design safety margins.

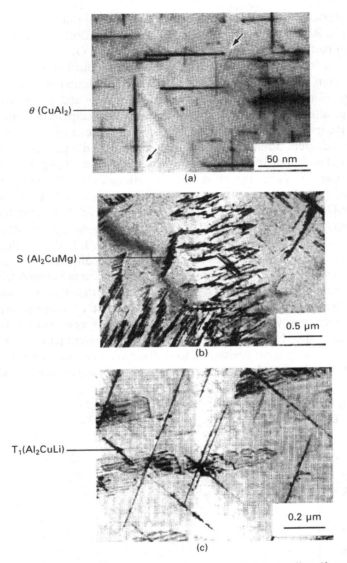

θ (CuAl₂)

50 nm

(a)

S (Al₂CuMg)

0.5 μm

(b)

T₁(Al₂CuLi)

0.2 μm

(c)

8.10 Precipitates in an age-hardenable aluminium alloy (from I. J. Polmear, *Light alloys*, Butterworth–Heinemann, 1995).

8.6.5   Properties of age-hardened aluminium

The mechanical properties of age-hardenable alloys are dependent on the temperature and time of the ageing operation. Figure 8.11 shows the typical effect of ageing temperature on the tensile strength of an aluminium alloy. The strength increases as the metal undergoes the transformations from a supersaturated solid solution to GP zones to intermediate (coherent) precipitates. At the stage when the precipitates transform from coherent to incoherent particles the maximum strength is reached. Over-ageing causes a deterioration in strength owing to coarsening of the incoherent particles.

Aluminium alloys are strengthened by a combination of mechanisms involving solid solution hardening, work hardening and grain boundary hardening, although the dominant mechanism is precipitation hardening. Without the extra strength provided by the precipitates many alloys would not have sufficient strength and toughness for use in lightweight aircraft structures. The initial improvement in strength shown in Fig. 8.11 results from GP zones resisting the movement of dislocations. GP zones generate an elastic strain in the surrounding matrix lattice that resists dislocation slip. Each GP zone provides only a small amount of resistance, but the very high density of GP zones (up to 10^{23} to 10^{24} m^{-3}) generates a sufficiently high internal strain to impede dislocation movement. It is the restriction of dislocation movement that causes the yield strength of aluminium to increase during the early stage of ageing. However, GP zones cannot completely stop the movement of dislocations. Dislocations can cut through the zones and continue to move through the aluminium matrix. The coherent precipitates cause a further improvement in strength because they generate a higher internal strain in the matrix lattice than GP zones. As the coherent particles grow in size they provide greater resistance to dislocation slip. As with GP zones, when a dislocation reaches a coherent precipitate it cuts through and then

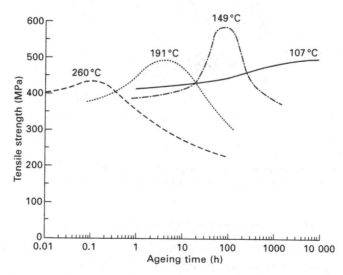

8.11 Effect of ageing temperature on the tensile strength of an aluminium alloy.

continues to move through the matrix. Maximum strength is achieved at the stage when the precipitates transform from coherent to incoherent particles. Dislocations are unable to cut through incoherent particles, and instead must move around them by the Orowan mechanism. The Orowan strengthening mechanism is described in chapter 4. This mechanism is very resistant to dislocation movement and, thereby, is extremely effective in raising the yield strength. High resistance to dislocation slip occurs when the precipitates are small and closely spaced, that is the situation when the particles are initially transformed into incoherent particles. Over-ageing beyond this stage causes the incoherent precipitates to coarsen and become more widely spaced, and this reduces the efficacy of the strengthening process.

The maximum strength that can be achieved by ageing is dependent on the temperature. Figure 8.12 shows the maximum strength and the heat-treatment time taken to reach the maximum strength for a range of ageing temperatures. The peak strength decreases with increasing temperature, although the time required to reach maximum strength increases rapidly with decreasing temperature. It is often not practical to heat treat a metal product over many days or weeks. A temperature should be selected that provides a compromise between high strength and short ageing time. In the production of aluminium aircraft structures, the ageing temperature is usually in the range of 150–200 °C, which provides a good balance between strength and process time.

Although ageing improves mechanical properties such as strength and fatigue resistance, the ageing process may degrade some other properties. Ageing lowers the ductility of aluminium, although the elongation-to-failure of many fully-aged alloys is above 5–10%. The resistance of aluminium alloys to stress corrosion cracking (SCC) may also be affected by age-hardening. The SCC process is described in chapter 21 and involves the growth of cracks under the combined effects of tension loads and corrosive fluids that lowers the fracture stress of the material. Figure 8.13 shows the effect of ageing time on the SCC resistance of an aluminium alloy, and it reaches a minimum level when the alloy is fully hardened. It is therefore necessary to protect age-hardened alloys against SCC when used in aircraft by using corrosion-resistant protective coatings. Several types of coatings are used for aircraft, including cladding and anodised films, and these are described in chapter 21.

8.7   High-temperature strength of aluminium

An important consideration when using aluminium alloys (and other materials) in aircraft structures is softening that occurs at elevated temperature. Care must be taken when selecting materials for supersonic aircraft to ensure structural weakening does not occur owing to excessive heating. Material properties

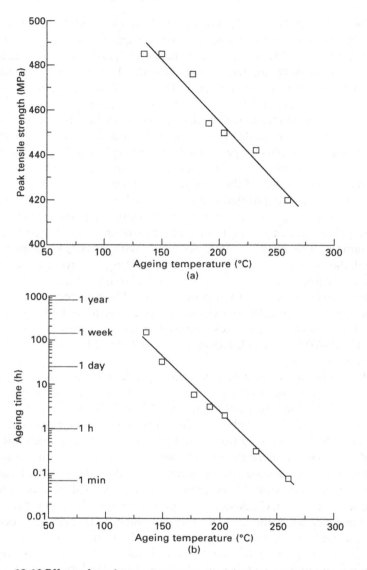

18.12 Effect of ageing temperature on (a) maximum tensile strength and (b) time to reach maximum strength of an age-hardenable aluminium alloy.

such as stiffness, strength, fatigue resistance and toughness are degraded at high temperature. The loss in stiffness and strength of an aluminium alloy with increasing temperature is shown in Fig. 8.14. The sensitivity of the engineering properties to temperature differs between alloy types, but they

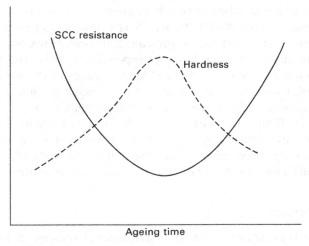

8.13 Effect of ageing time on the resistance of aluminium to stress corrosion cracking (SCC).

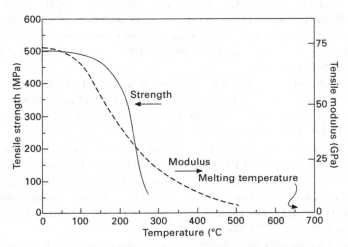

8.14 Effect of temperature on the tensile properties of an aluminium alloy.

all experience a large reduction in their mechanical properties when heated above 100–150 °C.

Thermal softening is a key factor in the selection of aluminium alloys for aircraft structures that experience a large temperature rise as a result of frictional skin heating. The nose section, leading edges and skins of aircraft are heated owing to friction caused by molecules in the atmosphere moving across the aircraft surfaces at high speed. Frictional heating is generally not a problem when flying at subsonic speeds because the surface temperature does not rise much above 80–90 °C, which is below the temperature at which aircraft materials experience significant softening. However, higher temperatures are generated when flying at supersonic speeds. The hottest parts of the aircraft are the nose cone and leading edges of the wings and tail sections, with temperatures in excess of 150 °C being reached at Mach 2. As Fig. 8.14 shows, at 150 °C the modulus and strength is 10–20% below room temperature. Temperatures exceeding 1000 °C are experienced during re-entry of spacecraft such as the space shuttle and temperatures above 500 °C can occur in ultra-high speed aircraft. At these temperatures, the aluminium must be thermally protected by a heat shield, such as ceramic tiles.

8.8   Summary

Aluminium is a popular aerospace structural material because of its light weight, moderate cost (for both the raw material and fabrication processing), and good mechanical performance including specific stiffness, specific strength, ductility and fracture toughness. For these reasons, aluminium (along with carbon-fibre composite) is the most common structural material for aircraft and helicopters. Aluminium typically accounts for 60–80% of the structural weight of modern passenger airliners and 40–60% for fighter aircraft and helicopters.

Potential problems with using aluminium in aircraft structures include stress corrosion cracking; corrosion when in direct contact with carbon–epoxy composite; difficulty with welding high-strength connections; and softening at relatively low temperatures (above about 150 °C).

There are many grades of aluminium alloy, although aircraft structures are fabricated using 2000 (Al–Cu), 7000 (Al–Zn–Mg) and, in fewer cases, 8000 (Al–Li) alloys. The development of aluminium alloys has proceeded in two distinct directions: one of the tension-dominated sections of the airframe that use primarily the 2000 alloys, and the other for the compression-dominated sections that use the 7000 alloys.

2000 alloys generally have better fatigue resistance and fracture toughness than 7000 alloys, and therefore are used where these properties are important such as fuselage skins, lower wing panels and control surfaces.

7000 alloys have higher strength than 2000 alloys and, therefore, are used in aircraft structures that are required to carry heavier loads. 7000 alloys are used in upper wing surfaces, spars, stringers, pressure bulkheads, and fuselage frames.

8000 series alloys are characterised by high specific stiffness owing to the ability of lithium to lower the density and increase the elastic modulus of aluminium. Many 8000 alloys also have superior fatigue resistance compared with 2000 and 7000 alloys. However, 8000 alloys are expensive and therefore used sparingly as a structural material. The use of 8000 alloys is currently restricted to a small number of aircraft and helicopter components where high specific stiffness and excellent fatigue performance is critical.

Aircraft alloys are heat treated by age-hardening to maximise their strength properties. The alloys are solution treated, quenched, and then thermally aged to increase the strength by precipitation hardening. Other important strengthening processes in aluminium are solid solution hardening, work-hardening and grain-size refinement.

Aerospace applications are small for aluminium alloys that cannot be strengthened by age-hardening. Alloys from the 1000, 3000, 4000 and 5000 series cannot be heat treated to induce precipitation hardening and, therefore, lack the strength and fatigue resistance needed in weight-efficient structures.

The susceptibility of aluminium to stress corrosion cracking increases with the improvement in strength gained by thermal ageing. Special ageing treatments have been developed to minimise the risk of stress corrosion damage in aluminium structures.

8.9   Further reading and research

Campbell, F. C., *Manufacturing technology for aerospace structural materials*, Elsevier, Amsterdam, 2006.

Gregson, P. J., 'Aluminium alloys: physical metallurgy, processing and properties', in *High performance materials in aerospace*, ed. H. M. Flower, Chapman & Hall, London, 1195, pp. 49–83.

Polmear, I. J. *Light alloys: metallurgy of light metals*, Hodder and Stoughton, London, 1995.

Smith, A., 'Aluminium–lithium alloys in helicopter airframes', in *Aerospace materials*, edited B. Cantor, H. Assender and P. Grant, Institute of Physics Publishing, Bristol, 2001, pp. 38–46.

Starke, E. A. and Staley, J. T., 'Application of modern aluminum alloys to aircraft', *Progress in aerospace sciences*, **32** (1996), 131–172.

9

Titanium alloys for aerospace structures and engines

9.1   Introduction

Titanium alloys are used in airframe structures, landing gear components and jet engine parts for their unique combination of properties: moderate density, high strength, long fatigue life, fracture toughness, creep strength, and excellent resistance to corrosion and oxidation. Titanium alloys also have good mechanical performance at high temperature (up to 500–600 °C), which is well above the operating temperature limit of lightweight aerospace materials such as aluminium alloys, magnesium alloys and fibre–polymer composites. For this reason, when titanium alloys were originally used in aircraft in the early 1950s it was for high-temperature applications. The earliest use of titanium was in compressor discs and fan blades for gas turbine engines, which require excellent creep resistance at high operating temperature. The use of titanium was important in the early development of jet engines, which were originally built using heat-resistant steels and nickel alloys. Both steels and nickel alloys are 'heavy materials', and their replacement with titanium in discs and blades reduced the weight of early jet engines by more than 200 kg.

Titanium has been an important engineering material in gas turbine engines for more than fifty years, and currently it accounts for 25–30% of the weight of most modern engines. Titanium can be used in engine components required to operate at temperatures up to 500–600 °C. The engine components made using titanium are fan blades, shafts and casings in the inlet region; low-pressure compressors; and plug and nozzle assemblies in the exhaust section. Titanium is also used in the engine frames, casings, manifolds, ducts and tubes. It is not possible to use titanium in all parts of the engine, and it is unsuitable within the combustion chamber and other sections where the temperature exceeds 600 °C. Above this temperature, titanium rapidly softens, creeps and oxidises, and more heat-resistant materials such as nickel alloys are required. The application and metallurgical properties of nickel alloys for jet engines is covered in chapter 12.

Titanium is also an important material for heavily-loaded airframe structures. Titanium is used in a wide variety of structures on commercial aircraft, including wing boxes, wings and undercarriage parts. Titanium is

202

used when its high strength (compared with aluminium) allows the same load to be carried by a physically smaller structural part, even though there is no weight advantage because of the higher density. Forged titanium is used in airframes requiring high strength and high toughness or when there is too little airframe space for aluminium alloys. For many years, steel has been used in aircraft landing gear because of its high stiffness, strength, fatigue resistance and toughness. However, high-strength steels are susceptible to corrosion and hydrogen embrittlement, which is the phenomenon whereby steel becomes very brittle owing to the absorption of hydrogen. Titanium is used as a replacement material for steel in landing gear to eliminate the problems of corrosion and hydrogen embrittlement as well as to achieve a significant weight saving. The use of titanium in large passenger aircraft designed before 1980 was relatively modest at 3–5% of the structural weight, but more recent aircraft use a greater percentage (Fig. 2.9). For example, titanium accounts for 9% of the structural weight of the A380 and 10% of the B777.

Aerospace is the single largest market for titanium products; with the industry consuming about 80% of the global production of the metal. The aerospace applications of titanium in the USA are approximately:

- jet engines for commercial aircraft: 37%;
- jet engines for military aircraft: 24%;
- airframes for commercial aircraft: 18%;
- airframes for military aircraft: 12%;
- rockets and spacecraft: 8%; and
- helicopters and armaments: 1%.

Titanium accounts for a higher percentage of the structural mass of military aircraft compared with commercial airliners in order to withstand the higher airframe loads generated by extreme manoeuvres during combat operations. The structural mass of fighter aircraft made using titanium is often in the range of 10–30%. For example, Fig. 9.1 shows that titanium alloys account for 26% (or 7000 kg) of the structural weight of the F-14 *Tomcat* fighter aircraft. The components made using titanium range in size from highly stressed wing structures, landing gear parts and tail sections down to small fasteners, springs and hydraulic tubing. Other examples of the use of titanium in fighter aircraft are the F/A-18 *Hornet* and F-15 *Eagle*. About 26% of the airframe of the F-15 is made of titanium, including the fuselage skin, wing torque box, wing spars and bulkheads, engine pod frames, and firewalls that separate the engines from the main frame. Titanium is also used in helicopters for the main rotor hub, tail rotor hub, pivots, clamps, and blade tips which require high strength and fracture toughness. Titanium alloys are also used in solid-fuel and liquid-fuel engines, high-pressure gas and fuel storage tanks and, in some cases, the skin of rockets.

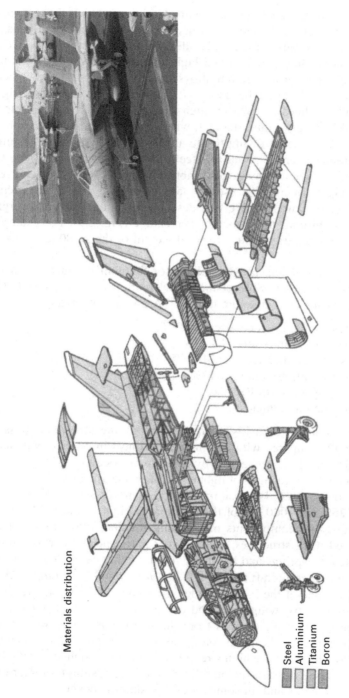

Materials distribution

Steel
Aluminium
Titanium
Boron

9.1 Use of titanium alloys in the F-14 *Tomcat* (reproduced with permission from M. Spick, *The great book of modern warplanes*, Salamander Books Ltd, 2000).

In this chapter, we discuss the metallurgical and structural properties of titanium alloys and their applications in aerospace structures and engines. The benefits and drawbacks of using titanium in aircraft are examined and the different types of titanium alloys, and how their properties are controlled by alloying, forging and heat-treatment are studied. New types of titanium-based materials with potential aerospace applications are also reviewed. The metallurgical and mechanical properties of titanium aluminides and titanium shape-memory alloys are described, together with their potential applications in future aircraft.

9.2 Titanium alloys: advantages and disadvantages for aerospace applications

9.2.1 Advantages

There are many benefits to using titanium in aircraft structures and engines, although (as with any material) there are also disadvantages. Table 9.1 compares the stiffness and strength of several types of titanium with that of other aerospace structural materials. There are different classes of titanium which are called commercially pure titanium, α-titanium, β-titanium and $\alpha+\beta$-titanium. The specific stiffness of titanium alloys is slightly lower than that of other aerospace materials, so there is no benefit in using them in aircraft structures designed for high stiffness. The strength of the titanium alloys varies over a wide range. This is because the properties are dependent on the alloy composition and heat treatment. However, the specific strength properties of titanium alloys are superior to the other materials (except carbon–epoxy composite), and for this reason they are good materials to

Table 9.1 Comparison of mechanical properties of titanium alloys with other aerospace structural materials

Materials	Average specific gravity (g cm^{-3})	Young's modulus (GPa)	Specific modulus (MPa m^3 kg^{-1})	Yield strength (MPa)	Specific strength (kPa m^3 kg^{-1})
Pure titanium	4.6	103	22.4	170–480	37–104
α-Ti alloys	4.6	100–115	21.7–25.1	800–1000	174–217
β-Ti alloys	4.6	100–115	21.7–25.1	1150–1300	180–282
$\alpha+\beta$-Ti alloys	4.6	100–115	21.7–25.1	830–1300	180–282
Aluminium (2024-T6)	2.7	70	25.9	385	142
Aluminium (7075-T76)	2.7	70	25.9	470	174
Magnesium	1.7	45	26.5	200	115
High-strength low-alloy steel	7.8	210	26.9	1000	128
Carbon–epoxy composite*	1.7	50	29.4	760	450

* [0/±45/90] Carbon–epoxy; fibre volume content = 60%.

use in aircraft structures required to carry high loads, such as airframe components, undercarriage parts, and wing boxes.

Other benefits of using titanium include high strength, good fatigue resistance, creep resistance at high temperature, and excellent oxidation resistance up to 600 °C. Certain types of titanium alloys can be joined by welding or diffusion bonding; thereby reducing the need for mechanical fasteners (bolts, screws, rivets) and adhesive bonding that is a requirement for age-hardened aluminium assemblies. Titanium has better resistance to corrosion than high-strength aluminium alloys, including the most damaging forms such as stress corrosion cracking and exfoliation. Titanium has the ability to form a thin oxide surface layer, which is resistant and impervious to most corrosive agents and which provides the material with excellent corrosion resistance. Titanium is used as a replacement material for aluminium in aircraft structures when corrosion resistance is the prime consideration.

An important advantage of titanium alloys over many other aerospace materials, particularly aluminium alloys and fibre–polymer composites, is high strength at elevated temperature. Titanium is often selected for use at temperatures too high for aluminium or composite material but where the temperatures/loads do not dictate the use of steel or nickel superalloys if weight is a key consideration. Figure 9.2 shows the effect of temperature on the yield strength of two titanium alloys used in gas turbine engines. The loss in strength with increasing temperature for an aluminium alloy (7075 Al-T6) is shown for comparison. The strength of titanium drops gradually with increasing temperature, and adequate strength is retained to 500–600 °C.

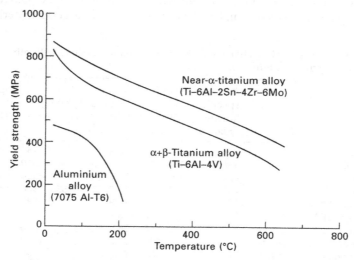

9.2 Effect of temperature on the yield strength of titanium and aluminium alloys.

It is for this reason, together with excellent resistance to creep and oxidation, that titanium is used in gas turbine engines and other high-temperature applications.

9.2.2   Disadvantages

Disadvantages of using titanium include relatively high density (4.5 g cm^{-3}) compared with aluminium alloys (2.7 g cm^{-3}) and carbon–epoxy composites (1.5–2.0 g cm^{-3}). However, it is lighter than nickel superalloys (8.7 g cm^{-3}) when used in jet engines. A major disadvantage of titanium is the high cost, which varies with the usual price fluctuations in the global metals commodity market; the raw material cost is typically \$10 000–12 000 per tonne (June 2010). The high cost is the result of the expensive process used to extract titanium from its ore together with the costly processes used in fabricating and shaping the metal into an aerospace component. Titanium is difficult to machine and requires specialist material removal processes (such as laser-assisted machining) to produce aircraft components free of machining damage.

9.3   Types of titanium alloy

9.3.1   Phases of titanium

A metallurgical property of titanium alloys that is important in their use in aircraft is allotropy, which is defined as different physical forms of the same material that are chemically very similar. Many metals, including the aluminium, magnesium and nickel alloys used in aircraft, can only occur in one crystalline phase at 20 °C. For instance, aluminium can only have a face-centred-cubic crystal structure at room temperature, and no other. As another example, the crystal structure of magnesium is always hexagonal close packed at 20 °C. Titanium alloys, on the other hand, can have a hexagonal-close-packed (hcp) crystal structure, which is known as α-Ti, and a body-centred-cubic (bcc) structure, which is called β-Ti, at room temperature.

The crystal structure and properties of the α-Ti and β-Ti phases are given in Fig. 9.3. The chemical composition of both phases is virtually identical, but their properties are different because of the different crystal structures. From a practical viewpoint, this means that α-Ti and β-Ti have different uses on aircraft: α-Ti has better creep resistance and ductility at high temperature than β-Ti, which makes it more suited for aeroengine applications. β-Ti has higher tensile strength and fatigue resistance than α-Ti owing to fewer slip systems in the bcc crystal, which makes it better suited for highly-loaded aircraft structures. It is also possible for titanium alloys to consist of a mixture of the α and β phases. α+β-Ti occurs when the metal contains both α-Ti and

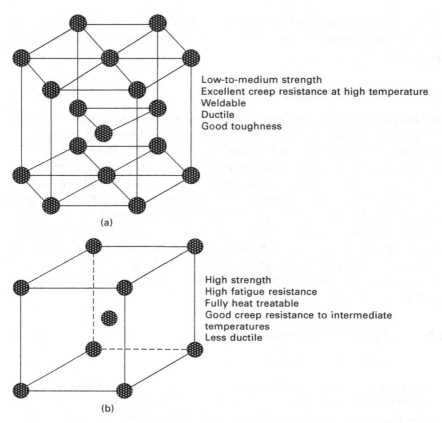

Low-to-medium strength
Excellent creep resistance at high temperature
Weldable
Ductile
Good toughness

(a)

High strength
High fatigue resistance
Fully heat treatable
Good creep resistance to intermediate
temperatures
Less ductile

(b)

9.3 Crystal structure and properties of (a) α-titanium and (b) β-titanium.

β-Ti grains, as shown in Fig. 9.4. The properties of α+β-Ti is somewhere between pure α-Ti and pure β-Ti. The α+β-Ti alloys generally have better strength than α-Ti alloys and higher creep strength and ductility than β-Ti alloys, which makes them useful for both aircraft engines and structures.

Pure titanium metal undergoes an allotropic transformation that changes the crystal structure from the α-phase to the β-phase at about 885 °C. An allotropic transformation simply means the crystal structure changes when the material is heated above or cooled below a critical temperature called the transus temperature. In pure titanium, the β-phase can only exist above the transus temperature, and the α-phase only occurs below the transus temperature. However, it is possible for β-Ti to be stable at room temperature by the addition of certain alloying elements. The ability of alloying elements to stabilise the formation of the α or β phases is determined by their electron valence number. This number describes the number of electrons in the outer

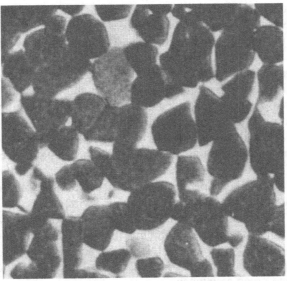

9.4 Microstructure of α+β-Ti alloy. The α-phase is the dark region and the β-phase the light region (reproduced with permission from R. M. Brick, A. W. Pense and R. B. Gordon, *Structure and properties of engineering materials*, McGraw–Hill Book Company, 1977).

orbital shell of an atom. Titanium has an electron valence number of two, three or four, and any element with a different valence number can promote the formation of the α- or β-phases. Elements with a lower valence number than titanium promote the formation of α-Ti, and these are called alpha-stabilisers. The most common α-stabilising element is aluminium, although tin is also regularly used. Elements with a higher electron valence number promote the formation of the β-phase. The elements stabilise the β-phase when the metal is cooled rapidly from the transus temperature to room temperature. These elements are the so-called beta-stabilisers, and include vanadium, molybdenum, chromium and copper. Alloying elements with the same valence number as titanium are called neutral elements because they do not stabilise either the α- or β-phases. Neutral elements are used to increase strength or improve some other property.

9.3.2   Commercially pure titanium

Commercially pure titanium (>99% Ti) is a low-to-moderate strength metal that is not well suited for aircraft structures or engines. The yield strength of high-purity titanium is within the range of 170–480 MPa, which is too low for heavily loaded aerostructures. The composition and properties of several commercially pure titanium alloys are given in Table 9.2. In the USA, the commercially pure forms of titanium are classified according to the ASTM standard, although not all countries have adopted this system. The standard simply classifies the metal types according to a numbering system, e.g. Grade 1, Grade 2 and so on.

Table 9.2 shows that some grades of pure titanium have a tensile strength of more than 450 MPa, which is similar to the 2000 aluminium alloys used in aircraft structures. However, the specific strength of pure titanium is not as high as the aluminium alloys because of its higher density. Commercially pure titanium contains a low concentration of impurities which remain in the metal after refining and processing from the rutile ore. Titanium contains trace amounts of impurities such as iron and atomic oxygen, and they have the beneficial effect of increasing strength and hardness by solid solution hardening. For example, ultra-high purity titanium (with an oxygen content of under 0.01%) has an ultimate tensile strength of about 250 MPa. In comparison, the ultimate strength of titanium with a small amount of oxygen (0.2–0.4%) is 300–450 MPa. However, it is not advantageous to strengthen titanium using impurity elements because there is a large loss in ductility, thermal stability and creep resistance.

Pure titanium is rarely, if ever, used in aircraft. However, one important engineering property of titanium is the ability to retain strength and ductility at very low temperatures, and therefore is useful for cryogenic applications. An aerospace application of commercially pure titanium is in fuel storage tanks containing liquid hydrogen for space vehicles. Liquid hydrogen must be stored below −210 °C under normal atmospheric conditions, at which temperature commercially pure titanium has good strength and toughness.

9.3.3   Alpha titanium alloys

There are two major groups of alpha titanium alloys known as super-alpha and near-alpha. Super-alpha alloys contain a large amount of α-stabilising alloying elements (>5 wt%) and are composed entirely of α-Ti grains. Near-alpha alloys contain a large amount of α-stabilisers with a smaller quantity of β-stabilising elements (<2 wt%). The microstructure of near-alpha alloys consists of a small volume fraction of β-Ti grains dispersed between the much greater volume fraction of α-Ti grains.

Near-alpha alloys have higher strength properties than super-alpha alloys (owing to the small amount of the hard β-Ti phase) and also have excellent creep resistance at high temperature. For this reason, near-alpha alloys are preferred over super-alpha alloys in components for gas turbine engines and rocket propulsion systems required to operate for long times at 500–600 °C. Above this temperature, the alloys soften and creep, and other materials with better thermal stability (such as nickel superalloys) are required.

Table 9.2 Composition and tensile properties of commercially pure titanium alloys

Type	Maximum impurity limits (wt%)					Young's modulus (GPa)	0.2% Yield strength (MPa)	Tensile strength (MPa)	Elongation (%)
	N	C	H	Fe	O				
ASTM Grade 1	0.03	0.10	0.015	0.20	0.18	103	170	240	25
ASTM Grade 2	0.03	0.10	0.015	0.30	0.25	103	280	340	20
ASTM Grade 3	0.05	0.10	0.015	0.30	0.35	103	380	450	18
ASTM Grade 4	0.05	0.10	0.015	0.50	0.40	103	480	550	15

The composition and properties of several super-alpha and near-alpha alloys are given in Table 9.3. There is no standardised system for classifying titanium alloys based on their composition. The designation system that is most often used simply lists the weight percentages of the principal alloying additions. As examples, Ti–8Al–1Mo–1V contains 8% aluminium, 1% molybdenum and 1% vanadium and Ti–6Al–4V means 6% aluminium and 4% vanadium. In addition, the IMI numbering system is used for some, but not all, titanium alloys. For example, the alloy Ti–5Al–2.5Sn is known as IMI317 whereas Ti–2.25Al–11Sn–5Zr–1Mo is known as IMI679. Although commonly used, the IMI number gives no clue to the types and amounts of alloying elements.

Near-alpha alloys are used more often than super-alpha alloys in aircraft because of their superior high-temperature properties, particularly strength and creep resistance. In fact, only one super-alpha alloy is used in significant amounts. Ti–5Al–2.5Sn (IMI317) is used in cryogenic applications for its ability to retain high ductility and fracture toughness at very low temperatures. Of the α-Ti alloys listed in Table 9.3, the near-alpha alloys IMI 685, IMI 829 and IMI 834 are used in jet engines. The yield strength of these alloys is within 800 to 1000 MPa, which is nearly twice that that for age-hardened aluminium alloys.

Strengthening of α-Ti alloys is achieved by work hardening, solid solution hardening and grain-size refinement. Work hardening by plastic forming processes such as rolling or extrusion can more than double the tensile strength from about 350 to 800 MPa. Solid solution hardening increases the tensile strength between 35 and 70 MPa for every 1% of alloying element. Aluminium is the main alloying element and is used to stabilise the α-phase and to increase the creep and tensile strengths. Figure 9.5 shows the effect of aluminium content on the yield strength and ductility of α-Ti, and the strength increases rapidly owing to solid solution hardening whereas the ductility drops owing to embrittlement. Adding more than about 9% Al promotes the formation of brittle titanium aluminide (Ti_3Al) precipitates that reduce the fracture toughness and ductility, and for this reason most

Table 9.3 Composition and properties of alpha alloys used in gas turbine engines

Alloy type	Composition	Young's modulus (GPa)	0.2% Yield strength (MPa)	Tensile strength (MPa)
Super-α	Ti–5Al–2.5Sn (IMI317)	103	760	790
Near-α	Ti–6Al–2Sn–4Zr–6Mo	114	862	930
	Ti–5.5Al–3.5Sn–3Zr–1Nb (IMI829)	120	860	960
	Ti–5.8Al–4Sn–3.5Zr–0.7Nb (IMI834)	120	910	1030
	Ti–2.25Al–11Sn–5Zr–1Mo (IMI679)	115	900	1000
	Ti–6Al–4Zr–2Mo (IMI685)	115	960	1030

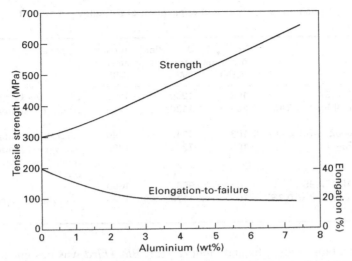

9.5 Effect of aluminium content on the tensile properties of α-Ti.

α-Ti alloys have an aluminium content under 6–7%. A class of aerospace titanium alloys known as titanium aluminides are produced by the addition of a high concentration of aluminium. Titanium aluminides are composed entirely of Ti_3Al, and are described later in this chapter. The strength of α-Ti alloys cannot be raised by heat treatment, which is one reason for their use in high-temperature applications. The thermal stability and resistance to thermal ageing of near-alpha alloys ensures that their mechanical properties are not changed appreciably when operating at high temperatures for long periods.

9.3.4   Beta titanium alloys

Beta-titanium alloys are produced by the addition of β-stabilising elements that retain the β-phase when the metal is cooled rapidly from the transus temperature. A large number of alloying elements can be used as β-stabilisers, although only V, Mo, Nb, Fe and Cr are used in significant amounts (typically 10–20 wt%). The composition, tensile properties and aircraft applications of several β-Ti alloys are given in Table 9.4. The strength and fatigue resistance of β-Ti alloys is generally higher than the α-Ti alloys. However, the use of β-Ti alloys is very low; accounting for less than a few percent of all the titanium used by the aerospace industry owing to their low creep resistance at high temperature.

The first β-Ti alloy to be used in commercial quantities was Ti–13V–11Cr–3Al, which was used in the airframe of the SR-71 *Blackbird* military aircraft. The alloy was used in the fuselage frame, wing and body skins,

Table 9.4 Composition, properties and application of β-Ti alloys in aircraft structures

Composition	Young's modulus (GPa)	0.2% Yield strength (MPa)	Tensile strength (MPa)	Application
Ti–13V–11Cr–3Al	103	1200	1280	SR-71 *Blackbird*
Ti–8V–6Cr–4Mo–4Zr–3Al (Beta C)	103	1130	1225	Aircraft fasteners
Ti–11.5Mo–6Zr–4Sn (Beta III)	103	1315	1390	Aircraft fasteners
Ti–10V–2Fe–3Al	103	1250	1320	Airframes, landing gear, helicopter rotor blades
Ti–15V–3Al–3Cr–3Sn	103	966	1000	Airframes
Ti–15Mo–2.7Nb–3Al–0.2Si (Timetal 21S)	103	1170	1240	Jet engine nacelles

longerons, bulkheads, ribs and landing gear. *Blackbird* was designed to fly at Mach 2.5 (3200 km h^{-1}) and, at this speed, the skins are heated to nearly 300 °C by frictional aerodynamic drag. Aluminium alloys soften at these temperatures, and therefore titanium alloy was used because of its superior strength at high temperature. The Boeing 777 uses β-Ti alloys in many components, including landing gear, exhaust plugs, nozzles, and sections of the nacelle. The β-Ti alloy used in the B777 landing gear is Ti–10V–2Fe–3Al, which is used as a replacement to high strength steel. β-Ti alloy is used to eliminate the risk of hydrogen embrittlement experienced with steel as well as providing a weight saving of 270 kg. β-alloys are also used in the fan disc of gas turbine engines to save weight. The two most common β-Ti alloys for fan discs are Ti–5Al–2Zr–2Sn–4Cr–4Mo and Ti–6Al–2Sn–4Zr–6Mo, and they provide weight reductions of more than 25% compared with the use of high-temperature steel or superalloy.

The yield strength of most β-Ti alloys is between 1150 and 1300 MPa, which is greater than the strength of α-Ti alloys (750–1000 MPa). The higher strength is the result of greater solid solution hardening together with precipitation hardening. Solution treatment and thermal ageing can increase the strength by 30–50% or more over the annealed condition. The heat treatment conditions used to strengthen β-Ti alloys are more extreme than the age-hardening of aluminium alloys described in chapter 8. A typical treatment for β-Ti alloys involves solution treatment at about 750 °C, rapid quench to room temperature, and then thermal ageing at 450–650 °C for several hours. The ageing causes some of the β-phase to transform into α-Ti particles. These particles are finely dispersed through the β-matrix, and are often found at grain boundaries and dislocations. The particles increase the yield strength of β-Ti by the precipitation hardening mechanism. In addition, the ageing process produces ω-phase precipitate particles that also strengthen the β-Ti alloy. The tensile strength of β-Ti alloys increases with the volume fraction of α- and ω-particles to 1400 MPa. The strength of β-Ti alloys can also be increased by work-hardening. Sheets of β-Ti alloy can be strengthened up to 1500 MPa by cold deformation by 50–70%. However, the increase in strength by cold working is accompanied by losses in ductility and toughness and, therefore, cold deformation is not widely used for raising the strength of aircraft titanium alloys.

9.3.5   Alpha+beta titanium alloys

α+β-Ti alloys are by far the most important group of titanium alloys used in aircraft. These alloys are produced by the addition of α-stabilisers and β-stabilisers to promote the formation of both α-Ti and β-Ti grains at room temperature. The popularity of α+β-Ti alloys stems from their excellent high temperature creep strength, ductility and toughness (from the α-Ti phase) and high tensile strength and fatigue resistance (from the β-Ti phase). The composition and tensile properties of several α+β-Ti alloys are given in Table 9.5, and the amounts of α-stabilisers and β-stabilisers are typically in the range of 2–6% and 6–10%, respectively.

Of the many types of α+β-Ti alloys, the most important is Ti–6Al–4V (IMI318), which makes up more than one-half the sales of titanium in the United States and Europe, and is the most used titanium alloy in aircraft. Ti–6Al–4V is used in both jet engines and airframes; it accounts for about 60% of the titanium used in jet engines and up to 80–90% for airframes. Figure 9.6 shows a Ti–6Al–4V engine blisk for the F-35 *Lightning II* fighter. The maximum operating temperature limit of Ti–6Al–4V under creep conditions is 300–450 °C and, therefore, this alloy is used for the fan and cooler parts of the engine compressor, whereas α-Ti alloys are used in hotter engine components where higher temperature creep strength is required. Ti–6Al–4V is the most commonly used titanium alloy in airframes, and is found in highly-loaded structures such as wing boxes, stiffeners, spars and

Table 9.5 Composition and tensile properties of α+β alloys

Composition	Young's modulus (GPa)	0.2% Proof strength (MPa)	Tensile strength (MPa)	Elongation (%)
Ti–6Al–4V (IMI318)	114	830	900	10
Ti–6Al–2Sn–4Zr–6Mo	114	1100	1170	10
Ti–5Al–2Sn–2Zr–4Mo–4Cr	114	1055	1125	–
Ti–10V–2Fe–3Al	103	1100	1170	–
Ti–2Al–2Sn–4Mo–0.5Si	114	1000	1100	13
Ti–6Al–6V–2Sn	114	1170	1275	10

9.6 Ti–6Al–4V blisk for a jet engine in the Joint Strike Fighter project. This photograph is reproduced with the permission of Rolls-Royce plc, copyright © Rolls-Royce plc 2010.

skin panels. For example, most of the titanium structures in the F-14 *Tomcat* shown in Fig. 9.1 are made using Ti–6Al–4V.

The mechanical properties of α+β-Ti alloys are often between those of α-Ti and β-Ti alloys. For example, the yield strength of annealed Ti–6Al–4V is about 925 MPa, which is higher than most near-alpha Ti alloys (~800 MPa) but lower than many β-Ti alloys (1150–1400 MPa). Similar comparisons can be made for other properties, including creep strength, fatigue resistance, fracture toughness, ductility and ultimate tensile strength. The strength of α+β-Ti alloys is derived from several hardening processes, including solid solution hardening, grain boundary strengthening and work hardening, although the most important is precipitation hardening of the β-Ti grains. As with β-Ti alloys, the thermal ageing of α+β-Ti alloys causes some of the β-phase to transform into α-Ti particles and ω precipitates which raise the strength. α+β-Ti alloys are strengthened following thermal ageing, normally at 480–650 °C, which increases the proof strength by 30–50% over the annealed alloy. Fully age-hardened α+β-Ti alloys have tensile strengths exceeding 1400 MPa.

9.4   Titanium aluminides

Titanium aluminides are a special class of titanium alloy emerging as an important high-temperature material for next-generation aeroengines. These materials have a similar or lower density and higher Young's modulus than conventional titanium alloys. The density of titanium aluminides is between 3.9 and 4.3 g cm^{-3} against 4.4 to 4.8 g cm^{-3} for titanium alloys. The Young's modulus of titanium alumimides is 145–175 GPa versus 90–120 GPa for titanium. Titanium aluminides also have higher oxidation resistance and strength retention at high temperature, which makes them better suited for use in gas turbine engines and rocket propulsion systems. Several types of titanium aluminide alloys retain strength to 750 °C, which is at least 150 °C higher than the operating temperature limit of conventional titanium alloys.

Titanium aluminides are classified as ordered intermetallic compounds, which means they form when atoms of two or more metals combine in a fixed ratio to produce a crystalline material different in structure from the individual metals. For titanium aluminides, the two metals are titanium (with a body-centred-cubic crystal structure) and aluminium (face-centred-cubic structure). The metals combine to produce Ti_3Al (hexagonal) and TiAl (face-centred tetragonal) compounds (Fig. 9.7). Ti_3Al is called α_2-aluminide and forms when the aluminium content of the alloy is about 25%. The composition of common α_2-aluminide alloys are Ti–25Al–5Nb, Ti–24Al–11Nb, Ti–25Al–10Nb–3V–1Mo and Ti–24.5Al–12.5Nb–1.5Mo. Niobium (Nb) is added to produce Ti_2Nb precipitates, which contribute to strength retention at high temperature and increased ductility at room temperature. The TiAl compound is known as γ-aluminide, and is formed when there are approximately equal amounts of Ti and Al. γ-aluminide alloys include Ti–48Al–1V, Ti–48Al–2Mn, Ti–48Al–2Cr–2Nb and Ti–47Al–2Cr–2Ni.

A benefit of using titanium aluminide as a structural material is high-temperature strength. Titanium aluminides have a small number of slip systems in their crystal structure which limits the movement of dislocations at high temperature, thereby retaining the strength. Figure 9.8 shows the effect of increasing temperature on the specific proof strengths of α_2- and γ-aluminides against a near-alpha titanium alloy (IMI834) that is currently used in jet engines. The α_2-alloy has superior specific strength compared with the near-alpha alloy up to 800 °C, and therefore Ti_3Al could be used as a lightweight replacement for conventional titanium alloys in jet engines. The γ-aluminide has a lower specific yield strength than the near-alpha alloy, but is able to retain its strength to higher temperatures. The maximum operating temperature of γ-aluminides is similar to nickel superalloys used in the hottest sections of jet engines. Titanium aluminides have potential application in high-pressure compressor and turbine blades. However, γ-aluminides have more than one-half the density (3.9 g cm^{-3}) of nickel alloys (8.7–8.9 g cm^{-3}) and therefore can be used as a lighter-weight engine material.

The outstanding high-temperature strength of titanium aluminides offers the possibility of increasing the maximum temperature limit of titanium-based alloys as well as being a lighter-weight substitute for nickel alloys in

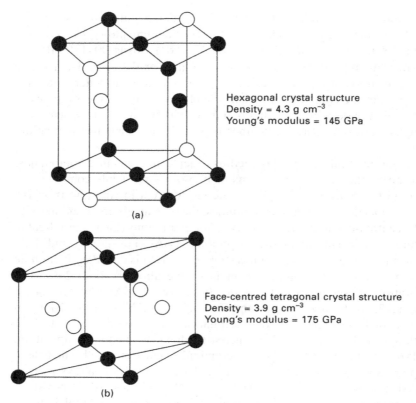

Hexagonal crystal structure
Density = 4.3 g cm^{-3}
Young's modulus = 145 GPa

(a)

Face-centred tetragonal crystal structure
Density = 3.9 g cm^{-3}
Young's modulus = 175 GPa

(b)

9.7 Crystal structure and properties of (a) Ti$_3$Al and (b) TiAl; the black and white circles represent titanium and aluminium atoms, respectively.

jet engines. However, titanium aluminides are susceptible to brittle fracture owing to their low ductility and poor fracture toughness, and this is a key factor stopping their use in aircraft engines. Typically, titanium aluminides have a room temperature ductility of 1–2% and a fracture toughness of about 15 MPa m$^{-0.5}$, which is lower than the ductility and toughness of conventional titanium alloys. Another problem is the high cost of processing engine components from titanium aluminides, being more expensive than for conventional titanium and nickel alloys. Aerospace companies are seeking ways to reduce the manufacturing costs and increase the ductility so titanium aluminides may eventually find use in jet engines.

9.5   Shape-memory titanium alloys

Shape-memory alloys are a unique class of material with the ability to 'remember' their shape after being plastically deformed. Shape-memory

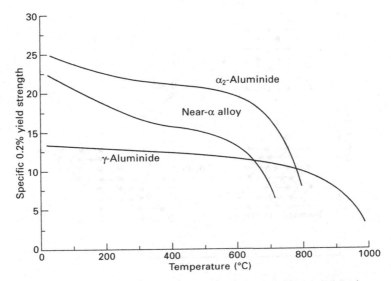

9.8 Effect of temperature on the average specific yield strengths of α_2-aluminide, γ-aluminide and a near-alpha alloy (data from J. Kumpfert and C. H. Ward, in *Advanced aerospace materials*, ed. H. Buhl, Springer Verlag, Berlin, 1992).

alloys revert back to their original shape by heating or some other external stimulus, provided the deformation they experience is within a recoverable range. The process of deformation and shape recovery can be repeated many times, and this offers the possibility of using shape-memory alloys in flight control systems. These alloys are being evaluated for a number of aircraft systems that would benefit from the shape-memory effect, such as control surfaces and hydraulic systems. Shape-memory alloys are not presently used in aircraft, although it is still worthwhile studying these materials because of their possible use in the future.

The most effective and common shape-memory alloy is nickel–titanium, where the titanium content is 45–50%. Nickel–titanium alloys that are commercially available are traded under the brand-name 'Nitinol'. Several non-titanium alloys also have shape-memory properties, most notably copper–zinc–aluminium and copper–aluminium–nickel alloys, but their shape-memory performance and mechanical properties are not as good as nickel–titanium.

The shape-memory effect is possible because a phase change or, in other words, a change to the crystal structure occurs in the material with the application of stress and heat, as shown schematically in Fig. 9.9. Nickel–titanium alloys being considered for use in aircraft exist as two phases: martensite and austenite. A shape-memory alloy in the undeformed

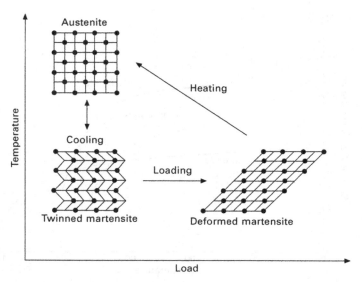

9.9 Effect of load and temperature on the shape-memory effect in nickel–titanium alloys.

(original) condition exists as twinned martensite. The microstructure of twinned martensite consists of grains separated by twin boundaries. These boundaries can be considered as planes of symmetry with a mirror-image of identical bonding and atomic configuration in both directions. When twinned martensite is deformed under an externally applied load the twin boundaries move to produce deformed martensite. This change is called a displacive transformation, which involves the co-operative rearrangement of atoms into a different, more stable crystal structure without a corresponding change in volume.

The shape-memory effect occurs because deformed martensite converts back to twinned martensite by heating the nickel–titanium alloy to a moderate temperature. Heat provides the energy needed to transform deformed martensite into austenite, which has a cubic crystal structure, which then transforms to twinned martensite upon cooling. When deformed martensite converts back into twinned martensite via the austenite phase the alloy recovers its original shape. This process of shape change under applied loading and then shape recovery by heating can be repeated many times. The temperature that shape-memory alloys must be heated to in order to induce the shape-memory effect is determined by their chemical composition, and for nickel–titanium alloys this is within the range of −100 and +100 °C.

It is worth noting that there is a strain limit to which shape-memory alloys can be deformed. Full shape recovery cannot be achieved when the material is deformed above this limit. The strain limit for nickel–titanium alloys is about 8.5%, which is high enough for many aerospace applications. There is a special class of material, known as ferromagnetic shape-memory alloys, which recover their shape under a strong magnetic field rather than heat.

Nickel–titanium alloys are being evaluated for a variety of aircraft components for their shape-memory properties. One of the most promising applications is control surfaces, such as flaps, rudders and ailerons. The movement of control surfaces on most modern aircraft is performed using hydraulic actuator systems, which are heavy, expensive and difficult to maintain. It is considered possible to replace hydraulic systems with shape-memory alloys embedded in the control surface. Figure 9.10 illustrates the design concept for a 'smart wing' that contains shape-memory alloy to control and manipulate a flexible surface. The wing contains wires of shape-memory alloy along the top and bottom surfaces. When the wing flexes down, the top wire is stretched thus transforming the alloy from twinned martensite to deformed martensite. Likewise, when the wing is bent upwards the bottom wire is stretched and thereby transformed to deformed martensite. The shape-memory effect in the wires is simply induced by heating them using an electric current, which causes the wing to recover its original shape. In this way, shape-memory alloys can be used to operate control surfaces without the need for conventional hydraulic systems.

9.6   Summary

Titanium alloys are used in heavily-loaded airframe sections, undercarriage parts, skins to high Mach speed aircraft, jet engines, and many other aircraft components requiring high strength, fracture toughness and resistance to fatigue, creep and corrosion. These mechanical properties are generally superior to lightweight structural materials, including aluminium alloys and magnesium alloys.

Titanium alloys occur in two allotropic forms at room temperature: α-Ti and β-Ti. α-Ti occurs when titanium contains a sufficient quantity of α-stabilising alloy elements, such as Al and Sn. β-Ti exists with the use of β-stabilising elements, such as V, Cr and Mo.

Near α-Ti is characterised by high creep resistance and, therefore, is used

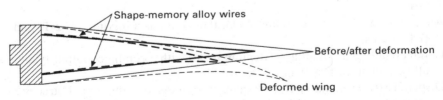

9.10 Schematic of a flexible smart aircraft wing containing shape-memory alloy.

in jet engine and rocket propulsion components operating up to 500–600 °C. α-Ti also has good fracture toughness and ductility.

β-Ti is characterised by high strength and fatigue resistance which makes it suitable for heavily loaded aircraft components.

Titanium alloys comprising both α-Ti and β-Ti are most often used in aircraft. Ti–6Al–4V is the most common α+β-Ti alloy, which possesses high creep resistance and toughness (from the α-Ti) and high strength and fatigue resistance (from the β-Ti). Ti–6Al–4V is used in airframes and jet engines.

Titanium aluminides (Ti_3Al, TiAl) may emerge as an important aerospace material in future jet engines because of their lower density, higher stiffness and superior high-temperature strength compared with titanium alloys. However, titanium aluminides are susceptible to brittle fracture owing to their low fracture toughness as well as being expensive to process and manufacture.

Shape-memory alloys based on nickel–titanium alloys possess the unique property of recovering their original shape following plastic deformation. These alloys undergo displacive transformation processes during deformation and then heating which allows the material to return to its original shape. Shape-memory alloys have potential aerospace applications where this property is useful, such as aircraft control surfaces.

9.7   Terminology

Allotropic transformation: Transformation process which changes the physical form (i.e. crystal structure) of a material.

Allotropy: Different physical forms (i.e. crystal structure) of the same material which are chemically very similar.

Alpha stabilisers: Alloying elements that promote the formation of the alpha phase of titanium.

Alpha titanium: Titanium metal with a hexagonal close packed (hcp) crystal structure.

Beta stabilisers: Alloying elements that promote the formation of the beta phase of titanium.

Beta titanium: Titanium metal with a body-centred-cubic (bcc) crystal structure.

Alpha + beta titanium: Titanium metal that contains a mixture of α-Ti and β-Ti grains.

Commercially pure titanium: Titanium metal with a combined impurity and alloy content of less than 1% by weight.

Displacive transformation: Rearrangement of atoms from one crystal structure to another structure without a change in the density (or volume) of the material.

Electron valence number: Number of electrons in the outer shell of an atom.

Near-alpha: Titanium metal composed mostly of α-Ti grains with a very small amount (<2% by volume) of β-Ti grains.

Neutral elements: Alloying elements that do not promote the formation of the α-Ti and β-Ti phases.

Ordered intermetallic compounds: Compounds formed from two or more metals that combine in a fixed ratio with a crystal different in structure from the individual metals.

Titanium aluminides: Ordered intermetallic compounds formed by the combination of titanium and aluminium. Common examples are TiAl and Ti_3Al. Characterised by good creep resistance and high strength at elevated temperature, but low ductility.

Shape-memory alloys: Metals alloys that can be permanently deformed and then reverted back to their original shape many times without being damaged. Shape recovery is driven by external stimuli, such as heat or electromagnetism.

Super-alpha: Titanium metal composed almost exclusively of α-Ti grains with a negligible amount of β-Ti grains.

Transus temperature: The temperature when an allotropic transformation occurs to a material.

Twin boundary: A grain boundary whose lattice structures are mirror images of each other in the plane of the boundary.

9.8   Further reading and research

Boyer, R. R., 'An overview on the use of titanium in the aerospace industry', *Materials Science and Engineering A*, **213**, (1996), 103–113.

Donanchie, M. J., *Titanium: a technical guide*, ASM International, 2000.

Loretto, M., 'TiAl-based alloys for aeroengine applications', in *Aerospace materials*, edited B. Cantor, H. Assender and P. Grant, Institute of Physics Publishing, Bristol, 2001, pp. 229–240.

Lütjering, G. and Williams, J. C., *Titanium* (2nd edition), Springer, 2007.

Polmear, I. J., *Light alloys: metallurgy of light metals*, Hodder and Stoughton, London, 1995.

Magnesium alloys for aerospace structures

10.1   Introduction

Magnesium is the lightest of all the metals used in aircraft. The density of magnesium is only 1.7 g cm^{-3}, which is much lower than the specific gravity of the other aerospace structural metals: aluminium (2.7 g cm^{-3}), titanium (4.6 g cm^{-3}), steel (7.8 g cm^{-3}). Only carbon-fibre composite material has a density (~1.7 g/cm^3) that is similar to magnesium. Magnesium alloys have lower stiffness and strength properties than the aerospace structural materials, but because of its low density the specific properties are similar. However, there are several problems with using magnesium alloys, including higher cost and lower strength, fatigue life, ductility, toughness and creep resistant properties compared with aluminium alloys. Poor resistance against corrosion is one of the greatest problems with magnesium. Magnesium and its alloys are one of the least corrosion-resistant metals, with its corrosion performance being vastly inferior to the other aerospace metals. There is a widespread belief that a serious safety concern with using magnesium is flammability. Magnesium burns when exposed to high temperature, and therefore may pose a major fire risk. However, there have been no cases of aircraft accidents caused by the ignition of magnesium. Magnesium meets all the aerospace standards for material flammability resistance. The main reason why magnesium is used sparingly in modern aircraft, typically less than 1% of the structural mass of large passenger aircraft, is poor corrosion resistance.

Although the use of magnesium in aerospace structures is now extremely limited, for many years it was used extensively in structural components in aircraft, helicopters and spacecraft because of its light weight. Magnesium was originally used in aircraft during the 1940s and for the next thirty years was a common structural material. Magnesium passed through a boom period between the 1950s and early 1970s when military and civilian aircraft were built using hundreds of kilograms of the material. Magnesium was used extensively in airframes, aviation instruments and low-temperature engine components for aircraft, especially fighters and military helicopters, and semistructural parts for spacecraft and missiles. Since the 1970s, however, the use of magnesium has declined owing to high cost, poor corrosion resistance and other factors, and it is now rarely used in aircraft, spacecraft and missiles. The use of magnesium alloys is now largely confined to engine parts, and

224

common applications are gearboxes and gearbox housings for aircraft and the main transmission housing for helicopters (Fig. 10.1). Magnesium has good damping capacity and therefore is often the material of choice in harsh vibration environments, such as helicopter gearboxes.

Despite the general decline in the use of magnesium, it remains an important material for specific aerospace applications. In this chapter, we study the metallurgy, mechanical properties and aerospace applications of magnesium. The following aspects of magnesium alloys are discussed: advantages and drawbacks of using magnesium in aircraft and helicopters; classification system for magnesium alloys; types of magnesium alloys used in aircraft and helicopters; and the engineering properties of magnesium alloys.

10.2   Metallurgy of magnesium alloys

10.2.1 Classification system for magnesium alloys

An international standardised system for classifying magnesium alloys, similar to the International Alloy Designation System for aluminium, does not exist. However the American Society for Testing and Materials (ASTM) has developed a coding system that is widely used by the magnesium industry. The system is a 'letter-letter-number-number' code. The letters indicate the two principal alloy elements listed in order of decreasing content. When the two alloying elements are present in an equal amount, then they are listed

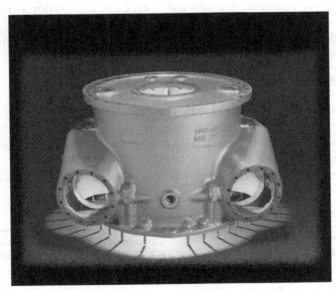

10.1 Helicopter gearbox casing made of magnesium alloy.

alphabetically. The code letters used to identify the alloying elements are given in Table 10.1. The two numbers specify the weight percent of the two principal alloying elements, rounded off to the nearest whole number and listed in the order of the two elements. For instance, the alloy AZ91 signifies that aluminium (A) and zinc (Z) are the two main alloy elements, and these are present in weight percentages of about 9 and 1%, respectively. As another example, WE43 identifies the alloy as containing 4% yttrium and 3% rare earth elements. In addition to the alloying elements specified in the code, magnesium alloys often contain other elements in lesser amounts. However, the coding system provides no information about these elements. In exceptional circumstances when an alloy contains three main alloying elements, then a 'three letter–three number' code system is used. For example, ZMC711 contains about 7% zinc, 1% manganese and 1% copper.

Magnesium alloys are classified as wrought or casting alloys. Wrought alloys account for only a small percentage (under 15%) of the total consumption of magnesium, and these alloys are not used in aircraft. A problem with wrought alloys is their low yield strength (typically less than 170 MPa). Most magnesium alloys that are used commercially, including those in aircraft and helicopters, are casting alloys. The casting alloys are often used in the tempered condition; that is heat-treated and work hardened, under conditions similar to the tempering of aluminium alloys. For this reason, the system used to describe the temper of aluminium alloys is also used for magnesium alloys (see Table 8.2). The temper conditions most often applied to magnesium are T5 (alloy is artificially aged after casting), T6 (alloy is solution treated, quenched and artificially aged), and T7 (alloy is solution treated only).

The major groups of magnesium alloys are based on the following compositions:

- magnesium–aluminium–manganese (Mg–Al–Mn)
- magnesium–aluminium–zinc (Mg–Al–Zn)
- magnesium–zinc–zirconium (Mg–Zn–Zr)
- magnesium–rare earth metal–zirconium (Mg–E–Zr)
- magnesium–rare earth metal–silver–zinc (with or without thorium) (Mg–E–Ag–Zn)
- magnesium–thorium–zirconium (with or without zinc) (Mg–Th–Zr)

Table 10.1 ASTM lettering system for magnesium alloys

A: Aluminium	B: Bismuth	C: Copper	D: Cadmium
E: Rare earths	F: Iron	H: Thorium	K: Zirconium
L: Lithium	M: Manganese	N: Nickel	P: Lead
Q: Silver	R: Chromium	S: Silicon	T: Tin
W: Yttrium	Z: Zinc		

10.2.2 Composition and properties of magnesium alloys

The composition and tensile properties of pure magnesium and several alloys used in aircraft are given in Table 10.2. Most of the magnesium alloys used in aircraft are from the Mg–Al–Zn, Mg–Al–Zr and Mg–E–Zr systems. The aerospace applications for magnesium alloys are listed in Table 10.3. As mentioned, the most common application is the transmission casings and main rotor gearbox of helicopters which utilise the low weight and good vibration damping properties of magnesium. Magnesium alloys are also used in aircraft engine and gearbox components.

10.2.3 Strengthening of magnesium alloys

Pure magnesium does not have sufficient strength or corrosion resistance to be suitable for use in aircraft. Magnesium has a hexagonal close packed (hcp) crystal structure. As mentioned in chapter 4, hcp crystals have few slip systems (three) along which dislocations can move during plastic deformation. As a result, it is not possible to greatly increase the strength of hcp metals by work-hardening. For example, annealed magnesium has a yield strength of about 90 MPa, and heavy cold-working of the metal only increases the strength to about 115 MPa. Another consequence of the hcp structure is the mechanical properties of wrought magnesium alloys are anisotropic and are

Table 10.2 Composition and properties of pure magnesium and its alloys used in aircraft and helicopters

Alloy	Composition	Yield strength (MPa)	Tensile strength (MPa)	Elongation (%)
Pure Mg (annealed)	>99.9% Mg	90	160	15
Pure Mg (cold-worked)	>99.9% Mg	115	180	10
WE43 (T6)	Mg–5.1Y–3.25Nb–0.5Zr	200	285	4
ZE41 (T5)	Mg–4.2Zn–1.3E–0.7Zr	135	180	2
QE21 (T6)	Mg–2.5Ag–1Th–1Nb–0.7Zr	185	240	2
AZ63 (T6)	Mg–6Al–3Zn–0.3Mn	110	230	3
AK61 (T6)	Mg–6Zn–0.7Zr	175	275	5

Table 10.3 Aerospace applications for magnesium alloys

Alloy	Application
RZ5	Helicopter transmission; aircraft gearbox casings; aircraft generator housing (e.g. A320, *Tornado, Concorde*)
WE43	Helicopter transmission (e.g. *Eurocopter* EC120, NH90; *Sikorsky* S92)
ZE41	Helicopter transmission
QE21	Aircraft gearbox casing; auxiliary gearbox (e.g. F-16, *Eurofighter, Tornado*)
ZW3	Aircraft wheels; helicopter gearbox (e.g. Westland *Sea King*)

different dependent on the loading direction. Owing to the anisotropy, the compressive yield strength of wrought magnesium alloys can be 30–60% lower than the tensile yield strength.

There are two broad classes of magnesium alloys that are strengthened by cold working or solid solution hardening combined with precipitation hardening. As mentioned, it is difficult to greatly increase the strength of magnesium by cold working owing to the hcp crystal structure, and therefore the majority of magnesium alloys used in aerospace applications are strengthened by the combination of solid solution and precipitation hardening. The strength properties of magnesium are improved by a large number of different alloying elements, and the main ones are aluminium and zinc. Other important alloying elements are zirconium and the rare earths. Rare earths are the thirty elements within the lanthanide and actinide series of the Periodic Table, with thorium (Th) and neodymium (Nd) being the most commonly used as alloying elements.

A problem with magnesium, however, is that the addition of alloying elements provides only a relatively small improvement to the strength properties. Compared with the annealed pure metal, the increase in yield strength of magnesium owing to alloying is typically in the range of 20 to 200%. In comparison, the alloying of annealed aluminium increases the yield strength by more than 1000% and the strength of annealed titanium is improved by up to 700%. The low response of magnesium to strengthening by alloying and work-hardening, together with low elastic modulus, ductility and corrosion resistance, are important reasons for the low use of this material in modern aircraft.

The majority of alloying elements used in magnesium increase the strength by solid-solution hardening and dispersion hardening. The alloying elements react with the magnesium to form fine intermetallic particles that increase the strength by dispersion hardening. The three most common intermetallic particles have the chemical composition: MgX (e.g. MgTl, MgCe, MgSn); MgX_2 (e.g. $MgCu_2$, $MgZn_2$); and Mg_2X (e.g. Mg_2Si, Mg_2Sn). These compounds are effective at increasing the strength by dispersion hardening, but they reduce the fracture toughness and ductility of magnesium. For example, the Mg–Al–Mn and Mg–Al–Zn alloys used in aircraft form particles ($Mg_{17}Al_{12}$) at the grain boundaries which lower the toughness and ductility.

Magnesium alloys must be heat treated before being used in aircraft to minimise the adverse effects of the intermetallic particles on toughness. This involves solution treating the magnesium at high temperature to dissolve the intermetallic particles in order to release the alloying elements into solid solution. The material is then thermally aged to maximise the tensile strength by precipitation hardening. A typical heat-treatment cycle involves solution treating at about 440 °C, quenching, and then thermally ageing at 180–200 °C for 16–20 h. These heat-treatment conditions are similar to those used to strengthen age-hardenable aluminium alloys. However, the response of magnesium to precipitation hardening is much less effective than aluminium. Only relatively small improvements to the tensile strength of magnesium alloys are gained by precipitation hardening. This is because the density and mobility of dislocations in magnesium is relatively low owing to the small number of slip systems in the hexagonal crystal structure.

The precipitation processes that occur in most magnesium alloys during thermal ageing are complex. In chapter 8, it is mentioned that aluminium alloys undergo the following transformation sequence in the age-hardening process: supersaturated solid solution → GP1 zones → GP2 zones → coherent intermetallic precipitates → incoherent intermetallic precipitates. Some magnesium alloys also undergo this sequence of transformations during ageing whereas other alloys form precipitates without the prior formation of GP zones. The types of precipitates that develop are obviously dependent on the composition and heat-treatment conditions. Precipitates in Mg–E–Zr alloys, such as WE43 used in helicopter transmissions, are $Mg_{11}NdY$ and/or $Mg_{12}NdY$ compounds. Precipitates in Mg–Zn alloys, such as ZE41 that is also used in helicopter transmissions, are coherent $MgZn_2$, semicoherent $MgZn_2$, and incoherent Mg_2Zn_3 particles. Several magnesium alloys used in aircraft contain aluminium, such as QE21 that is used in aircraft gearboxes. The main precipitate formed in Mg–Al alloys is $Mg_{17}Al_{12}$, which is effective at increasing strength. Figure 10.2 shows the effect of aluminium content on the tensile properties of a fully heat-treated Mg–Al–Zn alloy. The yield and ultimate tensile strengths increase with the aluminium content owing to

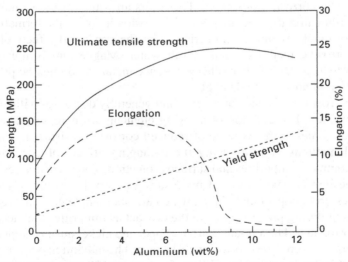

10.2 Effect of aluminium content on the tensile properties of a heat-treated magnesium alloy.

solid solution hardening and precipitation hardening. When the aluminium content exceeds 6–8%, the ductility is reduced owing to embrittlement of the grain boundaries by $Mg_{17}Al_{12}$ particles and, for this reason, the aluminium concentration is kept below this limit.

Two important alloying elements used in magnesium are zirconium and thorium. Zirconium is used for its ability to reduce the grain size. Cast magnesium has a coarse grain structure which results in low strength owing to the weak grain boundary hardening effect. Zirconium is used in small amounts (0.5% to 0.7%) to refine the grain structure and thereby increase the yield strength. In the past, thorium was often used to reduce the grain size and, for many years, magnesium–thorium alloys were used in components for missiles and spacecraft. However, thorium is a radioactive element that poses a health and environment hazard and, therefore, its use has been phased out over the past twenty years and it is now obsolete as an alloying element.

10.2.4 Corrosion properties of magnesium alloys

The biggest obstacle to the use of magnesium alloys is their poor corrosion resistance. Magnesium occupies one of the highest anodic positions in the galvanic series, and for this reason has a high potential for corrosion. There are many types of corrosion (as explained in chapter 21), and the most damaging forms to magnesium and its alloys are pitting corrosion and stress corrosion. Pitting corrosion, as the name implies, involves the formation of small pits over the metal surface where small amounts of material are dissolved by corrosion processes. These pits are sites for the formation of cracks within aircraft structures. Stress corrosion involves the formation and growth of cracks within the material under the combined effects of stress and a corrosive medium, such as salt water. Magnesium is much more susceptible to corrosion than other aerospace metals, and must be protected to avoid rapid and severe damage.

The corrosion resistance of magnesium generally decreases with alloying and impurities. The addition of alloying elements to increase the strength properties comes at the expense of reduced corrosion resistance. The most practical methods of minimising the damaging effects of corrosion are careful control of impurities and surface protection. The corrosion resistance of magnesium can be improved by ensuring a very low level of cathodic impurities. Iron, copper and nickel, which are not completely removed from magnesium during processing from the ore, act as impurities that accelerate the corrosion rate. The amount of iron, copper and nickel must be kept to very low levels to ensure good corrosion resistance. The maximum concentrations are only 1300 ppm (parts per million) for copper, 170 ppm for iron and and 5 ppm for nickel. In some aerospace magnesium alloys (e.g. AZ63), a small amount of manganese (<0.3%) is used to improve corrosion resistance. The manganese reacts with the impurities to form relatively harmless intermetallic compounds, some of which separate out during melting. The other way of protecting magnesium alloys from corrosion is surface treatments and coatings.

10.3   Summary

Magnesium used to be a popular aircraft structural material owing to its low density. However, the use of magnesium alloys has fallen from the boom period of the 1950s and 1960s when it was commonly used in aircraft, helicopters and missiles. The use in modern aircraft is limited mostly to gearboxes and gearbox housings for fixed-wing aeroplanes and main transmission housings on helicopters which require the high vibration damping properties afforded by magnesium.

The greatest problem with using magnesium in aircraft is poor corrosion resistance. The impurities content must be carefully controlled and the surface protected with treatments or coatings to avoid corrosion problems.

Other drawbacks include high cost and low stiffness, strength, toughness and creep resistance compared with other aerospace structural materials. It is difficult to increase the strength properties of magnesium owing to its low responses to cold-working and precipitation hardening.

The magnesium alloys most used in aerospace components are Mg–Al–Zn, Mg–Al–Zr and Mg–E–Zr alloys. These materials are strengthened predominantly by solid solution and precipitation hardening. However, the maximum strengths of magnesium alloys are much lower (50% or more) than high-strength aluminium alloys.

10.4   Further reading and research

Ostrovsky, I. and Henn, Y., 'Present state and future of magnesium application in aerospace industry', in *Proceedings of the International Conference on New Challenges in Aeronautics*, Moscow, 2007, pp. 1–5.

Polmear, I. J., *Light alloys: metallurgy of light metals*, Hodder and Stoughton, London, 1995.

Polmear, I. J., 'Magnesium alloys and applications', *Materials Science and Technology*, **10** (1994), 1–16.

11
Steels for aircraft structures

11.1 Introduction

Steel is an alloy of iron containing carbon and one or more other alloying elements. Carbon steel is the most common material used in structural engineering, with applications in virtually every industry sector including automotive, marine, rail and infrastructure. The world-wide consumption of steel is around 100 times greater than aluminium, which is the second most-used structural metal. Figure 11.1 shows the production of steel, aluminium, magnesium and composites over the course of the 20th century, and the usage of steel amounts to more than 90% of all metal consumed. Although steel is used extensively in many sectors, its usage in aerospace is small in comparison to aluminium and composite material. The use of steel in aircraft and helicopters is often limited to just 5–8% of the total airframe weight (or 3–5% by volume).

The use of steel in aircraft is usually confined to safety-critical structural components that require very high strength and where space is limited. In other words, steel is used when high specific strength is the most important

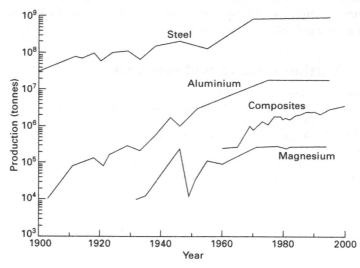

11.1 Production figures (approximate) for aluminium, magnesium, steel and composites.

232

criterion in materials selection. Steels used in aircraft have yield strengths above 1500–2000 MPa, which is higher than high-strength aluminium (500–650 MPa), α/β titanium (830–1300 MPa) or quasi-isotropic carbon–epoxy composite (750–1000 MPa). In addition to high strength, steels used in aircraft have high elastic modulus, fatigue resistance and fracture toughness, and retain their mechanical performance at high temperature (up to 300–450 °C). This combination of properties makes steel a material of choice for heavily-loaded aircraft structures. However, steel is not used in large quantities for several reasons, with the most important being its weight. The density of steel ($\rho = 7.2$ g cm^{-3}) is over 2.5 times higher than aluminium, 1.5 times greater than titanium, and more than 3.5 times heavier than carbon–epoxy composite. In addition to weight problems, most steels are susceptible to corrosion which causes surface pitting, stress corrosion cracking and other damage. High-strength steels are prone to damage by hydrogen embrittlement, which is a weakening process caused by the absorption of hydrogen. A very low concentration of hydrogen (as little as 0.0001%) within steel can cause cracking which may lead to brittle-type fracture at a stress level below the yield strength.

Aircraft structural components made using high-strength steel include undercarriage landing gear, wing-root attachments, engine pylons and slat track components (Fig. 11.2). The greatest usage of steel is in landing gear where it is important to minimise the volume of the undercarriage while having high load capacity. The main advantage of using steel in landing gear is high stiffness, strength and fatigue resistance, which provide the

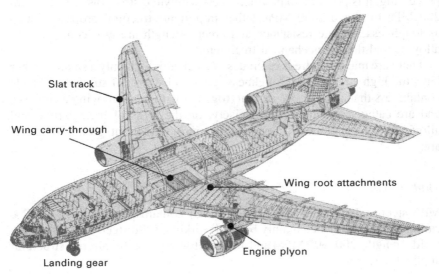

Slat track

Wing carry-through

Wing root attachments

Engine plyon

Landing gear

11.2 Aircraft applications of steel.

landing gear with the mechanical performance to withstand high impact loads on landing and support the aircraft weight during taxi and take-off. Owing to the high mechanical properties of steel, the load-bearing section of the landing gear can be made relatively small which allows storage within minimum space in the belly of an aircraft. Steel is also used in wing root attachments and, in some older aircraft, wing carry-through boxes owing to their high stiffness, strength, toughness and fatigue resistance. For similar reasons, steel is used in slat tracks which form part of the leading edge of aircraft wings.

This chapter presents an overview of the metallurgical and mechanical properties of steels for aircraft structural applications. The field of steel metallurgy is large and complex, although this chapter gives a short, simplified description of the basic metallurgy of steel. The chapter describes the steels used in landing gear and other high-strength components, namely, the maraging steels, medium-carbon low-alloy steels, and stainless steels.

11.2   Basic principles of steel metallurgy

11.2.1 Grades of steel

Iron is alloyed with carbon and other elements, forged and then heat-treated to produce high-strength steel. Pure iron is a soft metal having a yield strength under 100 MPa, but a tenfold or more increase in strength is achieved by the addition of carbon and other alloying elements followed by work hardening and heat treatment. By control of the alloy composition and thermomechanical processing, it is possible to produce steels with yield strengths ranging from 200 MPa to above 2000 MPa. Other important structural properties such as toughness, fatigue resistance and creep strength are also controlled by alloying and thermomechanical treatment.

There are many hundreds of grades of steel, although only a small number have the high strength and toughness required by aircraft structures. Steels contain less than about 1.5% carbon (together with other alloying elements), and are categorised, rather imprecisely, on the basis of their carbon and alloying element contents. Some of the most important groups of steels are:

Mild steels

Mild steels (also called low-carbon steels) contain less than about 0.2% carbon and are hardened mainly by cold working. Mild steels have moderate yield strength (200–300 MPa) and are therefore too soft for aircraft structural applications.

High-strength low-alloy steels

High-strength low-alloy (HSLA) steels contain a small amount of carbon (under 0.2%) like mild steels, and also contain small amounts of alloying elements such as copper, nickel, niobium, vanadium, chromium, molybdenum and zirconium. HSLA steels are referred to as micro-alloyed steels because they are alloyed at low concentrations compared with other types of steels. The yield strength of HSLA steels is 250–600 MPa and they are used in automobiles, trucks and bridges amongst other applications. The use of HSLA steels in aircraft is rare because of low specific strength and poor corrosion resistance.

Medium-carbon steels

Medium-carbon steels contain somewhere between 0.25 and 0.5% carbon and are hardened by thermomechanical treatment processes to strengths of 300–1000 MPa. This group of steels is used in the greatest quantities for structural applications, and they are found in motor cars, rail carriages, structural members of buildings and bridges, ships and offshore structures and, in small amounts, aircraft.

Medium-carbon low-alloy steels

Medium-carbon low-alloy steels also contain somewhere between 0.25 and 0.5% carbon but have a higher concentration of alloying elements to increase hardness and high-temperature strength. They contain elements such as nickel, chromium, molybdenum, vanadium and cobalt. Examples are given in Table 11.1 At the higher alloy contents these steels are used as tool steels (e.g. tool bits, drills, blades and machine parts) which require hardness and wear resistance at high temperature. Strength levels up to 2000 MPa can be achieved. These steels are used in aircraft, typically for undercarriage parts.

Maraging steels

Maraging steels also have a high alloy content, but with virtually no carbon (less than 0.03%). Alloying together with heat treatment (which, unlike that for the other steels described above, includes age-hardening) produces maraging steels with the unusual combination of high strength, ductility and fracture toughness. The strength of maraging steels is within the range of 1500–2300 MPa, which puts them amongst the strongest metallic materials. Maraging steel is used in heavily loaded aerospace components.

Table 11.1 Composition and properties of selected steels used in aircraft

Steel	Average composition	Yield strength (MPa)	Ultimate strength (MPa)	Elongation (%)
Maraging steels				
18Ni (200)	0.03 max C, 18 Ni, 8.5 Co, 3.3 Mo, 0.2 Ti, 0.1 Al	660	970	17
18Ni (250)	0.03 max C, 18 Ni, 8.5 Co, 5 Mo, 0.4 Ti, 0.1 Al	1700	1790	11
18Ni (300)	0.03 max C, 18 Ni, 9 Co, 5 Mo, 0.7 Ti, 0.1 Al	1950	2000	9
18Ni (350)	0.03 max C, 18 Ni, 12.5 Co, 4.2 Mo, 1.6 Ti, 0.1 Al	2300	2370	6
Medium-carbon low-alloy steels				
4130	0.3 C, 1.0 Cr, 0.5 Mn, 0.25 Si, 0.2 Mo	540	700	25
4340	0.4 C, 1.8 Ni, 0.8 Cr, 0.7 Mn, 0.25 Si, 0.25 Mo	410	750	22
300M	0.38 C, 1.8 Ni, 1.6 Si, 0.8 Cr, 0.8 Mn, 0.4 Mo, 0.05 min V	1590	1930	7
Aermet 100	0.25 C, 13.5 Co, 11 Ni, 3 Cr, 1.2 Mo	1720	1960	14
H11	0.35 C, 5.0 Cr, 1.5 Mo, 1.0 Si, 0.45 V, 0.4 Mn, 0.3 Ni	1650	2000	9
Precipitation-hardening stainless steels				
15-5 PH	0.07 C, 15 Cr, 4.5 Ni, 3.5 Cu, 1 Mn, 1 Si, 0.3 Nb	1400	1470	10
17-4 PH	0.07 C, 16 Cr, 4 Ni, 4 Cu, 1 Mn, 1 Si, 0.3 Nb	1150	1330	10

Stainless steels

Stainless steels are corrosion resistant materials that contain a small amount of carbon (usually 0.08–0.25%) and a high concentration of chromium (12–26%) and sometimes nickel (up to about 22%). There are several classes of stainless steels with various mechanical properties, and their yield strength covers a wide range (200–2000 MPa). Precipitation-hardening (PH) stainless steels are used increasingly in aerospace applications, particularly where both high strength and excellent corrosion resistance is required.

Of the many steels available, it is the medium carbon low-alloy steels, maraging steels and PH stainless steels that are most used in aircraft. The alloy composition and mechanical properties of several steels (but not all the grades) used in aircraft are given in Table 11.1.

11.2.2 Microstructural phases of steels

Steel is an allotropic material that can occur as several microstructural phases at room temperature depending on the alloy composition and heat treatment. The main phases of steel are called austenite, ferrite, pearlite, cementite, bainite and martensite. These phases have their own crystal structure and can all exist at room temperature. The high strength steels used in aircraft usually have a martensite microstructure, and steels with another microstructure are rarely used in highly-loaded aircraft components due to their lower strength. Although martensitic steels are the steel of choice for aircraft structural applications, it is still worth examining the other microstructural phases of steel to understand the reasons for selecting martensitic steels for aircraft.

The microstructural phases of steel can be understood from the phase diagram for iron–carbon, which is shown in Fig. 11.3. This diagram shows the equilibrium phases of iron that exist at various carbon contents and temperatures. This iron–carbon diagram is only valid when changes in temperature occur gradually (i.e. the steel is cooled or heated slowly) to allow sufficient time for the formation of stable (equilibrium) phases. The diagram is not valid when steel is cooled rapidly, which is a condition when metastable phases develop that are not represented in the diagram. Also, the phase diagram is only valid for the binary system of iron–carbon, and significant changes to the diagram may occur with the addition of alloying elements.

The phase diagram for iron–carbon is complex, although there are just a few simple things we need to understand from this diagram. The different phase regions of the diagram indicate different stable microstructures that exist for the range of temperatures and carbon contents bounded by the phase lines. The main phases in Fig. 11.3 are austenite (also called γ-iron), ferrite (α-iron) and cementite (Fe_3C).

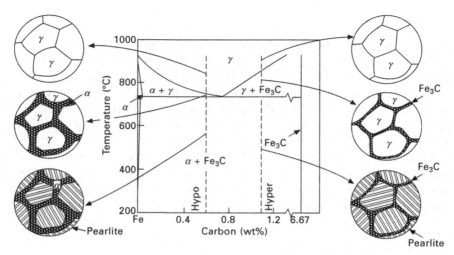

11.3 Iron–carbon phase diagram showing the microstructural phases during slow cooling (equilibrium conditions) of hypoeutectic and hypereutectic steels reproduced with permission (from D. R. Askeland, *The science and engineering of materials*, Stanley Thornes (Publishers) Ltd., 1996).

Austenite

Austenite is a materials science term for iron with a face-centred-cubic (fcc) crystal structure, and this phase occurs in the Fe–C system above the eutectoid temperature of 723 °C. The eutectoid temperature is the minimum temperature at which a material exists as a single solid solution phase or, in other words, when the alloying elements are completely soluble in the matrix phase. Alloying elements present in austenite are located at interstitial or substitutional sites in the iron fcc crystal structure, depending on their atomic size and valence number. Austenite becomes unstable when Fe–C is cooled below 723 °C, when it undergoes an allotropic transformation to ferrite and cementite (α-Fe + Fe$_3$C) during slow cooling. However, the addition of certain alloying elements, such as nickel and manganese, can stabilise the austenite phase at room temperature. On the other hand, elements such as silicon, molybdenum and chromium can make austenite unstable and raise the eutectoid temperature.

Ferrite

One phase formed during slow cooling of steel from the austenite phase is ferrite, which is a solid solution of body-centred-cubic (bcc) iron containing interstitial elements such as carbon and substitutional elements such as manganese and nickel. Carbon has a powerful hardening effect on ferrite by solid solution strengthening, but only a very small concentration can be dissolved into the interstitial lattice sites. The maximum solubility of carbon is about 0.02% at 723 °C, and the soluble concentration drops with the temperature to 0.005% at room temperature. Despite the low solubility of carbon, its presence in iron increases the strength at room temperature by more than five times. This is because the carbon atoms, which are about twice the diameter of the interstitial gaps in the ferrite crystal, induce a high elastic lattice strain which causes solid solution hardening. However, ferrite is soft and ductile compared with other phases of steel (i.e. cementite, bainite, martensite).

Cementite and pearlite

When the carbon content of steel exceeds the solubility limit of ferrite, then the excess carbon reacts with the iron to form cementite (or iron carbide) during slow cooling from the austenite phase region. Cementite is a hard, brittle compound with an orthorhombic crystal structure having the composition Fe_3C (Fig. 11.4). Upon slow cooling from the austenite phase, cementite and ferrite form as parallel plates into a two-phase microstructure called pearlite. The microstructure of pearlite is shown in Fig. 11.5, and it consists of thin plates of alternating phases of cementite and ferrite. The plates are very

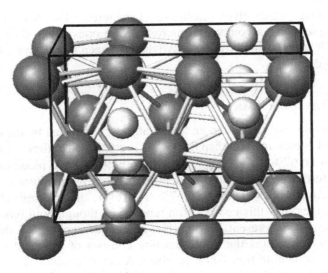

11.4 Orthorhombic Fe_3C; iron is shown as the dark atoms and carbon as the lighter atoms.

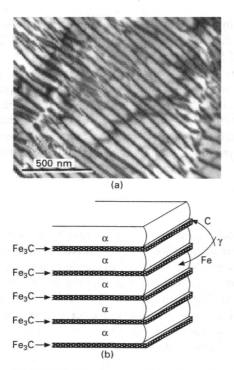

11.5 (a) Photograph and (b) schematic representation of pearlite (the photograph is from msm.cam.ac.uk).

narrow and long, which gives pearlite its characteristic lamellae structure. Pearlite is essentially a composite microstructure consisting of cementite layers (which are hard and brittle) sandwiched between ferrite layers (which are soft and ductile).

Hypo- and hypereutectic steels

The microstructure of steel at room temperature is controlled by the carbon content. Steel containing less than 0.005% C is composed entirely of ferrite, whereas above 6.67% C it is completely cementite. The Fe–C phase diagram shows that ferrite co-exists with cementite at room temperature over the carbon content range of 0.005 to 6.67%. However, the microstructure of the steel changes over this composition range even though both ferrite and cementite are always present. At a carbon content of about 0.8% the steel is eutectic, which means the microstructure consists of an equal amount of ferrite and cementite (i.e. 100% pearlite). A eutectic material has equal amounts of two (or more) phases and, for steel, it is composed of 50% ferrite and 50% cementite.

Steel containing more than 0.8% carbon is hypereutectic, which means the carbon content is greater than the eutectic composition. Hypereutectic steel contains grains of cementite and pearlite, with the volume fraction of cementite grains increasing with the carbon content above 0.8%. Steel is hypoeutectic when the carbon content is below 0.8%, and the microstructure consists of ferrite grains and pearlite grains (i.e. lamellae of ferrite and cementite within a single grain), as shown in Fig. 11.6. The volume fraction of pearlite grains increases, with a corresponding reduction in ferrite grains, when the carbon content is increased to the eutectic composition.

The majority of steels used in engineering structures, including aerospace, are hypoeutectic. Eutectic and hypereutectic steels lack sufficient ductility and toughness for most structural applications owing to the high volume fraction of brittle cementite. The mechanical properties of hypoeutectic steels are controlled by their carbon content. Figure 11.7 shows that raising the carbon content increases the strength properties and lowers the ductility and toughness. This is because the volume fraction of pearlite, which is hard and brittle, increases with the carbon content.

Martensite

The microstructure and mechanical properties of the hypoeutectic steels used in aircraft structures is controlled by heat-treatment as well as the carbon and alloy contents. Ferrite and cementite form when hypoeutectic steel

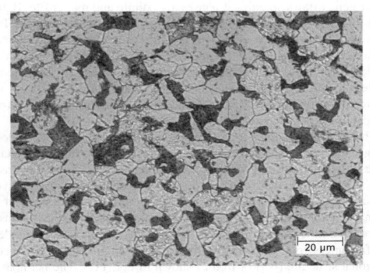

11.6 Hypoeutectic steel containing ferrite (light grey) and pearlite (dark grey) grains (from tms.org).

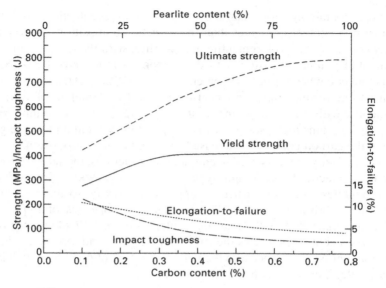

11.7 Effect of carbon content on the volume fraction of pearlite and mechanical properties for a hypoeutectic steel.

is cooled slowly from the austenite phase. Heat treatment processes that involve different cooling conditions can suppress the formation of ferrite and cementite and promote the formation of other phases which provide the steel with different mechanical properties such as higher hardness and strength. The principal transformational phases are bainite and martensite. Bainite is a metastable phase that exists in steel after controlled heat treatment. Bainite can form when steel is cooled from the austenite phase at an intermediate rate which is too rapid to allow the formation of ferrite and cementite but too slow to promote the formation of martensite. Bainite is generally harder and less ductile than ferrite and is tougher than martensite. Steels with a bainite microstructure are used in engineering structures, but not in aerospace applications.

When hot steel in the austenite-phase region is cooled rapidly it does not change into ferrite and cementite, but instead transforms to a metastable phase called martensite. Ferrite and cementite can only form from austenite when there is sufficient time for the iron and carbon atoms to move into the bcc crystal structure of α-iron and the orthorhombic structure of Fe_3C. When the cooling rate is rapid then the iron and carbon atoms in austenite do not have time to form these phases, and instead the material undergoes a diffusionless transformation to martensite, which has the body centred tetragonal structure (bct) of iron as shown in Fig. 11.8. The bct martensite structure is essentially the bcc austenite structure distorted by interstitial carbon atoms into a tetragonal lattice. The carbon atoms remain dissolved in the

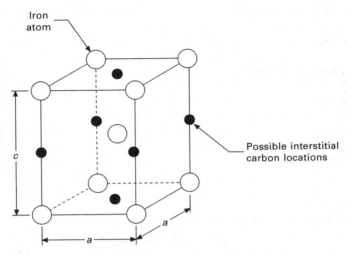

11.8 Body-centred-tetragonal structure of martensite.

crystal structure because there is insufficient time during quenching to form a carbon-rich second phase (Fe_3C). As a result, martensite is supersaturated with interstitial carbon atoms, which cause severe distortion and induce high strains in the crystalline lattice. Other alloying elements are dissolved in interstitial or substitutional sites and this also increases distortion of the lattice. This distortion increases the strength of martensite steel, with the yield stress being as high as 1800–2300 MPa. This is much higher than the strength of a steel with the same composition but consisting of ferrite and cementite (250–600 MPa). Unfortunately, the distortion of the crystal lattice makes martensite brittle and prone to cracking at low strain, and therefore as-quenched martensitic steel is not suitable for aircraft structures that require high toughness and damage tolerance.

Martensitic steels are tempered after quenching to increase ductility and toughness. Tempering involves heating the steel to a temperature below 650 °C to release some of the carbon trapped at the interstitial sites of the bct crystal and thereby relax the lattice strain. The freed carbon atoms react with the iron to produce iron carbides within the tempered martensite microstructure. However, tempering at too high a temperature causes the martensite to transform into ferrite and pearlite, and produces a large loss in strength. Figure 11.9 shows the effect of tempering temperature on the mechanical properties of a medium-carbon steel. The strength decreases whereas the ductility improves with increasing temperature owing to the lattice strain relaxation of the martensite. Tempered martensitic steels are used extensively in high hardness engineering structures, including many aircraft components.

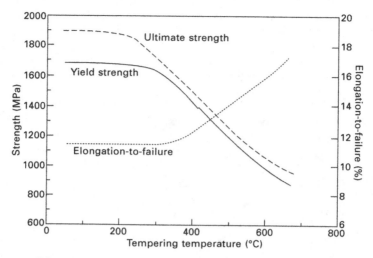

11.9 Effect of tempering temperature on the properties of 4340 medium-carbon low-alloy steel.

11.3   Maraging steel

Maraging steel is used in aircraft, with applications including landing gear, helicopter undercarriages, slat tracks and rocket motor cases – applications which require high strength-to-weight material. Maraging steel offers an unusual combination of high tensile strength and high fracture toughness. Most high-strength steels have low toughness, and the higher their strength the lower their toughness. The rare combination of high strength and toughness found with maraging steel makes it well suited for safety-critical aircraft structures that require high strength and damage tolerance. Maraging steel is strong, tough, low-carbon martensitic steel which contains hard precipitate particles formed by thermal ageing. The term 'maraging' is derived from the combination of the words martensite and age-hardening.

Maraging steel contains an extremely low amount of carbon (0.03% maximum) and a large amount of nickel (17–19%) together with lesser amounts of cobalt (8–12%), molybdenum (3–5%), titanium (0.2–1.8%) and aluminium (0.1–0.15%). Maraging steel is essentially free of carbon, which distinguishes it from other types of steel. The carbon content is kept very low to avoid the formation of titanium carbide (TiC) precipitates, which severely reduce the impact strength, ductility and toughness when present in high concentration. Because of the high alloy content, especially the cobalt addition, maraging steel is expensive.

Maraging steel is produced by heating the steel in the austenite phase region (at about 850 °C), called austenitising, followed by slow cooling in air to form a martensitic microstructure. The slow cooling of hypoeutectic steel from the austenite phase usually results in the formation of ferrite and pearlite; rapid cooling by quenching in water or oil is often necessary to form martensite. However, martensite forms in maraging steel upon slow cooling owing to the high nickel content which suppresses the formation of ferrite and pearlite. The martensitic microstructure in as-cooled maraging steel is soft compared with the martensite formed in plain carbon steels by quenching. However, this softness is an advantage because it results in high ductility and toughness without the need for tempering. The softness also allows maraging steel to be machined into structural components, unlike hard martensitic steels that must be tempered before machining to avoid cracking.

After quenching, maraging steel undergoes a final stage of strengthening involving thermal ageing before being used in aircraft components. Maraging steel is heat-treated at 480–500 °C for several hours to form a fine dispersion of hard precipitates within the soft martensite matrix. The main types of precipitates are Ni_3Mo, Ni_3Ti, Ni_3Al and Fe_2Mo, which occur in a high volume fraction because of the high alloy content. Carbide precipitation is practically eliminated owing to the low carbon composition. Cobalt is an important alloying element in maraging steel and serves several functions. Cobalt is used to reduce the solubility limit of molybdenum and thereby increase the volume fraction of Mo-rich precipitates (e.g. Ni_3Mo, Fe_2Mo). Cobalt also assists in the uniform dispersion of precipitates through the martensite matrix. Cobalt accelerates the precipitation process and thereby shortens the ageing time to reach maximum hardness. Newer grades of maraging steel contain complex $Ni_{50}(X,Y,Z)_{50}$ precipitates, where X, Y and Z are solute elements such as Mo, Ti and Al.

The precipitates in maraging steel are effective at restricting the movement of dislocations, and thereby promote strengthening by the precipitation hardening process. Figure 11.10 shows the effect of ageing temperature on the tensile strength and ductility of maraging steel. As with other age-hardening aerospace alloys such as the 2XXX Al, 7XXX Al, β-Ti and α/β-Ti alloys, there is an optimum temperature and heating time to achieve maximum strength in maraging steel. When age-hardened in the optimum temperature range of 480–500 °C for several hours it is possible to achieve a yield strength of around 2000 MPa while retaining good ductility and toughness. Over-ageing causes a loss in strength owing to precipitate coarsening and decomposition of the martensite with a reversion back to austenite. The strength of maraging steels is much greater than that found with most other aerospace structural materials, which combined with ductility and toughness, makes them the material of choice for heavily loaded structures that require high levels of damage tolerance and which must occupy a small space on aircraft.

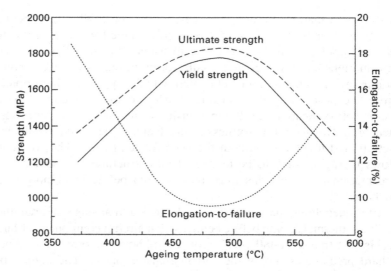

11.10 Effect of ageing temperature on the strength and ductility (percentage elongation to failure) of a maraging steel.

11.4   Medium-carbon low-alloy steel

Medium-carbon low-alloy steel contains 0.25–0.5% carbon and moderate concentrations of other alloying elements such as manganese, nickel, chromium, vanadium and boron. These steels are quenched from single-phase austenite condition, and then tempered to the desired strength level. Aircraft applications for this steel include landing gear components, shafts and other parts. There are numerous grades of medium-carbon low-alloy steel, and the most important for aerospace are Type 4340, 300M and H11 which cover the range from moderate-to-high strength, and provide impact toughness, creep strength and fatigue resistance.

11.5   Stainless steel

Aircraft components are made using stainless steel when both high strength and corrosion resistance are equally important. Stainless steel contains a large amount of chromium (12–26%) which forms a corrosion-resistant oxide layer. Chromium at the steel surface reacts with oxygen in the air to form a thin layer of chromium oxide (Cr_2O_3) which protects the underlying material from corrosive gases and liquids. The Cr_2O_3 layer is a very thin and impervious barrier which protects the underlying steel substrate.

There are several types of stainless steel: ferritic, austenitic, martensitic, duplex and precipitation-hardened stainless steels; although only the latter is used for structural aerospace applications because of its high tensile strength and toughness. Precipitation-hardened stainless steel used in aircraft has a tempered martensitic microstructure for high strength with a chromium oxide surface layer for corrosion protection. Like maraging steel, precipitation-hardened stainless steel is age-hardened by solution treatment, quenching and then thermal ageing at 425–550 °C. The most well known precipitation-hardened stainless steel is 17-4 PH (ASTM Grade A693), which contains a trace amount of carbon (0.07% max) and large quantity of chromium (15–17.5%) with lesser amounts of nickel (3–5%), copper (3–5%) and other alloying elements (Mo, V, Nb). The nickel is used to improve toughness and the other alloying elements promote strengthening by the formation of precipitate particles.

An early application of stainless steel was in the skins of super- and hypersonic aircraft, where temperature effects are considerable. Stainless steel was used in the skin of the Bristol 188, which was a Mach 1.6 experimental aircraft built in the 1950s to investigate kinetic heating effects. Stainless steel was also used in the American X-15 rocket aircraft capable of speeds in excess of Mach 6 (Fig. 11.11). The skin of these aircraft reaches high temperature owing to frictional heating, and this would cause softening of aluminium if it were used. Steel is heat resistant to 400–450 °C without any significant reduction in mechanical performance. Stainless steel is no longer used in super- and hypersonic aircraft owing to the development of other heat-resistant structural materials which are much lighter, such as titanium. Stainless steel is currently used in engine pylons and several other structural components which are prone to stress corrosion damage, although its use is limited.

11.6   Summary

There are many types of steels used in structural engineering, but only maraging steel, medium-carbon low-alloy steel and precipitation-hardened stainless steel are used in aircraft structures. Steel is used in structures requiring high strength and toughness but where space is limited, such as landing gear, track slats, wing carry-through boxes, and wing root attachments. The percentage of the airframe mass that is constructed of steel is typically 5–8%.

Advantages of using steel in highly-loaded aircraft structures include high stiffness, strength, fatigue resistance and fracture toughness. Stainless steel provides the added advantage of corrosion resistance. Problems encountered with steels include high weight, potential hydrogen embrittlement, and (when stainless steel is not used) stress corrosion cracking and other forms of corrosion.

Maraging steel is used in aircraft components because it combines very high strength (about 2000 MPa) with good toughness, thereby providing

(a)

(b)

11.11 Stainless steel was used in the (a) Bristol 188 and (b) X-15. (a) Photograph reproduced with permission from Aviation Archives. (b) Photograph reproduced with permission from NASA.

high levels of damage tolerance. These properties are achieved by the microstructure consisting of a ductile martensite matrix strengthened by hard precipitate particles formed by thermal ageing.

Medium-carbon low-alloy steel also combines high strength and toughness owing to a microstructure of tempered martensite containing hard carbide precipitates. This type of steel has similar mechanical properties to maraging steel, but is more susceptible to stress corrosion cracking. It is used widely in aerospace structural components.

Precipitation-hardened stainless steel is characterised by high strength and excellent corrosion resistance, and is used in aircraft structures prone to stress corrosion. The microstructure consists of martensite and precipitation particles. High corrosion resistance occurs by the formation of a chromium oxide surface layer which is impervious to corrosive gases and fluids.

11.7   Terminology

Austenite: Phase of iron or steel with a face-centred-cubic (fcc) crystal structure.

Austenitising: Heating steel to a temperature at which it transforms to austenite.

Bainite: Metastable phase of steel that occurs as a range of microstructures comprising fine carbide precipitates in an acicular (needle-shaped) ferrite matrix. The precise structure is a function of the temperature at which the carbides precipitate.

Cementite: Iron carbide (Fe_3C).

Eutectic: Mixture of two solid solutions, formed on solidification from a single liquid phase, at the eutectic temperature.

Eutectoid temperature: Lowest temperature at which austenite is stable.

Ferrite: Iron or steel having a body-centred-cubic (bcc) crystal structure.

Hypereutectic: Material with an alloy content greater than that of the eutectic composition, which is about 0.8% C for steel (Fe–C).

Hypoeutectic: Materials with an alloy content less than that of the eutectic composition, which is about 0.8% C for steel (Fe–C).

Hydrogen embrittlement: Loss in ductility and cracking of steel caused by the absorption of hydrogen.

Martensite: Microstructural phase of steel formed by a diffusionless shear mechanism when the material is cooled at a fast rate from the austenite phase. The fast cooling rate retains carbon in supersaturated solid solution of body-centred-tetragonal (bct) iron.

Pearlite: Steel consisting of alternating layered phases of ferrite and cementite.

11.8   Further reading and research

ASM handbook volume 1: Properties and selection: irons, steels, and high-performance alloys, ASM International, 1990.

Campbell, F. C., *Manufacturing technology for aerospace structural materials*, Elsevier, Amsterdam, 2006.

Davies, D. A., 'Structural steels' in *High-performance materials in aerospace*, edited H. M. Flower, Chapman and Hall, London, 1995, pp. 155–180.
Honeycomb, R. W. K. and Bhadeshia, H. K. D. H., *Steels: microstructure and properties*, Butterworth–Heinemann, Oxford, UK, 2006.

12
Superalloys for gas turbine engines

12.1 Introduction

Superalloys are a group of nickel, iron–nickel and cobalt alloys used in aircraft turbine engines for their exceptional heat-resistant properties. Materials used in jet engines must perform for long periods of time in a demanding environment involving high temperature, high stress and hot corrosive gas. Many materials simply cannot survive the severe conditions in the hottest sections of engines, where the temperatures reach ~1300 °C. Superalloys, on the other hand, possess many properties required by a jet-engine material such as high strength, long fatigue life, fracture toughness, creep resistance and stress-rupture resistance at high temperature. In addition, superalloys resist corrosion and oxidation at high temperatures, which cause the rapid deterioration of many other metallic materials. Superalloys can operate at temperatures up to 950–1300 °C for long periods, making them suitable materials for use in modern jet engines.

Superalloys have played a key role in the development of high thrust engines since the 1950s when the era of jet-powered civil aviation and rocketry began. Jet aircraft would fly at slower speeds and with less power without superalloys in their engines. The most effective way of increasing the thrust of jet engines is by increasing their operating temperature. This temperature is limited by the heat resistance of the engine materials, which must not distort, soften, creep, oxidise or corrode. Superalloys with their outstanding high-temperature properties are essential in the development of jet engines.

The important role of superalloys in raising the maximum operating temperature of jet engines is shown in Fig. 12.1. This figure shows the improvement in creep resistance using an industry benchmark of the maximum temperature that materials can withstand without failing when loaded at 137 MPa (20 ksi) for 1000 h. Over the era of jet aircraft, the maximum temperature has risen over 50%. The benefits that the increased operating temperature has provided in engine power have been enormous. Over the past 20 years the thrust of the gas turbine engine has increased by some 60% while over the same period the fuel consumption has fallen by 15–20%. The impressive achievements in engine power and fuel efficiency have been accomplished in part by improvements in the material durability in the hottest sections of the engine, and in particular the high-pressure turbine blades.

251

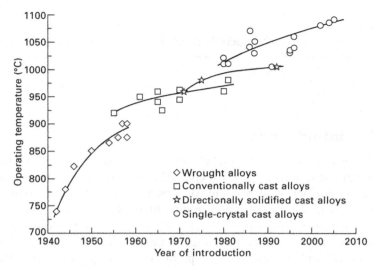

12.1 Improvement in the temperature limit of superalloys for aircraft turbine engines. The operating temperature is defined as the creep life of the material when loaded to 137 MPa (20 ksi) for 1000 h.

Engine durability has also improved dramatically owing to advances in engine design, propulsion technology and materials. Improved durability allows the operator to better utilise the aircraft by increasing engine life and reducing maintenance inspections and overhauls. When the Boeing 707 first entered service in 1957 the engines were removed for maintenance after about 500 h of operation. Today, a modern Boeing 747 engine can operate for more than 20 000 hours between maintenance operations. This remarkable improvement is the result of several factors, including the use of materials with improved high-temperature properties and durability.

The development of superalloys together with other advances in engine technology has pushed the operating limit to ~1300 °C, resulting in powerful engines for large civil aircraft and high thrust engines for supersonic military fighters. The processing methods used to fabricate engine components have been essential in raising the operating temperatures of engines. The development of advanced metal casting and processing methods has been important in increasing properties such as creep resistance at high-temperature.

Figure 12.2 shows the materials used in the main components of a modern jet engine. Superalloys account for over 50% of the total weight. Superalloys are used in the hottest components such as the turbine blades, discs, vanes and combustion chamber where the temperatures are 900–1300 °C. Superalloys are also used in the low-pressure turbine case, shafts, burner cans, afterburners and thrust reversers. In general, nickel-based superalloys are used in engine

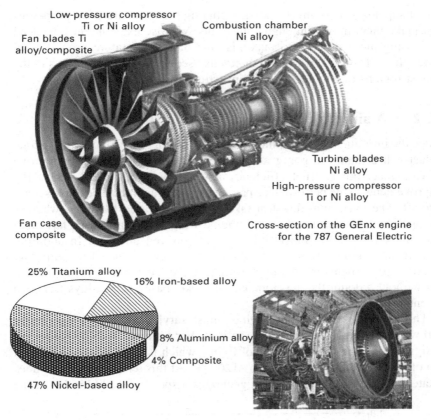

12.2 Material distribution in an aircraft turbine engine. The engine is the General Electric (CF6) used in the Boeing 787.

components that operate above 550 °C. A problem with superalloys is their high density of 8–9 g cm^{-3}, which is about twice as dense as titanium and three times denser than aluminium. Lighter materials are used whenever possible to reduce the engine weight. Titanium alloys are used to reduce the engine weight, but their use is restricted to components in the fan and compressor sections where the temperature is less than 550 °C. Titanium is used on the leading edges of carbon-fibre fan blades. Aircraft engines also contain aluminium alloys and fibre–polymer composites to reduce weight, although these materials can only be used in the coolest regions of the engine such as the fan blades and inlet casing, where the temperatures are less than 150 °C.

The use of superalloys in jet engines is explored in this chapter. The properties needed by materials used in engines are examined, such as high creep resistance and long fatigue life. The types of nickel, iron–nickel and cobalt superalloys are investigated, including their metallurgical properties that make them important engine materials. We also examine the advanced processing methods used to maximise the high-temperature properties of superalloys. The surface coating materials used to protect superalloys in the hottest regions of engines are also investigated.

12.2   A simple guide to jet engine technology

Since the introduction of commercial jet engines in the 1950s, the aerospace industry has made on-going advances in engine technology to improve performance, power, fuel efficiency and safety. Figure 12.3 shows the improvements in the performance parameters of jet engines used in passenger aircraft. The maximum thrust at take-off has increased by 300% whereas the specific fuel consumption and engine weight per unit thrust has more than halved since the introduction of jet-powered airliners. In addition, modern jet engines are more durable, reliable, quieter and less polluting. These improvements in engine performance are the result of many factors, but without a doubt the development of nickel-based superalloys has been essential to the success.

The materials used in jet engines must survive arduous temperatures and withstand high stress for long periods. For example, turbine blades are designed to last at least 10 000 h of flying, which is equivalent to 8 million km of flight, at temperatures up to ~1200 °C. At this temperature the blades rotate at more 10 000 rpm which generates a speed of 1200 km h^{-1} at the

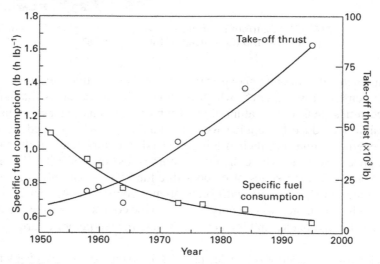

12.3 Performance improvements in jet engines (adapted from R. C. Reed, *The superalloys: fundamentals and applications*, Cambridge University Press, 2006).

blade tip and stress of about 180 MPa (or 20 tonne per square inch) at the blade root. To perform under such extreme conditions, materials used in the hot sections of jet engines must have some outstanding high temperature properties, which include:

- high yield stress and ultimate strength to prevent yield and failure;
- high ductility and fracture toughness to provide impact resistance and damage tolerance;
- high resistance to the initiation and growth of fatigue cracks to provide long operating life;
- high creep resistance and stress rupture strength;
- resistance against hot corrosive gases and oxidation;
- low thermal expansion to maintain close tolerances between rotating parts.

Jet engines are complex engineering systems made using various types of metals, ceramics and composites, although superalloys are the key material to their high thrust and long operating life. There are many types of jet engines, with the most common on large passenger aircraft and military fighter aircraft being turbojets or turbofans. The basic difference between turbojets and turbofans is the way air passes through the engine to generate thrust. Turbojets operate by drawing all the air into the core of the engine. With turbofans the air passes through the engine core as well as by-passing the core. Most modern jet engines are turbofans, which are more fuel efficient than turbojets during flights at subsonic speeds and, for this reason, their operation is described.

Figure 12.4 shows the main sections of a turbofan engine. The first inlet section is the fan, which consists of a large spinning fan system that draws air into the engine. Air flowing through the fan section is split into two streams: one stream continues through the main core of the engine whereas the second stream by-passes the engine core. The by-passed air flows through a duct system that surrounds the core to the back of the engine where it produces much of the thrust that propels the aircraft forward. The temperatures within the fan section are not high, which allows titanium or composite materials to be used for their high specific stiffness, strength and fatigue life. Other considerations in the selection of materials in the fan region include resistance to corrosion, erosion and impact damage (from bird strike), making titanium alloys and composites suitable.

The air in turbofan engines passes through the engine core and enters the compressor section where it is compressed to high pressure. The compressor is made up of stages, with each stage consisting of compressor blades and discs, which squeeze the air into progressively smaller regions. As air is forced through the compressor section its pressure and temperature rapidly increase, thus requiring the use of heat-resistant materials. Hot compressed

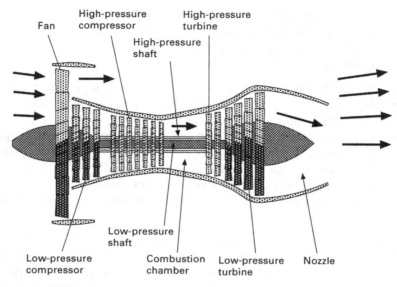

High-pressure
Fan   compressor

High-pressure
turbine

High-pressure
shaft

Low-pressure
shaft

Low-pressure
compressor

Combustion
chamber

Low-pressure
turbine

Nozzle

12.4 Main sections of a turbofan engine.

air then flows into the combustion chamber where it is mixed with jet fuel and ignited. This produces high-pressure gases that may reach a velocity of 1400 km h^{-1} and temperatures between 850 and 1500 °C. The combustion temperatures can exceed the melting point of the superalloys used in the combustion chamber. To survive, the superalloy engine components are protected with an insulating layer of ceramic material called a thermal barrier coating. The hot, high-pressure gases flow from the combustion chamber into the turbine section, which consists of bladed discs attached to shafts which run almost the entire length of the engine. The gases are allowed to expand through the turbine section which spins the blades. Power is extracted from the spinning turbine to drive the compressor and fan via the shafts. The turbine blades and discs are made using superalloys to withstand the hot gases coming from the combustion chamber. The gases that pass through the turbine are combined in the mixer section with the colder air that by-passed the engine core. The hot gases then flow into the rear-most section called the nozzle where high thrust is generated to propel the aircraft forward.

12.3   Nickel-based superalloys

12.3.1   Background

Nickel-based superalloy is the most used material in turbine engines because of its high strength and long fatigue life combined with good resistance to oxidation and corrosion at high temperature. Nickel-based superalloy is the material of choice for the hottest engine components that are required to operate above 800 °C. Without doubt, one of the most remarkable properties of nickel superalloys that is utilised in jet engines is their outstanding resistance against creep and stress rupture at high temperature. (The creep and stress rupture properties of materials are explained in chapter 22). Creep is an important material property in order to avoid seizure and failure of engine parts. Creep involves the plastic yielding and permanent distortion of materials when subjected to elastic loads. Most materials experience rapid creep at temperatures of 30–40% of their melting temperature. For example, aluminium and titanium alloys, which are used in the cooler regions of jet engines, creep rapidly above 150 and 350 °C, respectively. Nickel superalloys resist creep so well they can be used at 850 °C, which is over 70% of their melting temperature (T_m = 1280 °C). Very few other metallic materials possess excellent creep resistance at such high temperatures. The exceptional creep and stress rupture resistance of nickel superalloys means that engines can operate at higher temperatures to produce greater thrust. The outstanding creep and stress rupture resistance of nickel-based superalloys is shown in Fig. 12.5. Compared with the materials used in aircraft structures, aluminium, titanium and magnesium alloys, the stress rupture strength of nickel-based alloys is outstanding.

12.3.2 Composition of nickel superalloys

Nickel superalloys contain at least 50% by weight of nickel. Many of the superalloys contain more than ten types of alloying elements, including

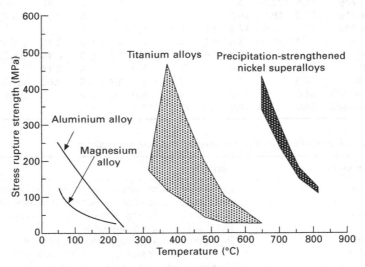

12.5 Stress rupture curves for aerospace materials.

high amounts of chromium (10–20%), aluminium and titanium (up to 8% combined), and cobalt (5–15%) together with small amounts of molybdenum, tungsten and carbon. Table 12.1 gives the composition of several nickel-based superalloys used in jet engines.

The functions of the alloying elements are summarised in Table 12.2. The elements serve several important functions, which are to:

- strengthen the nickel by solid solution hardening with the addition of elements such as molybdenum, chromium, cobalt and tungsten;
- strengthen the nickel by hard intermetallic precipitates and carbides with the addition of aluminium, titanium, carbon; and
- create a surface film of chromium oxide (Cr_2O_3) to protect the nickel from oxidation and hot corrosion.

Superalloys are not named or numbered according to any system; they are usually given their name by the company that developed or commercialised the alloy. Of the many alloys, the most important for aerospace is Inconel

Table 12.1 Average composition of nickel superalloys

Alloy	Composition									
	Ni	Fe	Cr	Mo	W	Co	Nb	Al	C	Other
Astroloy	55.0	–	15.0	5.3		17.0	–	4.0	0.06	
Hastelloy X	49.0	18.5	22.0	9.0	0.6	1.5	3.6	2.0	0.1	
Inconel 625	61.0	2.5	21.5	9.0	–	–	–	0.2	0.15	<0.25 Cu
Nimonic 75	75.0	2.5	19.5	–	–	–	–	0.15	<0.08	1 V
Inconel 100	60.0	<0.6	10.0	3.0	–	15.0	–	5.5	<0.08	2.9 (Nb+Ta)
Inconel 706	41.5	37.5	16.0	–	–	–	5.1	0.2	0.12	<0.15 Cu
Inconel 716	52.5	18.5	19.0	3.0	–	–	5.2	0.5	0.05	0.1 Zr
Inconel 792	61.0	3.5	12.4	1.9	3.8	9.0	–	3.5	0.04	
Inconel 901	42.7	34	13.5	6.2	–	–	–	0.2	0.16	0.3 V
Discaloy	26.0	55	13.5	2.9	–	–	3.5	0.2	0.15	0.5 Zr
Rene 95	61.0	<0.3	14.0	3.5	3.5	8.0	–	3.5	0.14	
Rene 104	52.0	–	13.1	3.8	1.9	182	1.4	3.5	0.03	2.7 Ta
SX PWA1480	64.0	–	10.0	–	4.0	5.0	–	5.0		2 Hf
DS PWA1422	60.0	–	10.0	–	12.5	10.0		5.0		

Table 12.2 Functions of alloying elements in nickel superalloys

Alloying element	Function
Chromium	Solid solution strengthening; corrosion resistance
Molybdenum	Solid solution strengthening; creep resistance
Tungsten	Solid solution strengthening; creep resistance
Cobalt	Solid solution strengthening
Niobium	Precipitation hardening; creep resistance
Aluminium	Precipitation hardening; creep resistance
Carbon	Carbide hardening; creep resistance

718 which accounts for most of the nickel superalloy used in jet engines. For example, superalloy 718 accounts for 34% of the material in the General Electric CF6 engine (shown in Fig. 12.2) used on the Boeing 787. Alloy 718 is a high-strength, corrosion-resistant alloy that is used at temperatures up to about 750 °C. Hastelloy X and Inconel 625 are often used in combustion cans and Inconel 901, Rene 95 and Discaloy are used in turbine discs. Nickel-based superalloys are available in extruded, forged and rolled forms. The higher strength forms are generally only found in the cast condition, such as directional and single crystal castings. The alloys PWA1480 and PWA1422 are special types of superalloys used in turbine blades that are produced by single crystal (SX) and directional solidification (DS) methods, respectively. These casting methods for producing blades are explained in chapter 6.

12.3.3 Properties of nickel superalloys

Nickel superalloys derive their strength by solid solution hardening or the combination of solid solution and precipitation hardening. Superalloys that harden predominantly by solid solution strengthening contain potent substitutional strengthening elements, such as molybdenum and tungsten. These two alloying elements have the added benefit of having low atomic diffusion rates in nickel, and move very slowly through the lattice structure at high temperatures. Creep is controlled by atomic diffusion processes, with the creep rate increasing with the diffusion rate of the alloying elements. Therefore the slow movement of alloying elements impedes the creep of nickel.

Solid solution-hardened nickel alloys have good corrosion resistance, although their high-temperature properties are inferior to precipitation hardened alloys. Figure 12.6 shows the creep-rupture strength of nickel-based superalloys hardened by means of solid solution strengthening or precipitation strengthening. Here, the stress necessary to cause tensile rupture in 100 h is given over a range of temperatures. Nickel superalloys hardened by precipitation strengthening provide the best high-temperature performance and, for this reason, are used in jet-engine parts such as blades, discs, rings, shafts, and various compressor components. Precipitation-hardened nickel alloys are also used in rocket motor engines.

Precipitation-hardened nickel alloys contain aluminium, titanium, tantalum and/or niobium that react with the nickel during heat treatment to form a fine dispersion of hard intermetallic precipitates. The most important precipitate is the so-called gamma prime phase (γ') which occurs as Ni_3Al, Ni_3Ti, or $Ni_3(Al,Ti)$ compounds. These precipitates have excellent long-term thermal stability and thereby provide strength and creep resistance at high temperature (Fig. 12.7). A high volume fraction of γ' precipitates is needed for high strength, fatigue resistance and creep properties (typically above 50%).

The γ' precipitates are formed by a heat-treatment process that involves

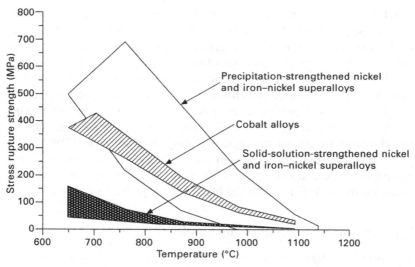

12.6 Stress rupture properties of nickel superalloys hardened by solid solution strengthening or precipitation strengthening.

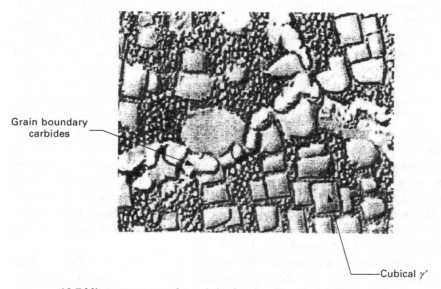

12.7 Microstructure of precipitation-hardened nickel superalloy showing γ' precipitates and carbides which provide high-temperature creep properties.

solution treatment followed by thermal ageing to dissolve the alloying elements into solid solution. Solution treatment is performed at 980–1230 °C. Single-crystal nickel alloys are solution treated at higher temperatures (up to 1320 °C). Following solution treatment and quenching, the nickel alloy is aged at 800–1000 °C for 4–32 h to form the γ' precipitates. The ageing temperature and time is determined by the application of the superalloy. Lower ageing temperatures and shorter times produce fine γ' precipitates for engine parts requiring strength and fatigue resistance, such as discs. Higher temperatures produce coarse γ' precipitates desirable for creep and stress rupture applications, such as turbine blades.

Precipitation-hardened nickel alloys contain other thermally stable compounds (in addition to γ') which also contribute to the high-temperature properties. Titanium, tantalum, niobium and tungsten react with carbon to form several types of hard carbide precipitates: MC, $M_{23}C_6$, M_6C and M_7C_3 where M stands for the alloying element. These carbides perform three important functions:

- prevent or slow grain boundary sliding that causes creep;
- increase tensile strength; and
- react with other elements that would otherwise promote thermal instability during service.

Small amounts of boron, hafnium and zirconium are often used as alloying elements in nickel. These elements combine with other elements to pin grain boundaries, thereby reducing their tendency to slide under load and thereby increasing the creep strength. Niobium is an important alloying element for the precipitation hardening of nickel alloys. The most commercially important superalloy is Inconel 718, and it is strengthened mainly by niobium precipitates (Ni_3Nb).

The newest types of superalloys contain rare earth elements (such as yttrium or cerium) at concentrations of 2.5 to 6% to increase the high-temperature creep strength by precipitation and solid solution hardening. However, rare earth elements are expensive to use, they increase the density of the superalloy, and they can produce casting defects. Ruthenium is increasingly being used instead of, or in combination with, the rare earth elements to achieve similar improvements in high-temperature mechanical performance without the problem of casting defects. The platinum group of elements (platinum, iridium, rhodium, palladium) are also being used to increase the operating temperature limit to 1100–1150 °C, but these are expensive.

Some grades of nickel superalloys contain submicrometre-sized oxide particles, such as ThO_2 or Y_2O_3, to promote higher elevated temperature tensile and stress rupture properties. For example, superalloy MA754 is produced by powder metallurgy involving mechanical alloying to introduce about 1% by volume of a fine dispersion of nano-sized oxide particles.

A problem with using metals at high temperature is rapid oxidation which quickly breaks down and degrades the material. Nickel alloys also contain 10–20% chromium to provide oxidation resistance through the formation of a protective surface oxide film composed of Cr_2O_3 or $NiCr_2O_4$. Nickel alloys rapidly degrade by oxidation without this surface film, which must be stable at the high operating temperatures of gas turbine engines.

12.4   Iron–nickel superalloys

Iron–nickel superalloys are used in gas turbine engines for their structural properties and low thermal expansion at high temperature. Iron–nickel alloys expand less than nickel or cobalt superalloys at high temperature, which is an important material property for engine components requiring closely controlled clearances between rotating parts. Iron–nickel alloys are generally less expensive than nickel- and cobalt-based superalloys, which is another advantage. The main uses for iron–nickel alloys in jet engines are blades, discs and casings.

The composition of several iron–nickel alloys used in jet engines is given in Table 12.3, and most contain 15–60% iron and 25–45% nickel. Iron–nickel superalloys are hardened by solid solution strengthening and precipitation strengthening. Aluminium, niobium and carbon are used as alloying elements to promote the formation of hard intermetallic precipitates or carbides that are stable at high temperature. The precipitates are similar to those present in nickel-based superalloys, and include γ' $Ni_3(Al,Ti)$, γ'' (Ni_3Nb) and various types of carbides. The precipitates provide iron–nickel alloys with good resistance against creep and stress rupture at elevated temperature. Chromium is used to form an oxide surface layer to protect the metal from hot corrosive gases and oxidation.

12.5   Cobalt superalloys

Cobalt superalloys possess several properties which make them useful materials for gas turbine engines, although they are more expensive than nickel superalloys. Cobalt alloys generally have better hot-corrosion resistance than nickel-based and iron–nickel alloys in hot atmospheres containing lead

Table 12.3 Composition of iron–nickel superalloys

Alloy	Composition (%)									
	Fe	Ni	Cr	Mo	W	Co	Nb	Al	C	Other
Solid solution-hardened alloys										
Haynes 556	29.0	21.0	22.0	3.0	2.5	20.0	0.1	0.3	0.1	0.5 Ta
Incoloy	44.8	32.5	21.0	–	–	–	–	0.6	0.36	
Precipitation-hardened alloys										
A-286	55.2	26.0	15.0	1.25	–	–	–	0.2	0.04	0.3 V
Incoloy 903	41.0	38.0	<0.1	0.1	–	15.0	3.0	0.7	0.04	

oxides, sulfur and other compounds produced from the combustion of jet fuel. Cobalt alloys have good resistance against attack from hot corrosive gases, which increases the operating life and reduces the maintenance of engine parts. However, comparison between nickel and cobalt alloys must be treated with some caution because there are wide differences in hot corrosion resistance within each group of superalloys. That is, certain nickel superalloys also have excellent resistance against hot corrosion. Cobalt alloys also have good stress rupture properties, although not as good as precipitation-hardened nickel-based alloys (Fig. 12.6).

Cobalt superalloys contain about 30–60% cobalt, 10–35% nickel, 20–30% chromium, 5–10% tungsten, and less than 1% carbon. The composition of some cobalt alloys used in jet engine components is given in Table 12.4. The main functions of the alloying elements are to harden the cobalt by solid solution or precipitation strengthening. The precipitates that form in cobalt alloys do not provide the same large improvement in high-temperature strength as nickel alloys and, for this reason, the resistance of cobalt alloys against creep and stress rupture is inferior to precipitation-hardened nickel-based and iron-nickel alloys. Cobalt alloys are generally used in components that operate under low stresses and need excellent hot-corrosion resistance.

12.6   Thermal barrier coatings for jet engine alloys

Turbine blades contain rows of hollow aerofoils for cooling to increase the engine operating temperature. Cool air flows through the holes, which are located just below the surface, to remove heat from the superalloy. The aerofoils are remarkably effective at cooling, which allows increased operating temperature and associated improvements in engine efficiency. To increase the operating temperature even further, the hottest engine parts are coated with a thin ceramic film to reduce heat flow into the superalloy. The film is called a thermal barrier coating, which has higher thermal stability and lower thermal conductivity (1 W m^{-1}K^{-1}) than nickel superalloy (50 W m^{-1}K^{-1}). The use of the coating allows higher operating temperatures (typically at least 170 °C) in the turbine section. The coating provides heat insulation and this lowers the temperature of the superalloy engine component. Thermal barrier coatings can survive temperatures well in excess of the melting

Table 12.4 Composition of cobalt-based superalloys

Alloy	Composition (%)									
	Co	Fe	Ni	Cr	Mo	W	Nb	Al	C	Other
Haynes 25	50.0	3.0	10.0	20.0	–	15.0	–	–	0.1	1.5 Mn
Haynes 188	37.0	<3.0	22.0	22.0	–	14.5	–	–	0.1	0.9 La
MP35-N	35.0	–	35.0	20.0	10.0	–	–	–	–	

temperature of the superalloy itself, and also provide protection from the effects of thermal fatigue and creep and the oxidising effect of sulfates and other oxygen-containing compounds in the combustion gases.

Thermal barrier coatings are complex multilayered structures, as shown in Fig. 12.8. The coating can be applied using various deposition methods, with the most common being electron-beam physical vapour deposition (EBPVD). In this process, a target anode material consisting of zirconium oxide (ZrO_2) and yttrium oxide (Y_2O_3) is bombarded with a high-energy electron beam under vacuum. The electron beam heats the anode material to high temperature which causes atoms and oxide molecules from the target to transform from the solid into gas phase. The gaseous atoms and molecules then precipitate as a thin solid layer of the anode material onto a substrate, such as a nickel superalloy component. The coating is deposited on the surface to a thickness of about 0.1–0.3 mm, which is sufficient to provide heat protection to the underlying metal.

The most common coating material is yttria-stabilised zirconia (YSZ), which is based on zirconia doped with 7% yttria. The YSZ is bonded to the surface via intermediate layers which improve the adhesion strength properties. An intermediate bond coat with a chemical composition MCrAlY (where M = Co, Ni or Co+Ni) or NiAl–Pt are often used. The bond coat also provides oxidation and corrosion resistance to the underlying superalloy component. Thermal barrier coatings are used on engine components in the combustion chamber and turbine sections, including high-pressure blades and nozzle guide vanes. However, YSZ TBCs are unsuitable for use on loaded rotating components because their low tensile strength and toughness causes them to crack and spall.

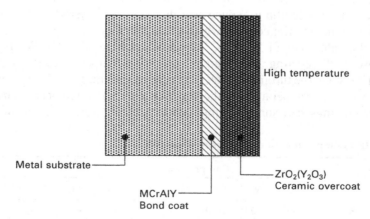

12.8 Through-thickness composition of a typical thermal barrier coating system.

12.7   Advanced materials for jet engines

The aerospace industry is continually developing new materials to increase the operating temperature limit of gas turbine engines. As mentioned in chapter 9, titanium aluminides are being developed for engine applications. A group of refractory intermetallics based on metal silicides (Mo_5Si_3, Nb_5Si_3, Ti_5Si_3) retain high strength to about 1300 °C, which is 200 °C higher than single-crystal nickel superalloys. Nb_5Si_3 is attracting the greatest interest of the silicides, although it is prone to oxidation at high temperature. Other types of advanced materials such as eutectic solidified ceramics, ceramic matrix composites (e.g. carbon/carbon), and silicon nitride are also being evaluated as high-temperature engine materials. For the foreseeable future, however, nickel-based superalloys are likely to be the dominant structural material for high-temperature components in jet engines.

12.8   Summary

Superalloys have played a central role in the development of jet engine technology. The development of superalloys with better high-temperature and hot-corrosion properties together with advances in engine design and propulsion technology has resulted in great improvements in engine performance. Over the past 20 years, the thrust of jet engines has increased by more than 60% whereas the fuel consumption has fallen by 15–20%, and these improvements are, in part, the result of improvements in the high-temperature properties of superalloys.

A variety of high-performance materials is used in modern jet engines. Aluminium and carbon-fibre composites are used in the coolest sections of engines (operating at temperatures below about 150 °C), such as the fan and inlet casing, to minimise weight. Titanium ($\alpha+\beta$ and β) alloys are used in engine components with operating temperatures below about 550 °C, which includes parts in the fan and compressor sections. Superalloys are used for components that operate above 550 °C, such as the blades, discs, vanes and other parts found in the combustion chamber and other high-temperature engine sections.

Materials used in the hottest engine components, such as high-pressure turbine blades and discs, must have high strength, fatigue life, fracture toughness, creep resistance, hot-corrosion resistance and low thermal expansion properties. Nickel-based superalloys are the material of choice of these engine components because of their capability to operate at temperatures up to 950–1200 °C for long periods of time.

Nickel-based superalloys used in jet engines have a high concentration of alloying elements (up to about 50% by weight) to provide strength, creep resistance, fatigue endurance and corrosion resistance at high temperature.

The types and concentration of alloying elements determines whether the superalloy is a solid solution-hardened or precipitation-hardened material. Precipitation-hardened superalloys are used in the hottest engine components, with their high-temperature strength and creep resistance improved by the presence of γ' [Ni_3Al, Ni_3Ti, $Ni_3(Al,Ti)$] and other precipitates that have high thermal stability.

The casting process is important in the production of heat-resistant superalloy engine components. The creep resistance of materials is improved by minimising or eliminating the presence of grain boundaries that are aligned transverse to the load direction. Superalloys are cast using directional solidification which produces a columnar grain structure with few transverse grain boundaries or single crystal casting which eliminates all grain boundaries.

Iron–nickel superalloys are used in jet engines for their high-temperature properties and low thermal expansion. These superalloys, which contain 15–60% iron and 25–45% nickel, are used in blades, discs and engine casings that require low thermal expansion properties.

Cobalt superalloys are used in jet engine components that require excellent corrosion resistance against hot combustion gases. The alloys contain 30–60% cobalt and high concentrations of nickel, chromium and tungsten which provide good resistance against lead oxides, sulfur oxides and other corrosive compounds in the combustion gas.

Thermal barrier coatings are a ceramic multilayer film applied to the superalloy surface to increase the operating temperature of the engine. The coating is an insulating layer that reduces the heat conducted into the superalloy. Yttria-stabilised zirconia (YSZ) is the most common coating material, and is used on engine components in the combustor chamber and turbine sections, including high-pressure blades and nozzle guide vanes.

12.9   Further reading and research

Campbell, F. C., *Manufacturing technology for aerospace structural materials*, Elsevier, Amsterdam, 2006.

Clarke, D. and Bold, S., 'Materials developments in aeroengine gas turbines', in *Aerospace Materials*, edited B. Cantor, H. Assender and P. Grant, Institute of Physics Publishing, Bristol, 2001, pp. 71–80.

Geddes, B., Leon, H. and Huang, X., *Superalloys: alloying and performance*, ASM International, 2010.

Grant, P., 'Thermal barrier coatings', in *Aerospace materials*, edited B. Cantor, H. Assender and P. Grant, Institute of Physics Publishing, Bristol, 2001, pp. 294–310.

Khan, T. and Bacos, M.-P., 'Blading materials and systems in advanced aeroengines', in *Aerospace materials*, edited B. Cantor, H. Assender and P. Grant, Institute of Physics Publishing, Bristol, 2001, pp. 81–88.

Reed, R. C., *The superalloys – fundamentals and applications*, Cambridge University Press, Cambridge, 2008.

Winston, M. R., Partridge, A. and Brooks, J. W., 'The contribution of advanced high-temperature materials for future aero-engines', *Proceedings of the Institute of Mechanical Engineers, Part L: Journal of Materials: Design and Applications*, **215** (2001), 63–73.

Polymers for aerospace structures

13.1 Introduction

Polymer is a generic term that covers a wide variety and large number of plastics, elastomers and adhesives. The three main groups of polymers are called thermoplastics, thermosetting polymers (or thermosets) and elastomers. The term 'plastic' is often used to describe both thermoplastics and thermosets, although there are important differences between the two. Elastomers are commonly called 'rubbers', although in the aerospace engineering community the former term is the correct one to use. Adhesives are an important sub-group of polymers, and can be thermoplastic, thermoset or elastomer.

The basic chemical properties used to distinguish between thermoplastics, thermosets and elastomers are shown in Fig. 13.1. Thermoplastics consist of long molecular chains made by joining together small organic molecules known as monomers. In effect, monomers are the basic building blocks that are joined end-to-end by chemical reactions to produce a long polymer chain. The property of thermoplastics that distinguishes them from other polymers is that strong covalent bonds join the atoms together along the length of the chain, but no covalent bonding occurs between the chains. Each thermoplastic chain is a discrete molecule. The molecular chains are entangled and intertwined in a thermoplastic, but the chains are not joined or connected. This structure provides thermoplastics with useful engineering properties, including high ductility, fracture toughness and impact resistance. The lack of bonding between the molecular chains causes thermoplastics to soften and melt when heated which allows them to be recycled.

Thermosetting polymers are also long molecular chains made by joining together smaller molecules, although the chemical reactions produce covalent bonds both along the chain and bridging across the chains. The bonding between the chains is known as crosslinking, and it produces a rigid three-dimensional molecular structure. It is the crosslinking of thermosets that distinguishes them from thermoplastics. The crosslinks generally provide thermosets with higher elastic modulus and tensile strength than thermoplastics, but they are more brittle and have lower toughness. Thermosets do not have a melting point because the crosslinks do not allow the chains to flow like a liquid at high temperature and, therefore, these polymers cannot be recycled.

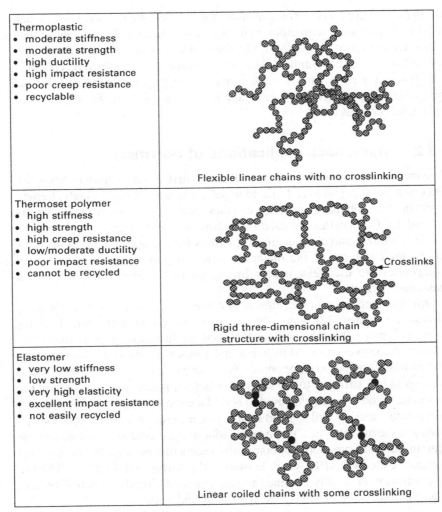

Thermoplastic
• moderate stiffness
• moderate strength
• high ductility
• high impact resistance
• poor creep resistance
• recyclable

Flexible linear chains with no crosslinking

Thermoset polymer
• high stiffness
• high strength
• high creep resistance
• low/moderate ductility
• poor impact resistance
• cannot be recycled

Crosslinks

Rigid three-dimensional chain structure with crosslinking

Elastomer
• very low stiffness
• low strength
• very high elasticity
• excellent impact resistance
• not easily recycled

Linear coiled chains with some crosslinking

13.1 Basic properties of thermoplastics, thermoset polymers and elastomers.

Elastomers are natural and synthetic rubbers with a molecular structure that is somewhere between thermoplastics and thermosets. Some crosslinking occurs between elastomer chains, but to a lesser amount than with thermosets. The elastomer chains are coiled somewhat like a spring, and this allows them to be stretched over many times their original length without being permanently deformed. Elastomers are characterised by low stiffness and low strength as well as good impact resistance and toughness.

In this chapter, the polymers, elastomers and structural adhesives used in aerospace structures are described. These materials lack the stiffness and strength to be used on their own in aircraft structures, but they are useful when used in combination with other materials such as fibre–polymer composites. The different types of polymers, viz. thermoplastics, thermosets, elastomers and adhesives, are examined. The polymerisation processes and chemical composition of polymers are studied. In addition, the mechanical and thermal properties of polymers, and their applications to aircraft, helicopters and spacecraft are discussed.

13.2   Aerospace applications of polymers

The most common use for polymers is the matrix phase of fibre composites. Polymers are the 'glue' used to hold together the high-stiffness, high-strength fibres in fibre–polymer composites, and these materials are described in chapters 14 and 15. Composites are used in the airframe and engine components of modern military and civilian aircraft, with polymers accounting for 40–45% of the total volume of the material. Moulded plastics and fibre–polymer composites are used extensively in the internal fittings and furniture of passenger aircraft.

Another important application of polymers is as an adhesive for joining aircraft components. It is possible to produce high strength, durable joints using polymer adhesives without the need for fasteners such as rivets and screws. Adhesives are used to join metal-to-metal, composite-to-composite and metal-to-composite components. For example, adhesives are used to bond ribs, spars and stringers to the skins of structural panels used throughout the airframe. Adhesives are also used to bond face sheets to the core of sandwich composite materials and to bond repairs to composite and metal components damaged during service. Thin layers of adhesive are used to bond together the aluminium and fibre–polymer composite sheets that produce the fibre–metal laminate called GLARE, which is used in the Airbus 380 fuselage. The use of elastomers is usually confined to nonstructural aircraft parts that require high flexibility and elasticity, such as seals and gaskets.

13.3   Advantages and disadvantages of polymers for aerospace applications

Polymers possess several properties that make them useful as aircraft materials, including low density (1.2–1.4 g cm^{-3}), moderate cost, excellent corrosion resistance, and high ductility (except thermosets). Some polymers are tough and transparent which makes them suitable for aircraft windows and canopies. However, polymers cannot be used on their own as structural materials because of their low stiffness, strength, creep properties and working temperature. Table 13.1 summarises the main advantages and disadvantages of thermoplastics, thermosets and elastomers whereas Table 13.2 gives the

Table 13.1 Comparison of the advantages and disadvantages of polymers for aircraft structural applications

Thermoplastic	Thermoset	Elastomer
Advantages		
• Non-reacting; no cure required • Rapid processing • High ductility • High fracture toughness • High impact resistance • Absorbs little moisture • Can be recycled	• Low processing temperature • Low viscosity • Good compression properties • Good fatigue resistance • Good creep resistance • Highly resistant to solvents • Good fibre wetting for composites	• Low processing temperature • High ductility and flexibility • High fracture toughness • High impact resistance
Disadvantages		
• Very high viscosity • High processing temperature (300–400 °C) • High processing pressures • Poor creep resistance	• Long processing time • Low ductility • Low fracture toughness • Low impact resistance • Absorb moisture • Limited shelf life • Cannot be recycled	• Long processing times • Poor creep resistance • Low Young's modulus • Low tensile strength

Table 13.2 Comparison of the typical properties of a structural polymer (epoxy resin) against aerospace structural materials

Materials	Average specific gravity (g cm^{-3})	Young's modulus (GPa)	Specific modulus (MPa m^3 kg^{-1})	Yield strength (MPa)	Specific strength (kPa m^3 kg^{-1})
Polymer (epoxy)	1.2	3	2.5	100	83
Aluminium (7075-T76)	2.7	70	25.9	470	174
Magnesium	1.7	45	26.5	200	115
α+β-Ti alloy	4.6	110	23.9	1000	217
Carbon/epoxy composite*	1.7	50	29.4	760	450

*[0/±45/90] carbon/epoxy; fibre volume content = 60%.

typical properties of polymers compared with several types of aerospace metals, and shows that the polymers are structurally inferior.

13.4   Polymerisation

13.4.1 Polymerisation processes

Polymerisation is the chemical process by which polymers are made into long chain-like molecules from smaller molecules. Polymerisation is the process by which small molecules, known as monomers, are joined to create macromolecules. The polymerisation process is used to make thermoplastics, thermosets and elastomers, although the reaction processes are different. The chemistry of polymerisation is a complex topic, and only the basics are described here.

The polymerisation process can be divided into two types of chemical reaction: addition polymerisation and condensation polymerisation. Addition polymerisation involves the linking of monomers into the polymer chain by a chemical reaction that does not produce molecular by-products. For example, the major steps in the addition polymerisation reaction for polyethylene is shown in Fig. 13.2, which involves the joining of ethylene monomers (C_2H_4) into a long polyethylene chain $(CH_2)_{2n}$ where n is the number of ethylene monomers in the chain. A catalyst, also known as the 'initiator' because it starts the reaction, is added to the monomers. The catalyst splits the double covalent bond within the ethylene monomer to produce unpaired electrons. Each unpaired electron needs to immediately pair with another electron, so it joins with an unpaired electron at the neighbouring monomer. The pairing forms a single covalent bond that binds the two monomer units, and the process is repeated many, many times to produce a long molecular chain. A polymer

13.2 Addition polymerisation.

chain is built-up from hundreds to many hundreds of thousands of monomer units, which are called mer units when joined into a chain. The addition of monomers to a chain continues until the supply of monomer units or catalyst is exhausted or a special chemical called a terminator is used to stop the reaction process, or there is self-termination when a chain end connects to the end of another chain growing independently. In addition to polyethylene, other examples of polymers produced by addition polymerisation are polyvinyl chloride (PVC), polystyrene and polytetrafluroethylene (Teflon).

Condensation polymerisation is a process involving two or more different types of molecules that react to produce a molecular chain made of combinations of the starting molecules. Figure 13.3 shows a simple example of a condensation reaction between carboxylic acid and alcohol. A feature of the condensation reaction is that small molecules are produced as a by-product. These by-product molecules do not form part of the polymer chain. Water is a by-product of many condensation reactions and, when possible, it should be extracted from the polymer before use in aircraft. Examples of aerospace polymers produced by condensation reactions are epoxy resin, which is used as the matrix phase of carbon-fibre composite structures and as a structural adhesive, and phenolic resin which is used inside aircraft cabins for fire resistance.

Polymerisation reactions for most of the polymers used in aircraft involve just one type of monomer. When the polymer chain is made using just one monomer type it is known as a homopolymer. Polymers can also be produced by the polymerisation of two types of monomers, which is called a copolymer. Although copolymers are not used extensively in plastic aircraft parts, they are sometimes used as elastomers in seals. The two monomer types can be arranged in various ways along the chain, as illustrated in Fig. 13.4, thus resulting in different material properties. When the two different monomers are distributed randomly along the chain it is called a random copolymer. Under controlled processing conditions, the two monomer types may alternate as single mer units along the chain, and this is known as a regular copolymer. If, however, a long sequence of one type of mer units is followed by a long sequence of another type of mer units, it is termed a block copolymer. This latter type is called a graft copolymer when the chains produced from one type of monomer are attached (or grafted) to the chain created from the other monomer.

Addition and condensation reactions are started using a chemical catalyst,

$$R-\overset{\overset{\displaystyle O}{\|}}{C}-OH + H-OR'' \longrightarrow R-\overset{\overset{\displaystyle O}{\|}}{C}-OR'' + H_2O$$

Carboxylic acid   Alcohol   Ester   Water

13.3 Condensation polymerisation.

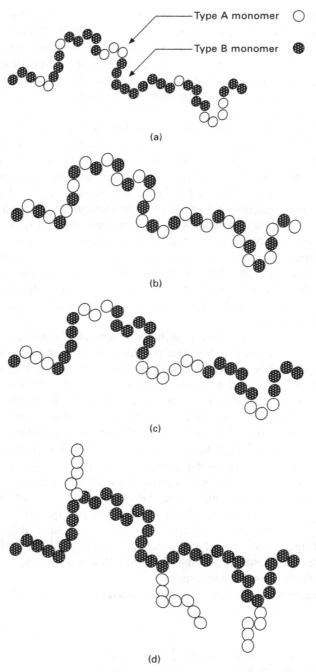

13.4 Structural types of copolymers: (a) random, (b) alternate, (c) block and (d) graft.

and the polymerisation rate is controlled by the amount of catalyst and the temperature. Increasing the catalyst content and temperature accelerates the reaction rate. However, other methods can be used to drive the polymerisation process, most notably electrons (e-beam) and ultraviolet (UV) radiation. E-beam curing is a process that involves irradiating the polymer with an electron beam to split the double covalent bonds in the monomer to produce unpaired electrons. Irradiating the polymer with UV radiation is another method for producing unpaired electrons. The unpaired electrons then bond with unpaired electrons from other monomer units to produce long polymer chains. These methods are gaining importance in the aircraft industry because they eliminate the need to use corrosive and toxic catalysts, although at the moment the use of catalysts remains the most common way of curing most types of polymer.

13.4.2 Polymer structure

The structural arrangement of the chain-like molecules has a major impact on the mechanical properties of a polymer. Thermoplastics can be polymerised by both addition and condensation reactions, although the monomers that make up thermoplastics are always bifunctional. This means they have two sites in the molecule where pairing can occur with other molecules. As a result, the polymer can only grow as a linear chain. Thermosetting polymers are also produced by addition and condensation reactions, but are formed from trifunctional monomers which have three reaction sites, and this allows the polymer to grow with covalent bonds along the chain and covalent (crosslink) bonds bridging across the chains. Elastomers contain a mixture of bifunctional and trifunctional monomers, which produces long linear segments in the chain from the bifunctional units, and widely spaced crosslinks from the trifunctional units.

Structural variations can occur within the chain during the polymerisation process that may significantly alter the engineering properties. The three major types of polymer structures are called atactic, isotactic and syndiotactic, and they are shown in Fig. 13.5. The chemical composition of the three types of polymer structure is the same; with the only difference being the location of atoms and small compounds ('R' groups) attached as side groups to the chain. When the side groups are located randomly along the chain, this is called an atactic structure. When the side groups are placed on the chain in the same location in each mer, it is called an isotactic structure. Syndiotactic refers to a structure with the R groups placed on the chain in a more or less regular pattern, but having alternate positions on either side of the chain.

The arrangement of the molecular chains at the end of the polymerisation process is complex, and simplified illustrations of the most common forms are shown in Fig. 13.6. The four most common structures are called linear,

13.5 Models of polymer chains with (a) atactic, (b) isotactic and (c) syndiotactic structures.

crosslinked, branched and ladder. Polymer chains are never straight, even over very short distances, and instead twist and turn along their length and become intertwined with other chains. Entanglement and intertwining of the chains is an important mechanism for providing stiffness and strength to polymers. This mechanism is particularly important for thermoplastics because they do not have crosslinking to resist sliding of the chains under an applied force. However, thermoplastics do have weak attraction forces between the chains. These are known as van der Waals bonds, also called 'secondary bonds', which is an electrostatic force between atoms owing to the nonsymmetric distribution of electrons around the atomic nucleus. These bonds are much weaker than the covalent bonds that form crosslinks between thermosets and, therefore, are less effective at resisting chain sliding.

13.5   Thermosetting polymers

Thermosetting polymers are crosslinked polymers that have a three-dimensional network structure with covalent bonds linking the chains. Thermosets are produced by mixing a liquid resin consisting of monomers and oligomers (several monomer units joined together) with a liquid hardener, which can be another resin or catalyst. The resin and hardener react in a process that joins the monomers and oligomers into long polymer chains and forms crosslinks between the chains. Heat is often used to accelerate the reaction process and pressure is applied to squeeze out volatile by-products. When the crosslinks are formed they resist rotation and sliding of the chains under load, and this provides thermosets with better strength, stiffness and hardness than thermoplastics.

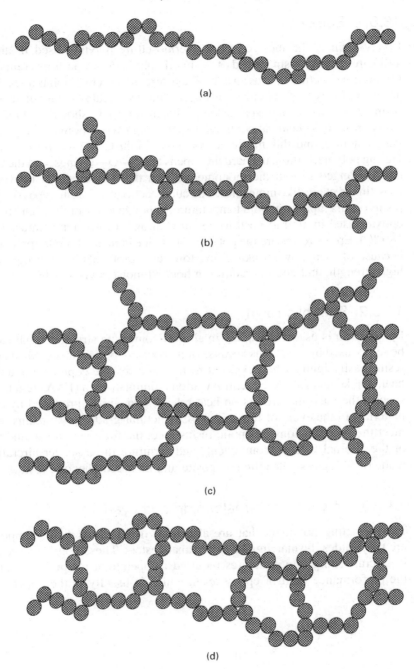

13.6 Schematic illustration of the molecular structure of (a) linear, (b) branched, (c) crosslinked, and (d) ladder polymers.

13.5.1 Epoxy resin

Epoxy resin is the most common thermosetting polymer used in aircraft structures. Epoxy resin is used as the matrix phase in carbon-fibre composites for aircraft structures and as an adhesive in aircraft structural joints and repairs. There are many types of epoxy resins, and the chemical structure of an epoxy resin often used in aerospace composite materials is shown in Fig. 13.7. Epoxy resin is a chemical compound containing two or more epoxide groups per monomer, and this molecule contains a tight $C-O-C$ ring structure. During polymerisation, the hardener opens the $C-O-C$ rings, and the bonds are rearranged to join the monomers into a three-dimensional network of crosslinked chain-like molecules. The cure reaction for certain types of epoxy resins occurs rapidly at room temperature, although many of the high-strength epoxies used in aircraft need to be cured at an elevated temperature (120–180 °C). Epoxy resins are the polymer of choice in many aircraft applications because of their low shrinkage and low release of volatiles during curing, high strength, and good durability in hot and moist environments.

13.5.2 Phenolic resin

Epoxy resin is used extensively in aircraft composite structures, but cannot be safely used inside cabins because of its poor fire performance. Most epoxy resins easily ignite when exposed to fire, and release copious amounts of heat, smoke and fumes. Federal Aviation Administration (FAA) regulations specify the maximum limits on heat release and smoke produced by cabin materials in the event of fire, and most structural-grade epoxy resins fail to meet the specifications. Phenolic resins meet the fire regulations, and most of the internal fittings, components and furniture in passenger aircraft are made of fibreglass–phenolic composite and moulded phenolic resin.

13.5.3 Polyimide, bismaleimide and cyanate

Thermosetting polymers that are also used in structural fibre composites are polyimides, bismaleimides and cyanate esters. These polymers are used in aircraft composite structures required to operate at temperatures above the performance limit of epoxy resin, which is usually in the upper range

13.7 Functional unit of epoxy resins used in aircraft composite materials.

of 160–180 °C. Polyimides can operate continuously at temperatures up to 175 °C and have an operating limit of about 300 °C. The polyimide called PMR-15 is the most common, and is used as the matrix phase of carbon-fibre composites in high-speed military aircraft and jet engine components. The down-side of using polyimides is their high cost. Bismaleimide (BMI) is also used in fibre composites required to operate at temperature, with an upper service temperature of about 180 °C. Carbon–BMI composites are used in the F-35 *Lightning II* fighter along with carbon–epoxy materials. Cyanate resins, which are also known as cyanate esters, cyanic esters or triazine resins, have good strength and toughness at high temperature, and their maximum operating temperature is approximately 200 °C. However, cyanate resins pose a safety risk because they produce poisonous hydrogen cyanide during the cure reaction process.

13.6   Thermoplastics

13.6.1 Aerospace thermoplastics

The use of thermoplastics in aircraft, whether as the matrix phase of fibre–polymer composites or as a structural adhesive, is small compared with the much greater use of thermosets. Some sectors of the aerospace industry are keen to increase the use of thermoplastics in composite materials, and the number of applications is gradually increasing. Thermoplastics provide several important advantages over thermosets when used in composite materials, most notably better impact damage resistance, higher fracture toughness and higher operating temperatures. However, thermoplastics must be processed at high temperature that makes them expensive to manufacture into aircraft composite components.

The group of thermoplastics that are most used in aircraft composite structures are called polyketones, and include polyether ketone (PEK), polyether ketone ketone (PEKK) and, the most common, polyether ether ketone (PEEK). The main types of thermoplastics used in aircraft are given in Fig. 13.8.

Several types of thermoplastics are transparent, tough and impact resistant which makes them well suited for aircraft windows and canopies. The thermoplastics most often used in aircraft windows are acrylic plastics and polycarbonates. Acrylic plastics are any polymer or copolymer of acrylic acid or variants thereof. An example of acrylic plastic used in aircraft windows is polymethyl methacrylate (PMMA), which is sold under commercial names such as Plexiglas and Perspex. Acrylic plastics are lighter, stronger and tougher than window glass. Polycarbonates get their name because they are polymers having functional groups linked together by carbonate groups ($-O-(C=O)-O-$) in the long molecular chain. Polycarbonates are stronger

Polyether ether ketone (PEEK)	
Polyphenylene sulfide (PPS)	
Polysulfone (PSU)	
Polyetherimide (PEI)	
Polycarbonate	

13.8 Thermoplastics used in aircraft.

and tougher than acrylic plastics and are used when high-impact resistance is needed, such as cockpit windows and canopies. In these applications, the material must have high impact resistance because of the risk of collision with birds. Although bird strikes do not occur at cruise altitudes, they present a serious risk at low altitudes, particularly during take-off and landing. Polycarbonate windscreens are also resistant to damage by large hailstones. Figures 13.9 and 13.10 show examples of damage caused to aircraft windows by bird strike or hail, respectively. Although the polycarbonate windows are damaged, they were impacted under severe conditions that would have caused most other polymer materials to rupture leading to cabin depressurisation. A large bird hit the window shown in Fig. 13.9 when the aircraft was flying at several hundred kilometres per hour and hailstones larger than golf balls caused the damage shown in Fig. 13.10. Had these windows been made with glass the bird and hailstones would almost certainly have punctured through and entered the cockpit. Polycarbonate windows offer good safety to the flight crew against severe impact events.

13.6.2 Crystallisation of thermoplastics

A unique property of thermoplastics is the ability of their polymer chains to be amorphous or crystalline (Fig. 13.11). Amorphous polymers consist of

13.9 Bird strike damage to a cockpit window. Photograph reproduced with permission from AirSafe.com.

13.10 Hail damage to a cockpit window.

long chains that are randomly shaped and disordered along their length. Each chain twists and turns along its length without order, and there is no pattern with the other chains. In comparison, crystalline polymers consist of chains with a well-ordered structure. Thermosets and elastomers always occur in the amorphous condition; crystallinity is a property unique to thermoplastics. The ability of thermoplastics to crystallise is important for their use in aircraft because crystalline polymers generally have better resistance to paint strippers, hydraulic fluids and aviation fuels, which can degrade amorphous polymers. A crystalline polymer also has higher elastic modulus and tensile strength than the same polymer in the amorphous condition.

Crystallisation occurs when a thermoplastic is cooled slowly from its melting temperature, which allows sufficient time for segments of the polymer chains to take an ordered structure. The folded chain model depicted in Fig. 13.11 is one representation of a crystallised polymer. A single chain folds back and forth over a distance of 50–200 carbon atoms, and the folded chains extend in three dimensions producing thin plates or lamellae. It is virtually impossible for thermoplastics to be completely crystalline because the entanglement and twisting of the chains stops them folding into an ordered pattern over large distances. Instead, thermoplastics can have both the amorphous and crystalline phases present together in the same material, and these are called semicrystalline thermoplastics. The crystalline phase occurs in tiny regions, typically only a few hundred angstroms wide, in which the chains are aligned. In some polymers these crystalline regions cluster into small groups known as spherulites. Surrounding these crystalline regions is the amorphous thermoplastic.

The percentage volume fraction of crystallised material within a thermoplastic typically varies from 30% to 90%, with the actual amount being determined by several factors. Crystallisation occurs more easily for linear polymers that have small side groups attached to the main chain. Thermoplastics that have large side groups or have extensive chain branching are more difficult to crystallise. For example, linear polyethylene can be crystallised to 90%

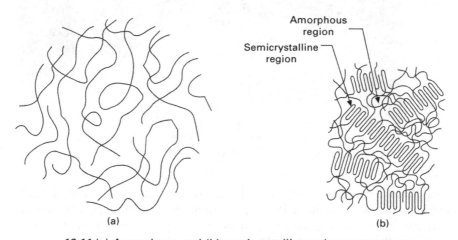

13.11 (a) Amorphous and (b) semicrystalline polymer structures.

by volume, whereas branched polyethylene can be crystallised to only 65%. This occurs because the branches interfere with the chain-folding process, thereby limiting the growth of crystalline zones. The length of the polymer chain also controls the amount of crystallisation. Crystallisation occurs more easily with shorter chains because the entanglement, which hinders crystallisation, is less.

The most common method to control the amount of crystallinity is careful management of the polymer processing conditions. Cooling a thermoplastic slowly from its melting temperature increases the amount of crystallisation because more time is given for the chains to move into an ordered pattern. Slowly deforming a thermoplastic at high temperature (but below the melting temperature) also increases crystallinity by allowing the chains to straighten in the direction of the applied stress. Lastly, the process of annealing, which involves holding the polymer at high temperature for a period of time, increases crystallinity. Often two or more of these methods are used to control the amount of crystallisation to ensure the thermoplastic has the properties best suited for a specific application.

13.7   Elastomers

13.7.1 Aerospace applications of elastomers

Elastomers are not suitable for use in aircraft structures because they lack stiffness and strength, but they do have exceptionally high elasticity with elongation values between one hundred and several thousand percent. This makes elastomers suitable when low stiffness and high elasticity is required, such as aircraft tyres, seals and gaskets. Many aircraft components that require a tight seal, such as window and door seals, use elastomers. These materials are used for their excellent elasticity; they can be easily compressed to make a tight seal without being damaged or permanently deformed. Although elastomers usually work well as seals and gaskets, they can gradually erode and degrade in harsh operating conditions, such as high temperatures. The most dramatic example of failure of an elastomer was the space shuttle *Challenger* accident. This accident is described in more detail in the case study in Section 13.15 at the end of the chapter.

13.7.2 Structure of elastomers

The molecular structure of elastomers is characterised by coiling of their chains. The coiled structure occurs because the chain is not balanced along its length owing to the bonding pattern and the location of side groups along the chain. Curvature of the chain is produced by the arrangement of atoms and side groups along the chain. An atom or side group R can be placed in either the *cis* or *trans* position along the carbon chain, as shown in Fig. 13.12. The unsaturated segment of the chain is where there are C=C double bonds. In the *trans* position, the unsaturated bonds are on opposite sides, whereas in the *cis* position the unsaturated bonds are on the same side of the chain and the R groups are on the opposite side. The location of the R groups on the same side of the chain causes it to become unbalanced, and this forces the chain to coil until it finds an equilibrium position. Elastomers usually have the unsaturated bonds in the *cis* position, which causes the chain to coil like a spring. It is the coiled structure of elastomers that distinguishes them from other polymers, and gives them their very high elasticity. Coiling allows the chains to stretch like a spring and, under an applied tensile force, they can elongate to many times their original length without permanently deforming. The chains spring back into their original coiled structure when the applied load is removed.

Elastomers have some crosslinking between the chains that provides a small amount of resistance to elastic stretching. The amount of crosslinking is much less than found in the heavily crosslinked thermosets. The crosslinks in elastomers are widely spaced along the coiled polymer chains. The crosslinks between the elastomer chains are created in a process called vulcanisation. The most common vulcanisation process involves heating rubber with sulfur to temperatures of about 120 to 180 °C. Without vulcanisation, elastomers behave like a very soft solid under load. When vulcanisation is carried too far, however, the chains are too tightly bound to one another by excessive crosslinking and the elastomer is very brittle. The crosslinks also improve the wear resistance and stiffness, which is important when elastomers are to be used in aircraft tyres. These properties improve with the amount of crosslinking, but the amount of sulfur present in the elastomer should not exceed about 5% otherwise the material becomes too brittle.

Not all elastomers are crosslinked; a special group known as thermoplastic elastomers are not crosslinked but still have the elasticity of a conventional

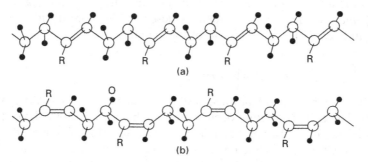

13.12 (a) *cis* and (b) *trans* structures of the polymer chain.

elastomer. As mentioned, the chain in a block copolymer is made using two types of monomers, with a long sequence of one type of monomer followed by a sequence of the other monomer type. An elastomer copolymer is made with thermoplastic and elastomer monomers, and the chains consist of long coiled lengths of the elastomer joined to shorter sections of thermoplastic. A common thermoplastic elastomer is styrene–butadiene–styrene (SBS), where the styrene is the thermoplastic-based monomer and butadiene the elastomer monomer. The SBS chains are made of short linear segments of styrene bonded to longer coiled segments of butadiene. The styrene units from a number of chains clump together, and this stops chains from sliding freely pass each other when the material is under load. The rubbery butadiene units uncoil under load providing SBS with high elasticity.

13.8   Structural adhesives

13.8.1 Aerospace applications of structural adhesives

A structural adhesive can be simply described a 'high-strength glue' that bonds together components in a load-bearing structure. Structural adhesives are used extensively in aircraft for bonding metal-to-metal, metal-to-composite and composite-to-composite parts. Adhesives are most commonly used in joints of thin aerostructures with a well-defined load path. Adhesives are also used to join thick airframe structures with complex, multidirectional load paths, although the joint is usually reinforced with fasteners such as bolts, rivets or screws for added strength and safety. Adhesives are also used to bond repair patches to damaged aircraft structures.

13.8.2 Performance of structural adhesives

Thermosets, thermoplastics and elastomers have adhesive properties, but specially formulated thermosets are used most often for bonding aircraft structures. There are many advantages of using adhesives for joining aircraft components, rather than relying solely on mechanical fasteners for strength. Adhesive bonding eliminates some, or all, of the cost and weight of fasteners in certain aircraft components. Another advantage of bonding is the reduced incidence of fatigue cracking in metal connections. Drilled holes for fasteners are potential sites for the start of fatigue cracks, and removing the need for holes by bonding reduces this particular problem. Adhesive bonding has the advantage of providing a more uniform stress distribution in the connection by eliminating the individual stress concentrations at fastener holes. Bonded joints are usually lighter than mechanically fastened joints and enable the design of smooth external surfaces. In addition to joining structural components, adhesives are used for bonding the skins and core in sandwich composites. Figure 13.13 shows the application of adhesively bonded sandwich composites in an aircraft.

The adhesive must have high elastic modulus, strength and toughness. Toughening agents (such as tiny rubber particles) are often blended into the adhesive to increase fracture toughness. Structural adhesives must have low shrinkage properties when cured to avoid the development of residual stresses in the joint. Adhesives must also be resistant to attack or degradation in the operating environment. Water and liquids such as solvents (paint strippers), hydraulic fluids and aviation fuel can attack the bond between the adhesive and substrate, and therefore durable adhesive systems must be used.

Structural adhesives with the best mechanical properties and durability are useless unless they can bond strongly to the substrate. The type of adhesive

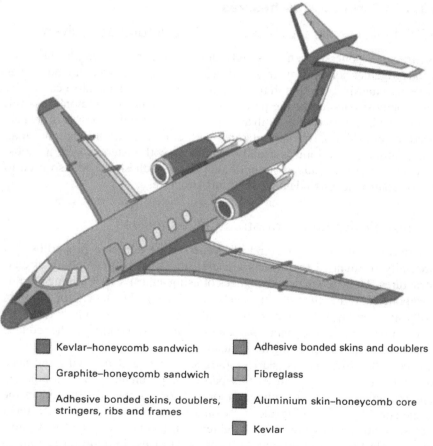

■ Kevlar–honeycomb sandwich   ■ Adhesive bonded skins and doublers

□ Graphite–honeycomb sandwich   ■ Fibreglass

■ Adhesive bonded skins, doublers,   ■ Aluminium skin–honeycomb core
  stringers, ribs and frames

                                      ■ Kevlar

13.13 Application of adhesively bonded sandwich composites in the *Citation* III aircraft.

used is critical to ensure strong adhesion. However, surface preparation of the substrate is the keystone upon which the adhesive bond is formed. The substrate must be grit blasted or lightly abraded, cleaned, dried, and then treated with a chemical coupling agent to ensure good adhesion. If the substrate surface is not suitably prepared, the adhesive fails owing to poor bonding.

13.8.3 Types of structural adhesives

The thermosetting adhesives used in aircraft structures are crosslinked polymers that are cured using heat, pressure or a combination of heat and pressure. Heat-curing adhesives are cured at temperatures close to (or preferably slightly over) the maximum-use temperature of the structure. Adhesives are available as films or pastes. Film adhesives are often used in bonding aircraft structures because they usually provide higher strength than paste adhesives. The film is simply placed between the substrates and then heated and pressurised to form a strong bond. Adhesive pastes can be a one-part system that is simply spread on the substrates and cured by heat and/or pressure. Alternatively, two-part systems consisting of resin and hardener are mixed together into an adhesive paste, which is then applied to the substrates.

The polymer most often used as a structural adhesive is epoxy resin because of its ability to adhere to most surfaces (including aircraft-grade aluminium alloys and fibre composites), high strength, and long-term durability over a wide range of temperatures and environments. Silicone is often used when a high toughness adhesive is needed, whereas bismaleimide and polyimide are used when a high-temperature adhesive is required. Other types of adhesives include urethanes, phenolic resins, acrylic resins and inorganic cements, although they are rarely used for bonding aircraft structures. Hot-melt adhesives are used occasionally, although not in highly-loaded aircraft structures. Hot-melt adhesives are thermoplastics or thermoplastic elastomers that melt when heated. On cooling the polymer solidifies and forms a bond with the substrate. Most commercial hot-melt adhesives soften at 80 to 110 °C, which makes them unsuitable as a high-temperature adhesive. Pressure sensitive adhesives (PSA) are usually elastomers or elastomer copolymers that are not crosslinked or are only slightly crosslinked and they bond strongly with a substrate. PSAs adhere simply by the application of pressure, and the most common (but obviously non-aerospace type) is Scotch® tape. Most PSAs are not suitable for joining aircraft structures because of their low strength at elevated temperature.

13.9   Mechanical properties of polymers

13.9.1 Deformation and failure of polymers

Polymers deform elastically and plastically when under load, much like metals. However, the deformation processes for polymers are different to those for metals. The elastic deformation of metals involves elastic stretching of the bonds in the crystal structure and plastic deformation involves dislocation slip. Polymers do not have crystal structures as found in metals nor do they contain dislocations, and so the processes responsible for deformation are different from those for metals.

Tensile stress–strain curves for a thermoplastic, thermosetting polymer and elastomer are presented in Fig. 13.14. These curves are representative of the stress response of these different types of polymers, where the elongation-to-failure increases in the order: thermoset, thermoplastic, elastomer. The curves can be divided into three regimes that are controlled by different deformation processes: elastic, plastic and fracture regimes.

Elastic deformation

When an elastic stress is applied to a polymer, the chains are stretched in the loading direction. Elastic stretching occurs in the covalent bonds between the atoms along the chain backbone. Stretching also involves some straightening of twisted segments of the chain. The stretching of the bonds and the twisted segments of the polymer chains increases with the applied stress. However,

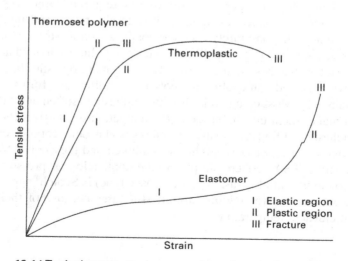

13.14 Typical stress–strain curves for a thermoplastic, thermoset polymer and elastomer.

when the load is removed the chains relax back into their original position. The slope of the stress–strain curve in the elastic regime is rectilinear for thermoplastics and thermosets, and is used to calculate material stiffness, i.e. Young's modulus.

The situation is different for many elastomers, which do not have a rectilinear strain–stress response in the elastic regime. Instead, the apparent elastic modulus for elastomers decreases with increasing strain as the chains begin to uncoil and stretch in the load direction. The elastomer chains are so tightly coiled that they can be stretched many times their original length before there is any significant resistance to loading. As the chains start to straighten, the stiffness begins to increase owing to elastic stretching of the bonds within the chain and the crosslinking bonds between the chains. Because elastomers are not heavily crosslinked, the chains can stretch without a large amount of resistance from the crosslinking bonds. At any point in the elastic regime, the elastomer reverts back to its original shape (in the absence of creep) when the load is removed. Elastic recovery occurs by the bonds along the chains and the crosslinking bonds between the chains relaxing into their original state, allowing the chains to return to their original coiled position.

Plastic deformation

An amorphous polymer plastically deforms when the applied stress exceeds the yield strength. The yielding mechanisms are complicated and involve the stretching, rotation, sliding and disentanglement of the chains under load deformation. Eventually the chains become almost aligned parallel and close together and, at this point, the polymer reaches its ultimate strength. Plastic deformation of thermosets is resisted by the crosslinks and, therefore, the yield strength of these polymers is often higher than thermoplastics. The lack of crosslinks in amorphous thermoplastics allows the chains to stretch, slide and disentangle more readily than in thermosets, which results in greater ductility and toughness.

The yielding process in semicrystalline thermoplastics is a combination of the deformation processes just described for amorphous polymers together with different processes for the crystalline phase. When a tensile load is applied, the crystalline lamellae slide past one another and begin to separate as the chains are stretched. The folds in the lamellae tilt and become aligned in the load direction; at this stage the ultimate strength is reached.

Fracture

The process leading to fracture of polymers starts with the breaking of the weakest or most highly strained segments of the polymer chains. For thermoplastics, this occurs along the backbone of the chains, whereas in thermosets and elastomers it can start by the breaking of crosslinks. The breaking of the chains continues under increasing strain until micrometre-sized voids develop at the most damaged regions, as shown in Fig. 13.15. These voids grow in size and coalesce into cracks until the polymer eventually breaks.

13.9.2 Engineering properties of polymers

Table 13.3 gives the mechanical properties of thermoplastics, thermosetting polymers and elastomers that have current or potential aerospace applications. The values are the average properties, which can vary owing to changes in the chemical state and processing conditions of the polymer. The Young's modulus of thermoplastics and thermosets is in the range of 2–6 GPa, which is much smaller than the elastic modulus of aluminium alloys (70 GPa), titanium alloys (110 GPa), steels (210 GPa) and other metals used in aircraft. Likewise, the tensile strength of most polymers is under 100 MPa, which is well below the strength of the aerospace alloys. It is because of the low mechanical properties that polymers are not used in heavily loaded aircraft structures.

The tensile strength of polymers can be controlled in several ways:

- Increasing the degree of polymerisation. The strength properties of polymers increase with the length of the chains, which is controlled by the degree of polymerisation. For example, the tensile strength of polyethylene is directly related to the chain length, which is defined by the molecular weight (MW). Low-density polyethylene (MW ~ 200 000

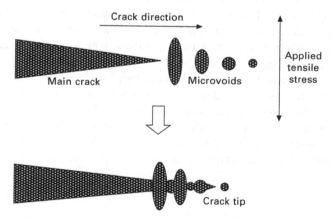

13.15 Schematic showing tensile failure of a polymer by the formation and coalescence of voids.

Table 13.3 Tensile properties of polymers used in aircraft

	Young's modulus (GPa)	Tensile strength (MPa)	Elongation-to-failure (%)
Thermoplastic			
Polyether ether ketone (PEEK)	3.7	96	50
Polyether ketone (PEK)	4.6	110	12
Polyphenylene sulfide (PPS)	6.0	110	12
Polysulfone (PSU)	2.4	68	40
Polyether sulfone (PES)	2.3	81	80
Polyaryl sulfone (PAS)	2.7	75	8
Polyetherimide (PEI)	3.7	100	42
Polycarbonate (PC)	2.7	73	130
Thermoset			
Epoxy	4.5	85	3.0
Phenolic	6	55	1.7
Bismaleimide (524C)	3.3	50	2.9
Polyimide (PMR15)	4.0	65	6.0
Cyanate ester	3.0	80	3.2
Elastomer			
Polyisoprene		20	800
Polybutadiene		23	
Polyisobutylene		27	350
Polychloroprene (Neoprene)		23	800
Butadiene–styrene		30	2000
Thermoplastic elastomer		33	1300
Silicone		7	700

monomer units) has a tensile strength of about 20 MPa, high-density polyethylene (MW ~ 500 000 monomer units) has the strength of 37 MPa, and ultra-high-density polyethylene (MW ~ 4 500 000 monomer units) the strength of 60 MPa. The strength increases with the chain length because they become more tangled and, therefore, a higher stress is needed to stretch and untangle them. The strength increases with the chain length up a limit beyond which further polymerisation does not add extra strength.

• Increasing the size of side groups attached to the main chain. Strength is increased by adding larger atoms or groups to the side of the chain backbone. This makes it more difficult for the chains to rotate, uncoil and disentangle when under load, thereby increasing strength. For example, polyetherimide (PEI) which has relatively large side groups has the tensile strength of about 100 MPa whereas polyphenylene sulfide (PPS) with no large side groups has the strength of only 60 MPa.

• Increasing the amount of chain branching. The strength is increased by extensive chain branching. A high density of branches along the main chain provides resistance against the stretching, sliding and disentanglement of neighbouring chains and, thereby, increases the strength. However, too much branching can prevent the close packing and crystallisation of the chains and, thereby, reduce the strength.

- Increasing the degree of crosslinking (for thermosets and elastomers). The strength increases with the number density of crosslinks between the chains.
- Increasing the degree of crystallisation (for thermoplastics). The strength increases with the volume percentage of thermoplastic that is crystallised. This is because a higher load is needed to permanently deform polymer chains when in a crystallised state rather than the amorphous condition.

Table 13.3 shows that thermosets have low ductility, with elongation-to-failure usually below 5%. This is because the high density of crosslinks between the chains resists extensive plastic deformation of the polymer. Many thermoplastics have high elongation-to-failure (above 50%), although their ductility decreases with increasing crystallinity. Elastomers have exceptionally high elongation-to-failure, in the range of several hundred to several thousand percent, because of their coiled chain structure.

13.9.3 Thermal properties of polymers

An important consideration in the use of polymers is softening that occurs owing to heating effects, such as the frictional heating of aircraft skins at fast flight speeds or the operating temperature of engine components. Polymers lose stiffness, strength and other properties at moderate temperature. Although other aerospace materials, including the metal alloys we have discussed, also lose mechanical performance when heated, the properties of polymers are reduced at much lower temperatures. The mechanical properties of most polymers drop sharply above 100–150 °C, and therefore their use in fibre composite airframe, bonded joints, engine components and other structural applications must be confined to lower temperatures. For high-temperature conditions, several polymer systems can be used, such as bismaleimides, polyimides and cyanates, but even here the maximum operating temperature is below 200–220 °C.

The effect of increasing temperature on the Young's modulus of polymers is shown in Fig. 13.16. At low temperature there is low internal energy within the polymer and, therefore, the chains are 'sluggish' and difficult to move under an applied force. In crystalline polymers, the sluggish behaviour makes it difficult for the chains to unfold and align themselves in the load direction. In amorphous polymers, the chains lack the energy to slide and move through the tangled network in order to align in the load direction. The ability of polymers to undergo large deformations is suppressed at low temperatures and they are more likely to resist the applied load, thus becoming

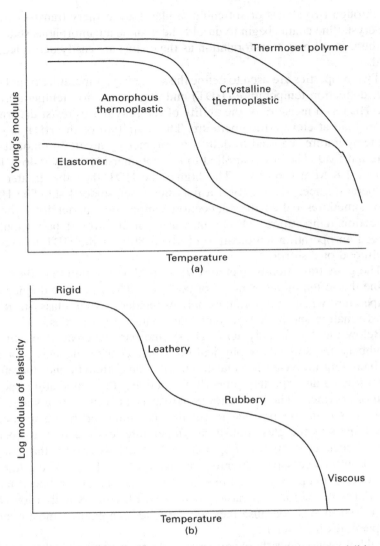

13.16 Variation of (a) Young's modulus with temperature and (b) deformability with temperature.

stiff, strong and brittle. In this condition, polymers are often described as being glassy. When the temperature is raised more energy is available to the polymer chains to move under load. This is observed as a reduction in the elastic modulus and strength and an increase in the ductility (elongation-to-failure). When an amorphous polymer transforms from a rigid, brittle material to a soft, ductile material with increasing temperature it is described as converting from a glassy state to a rubbery state. Crystalline materials go through two stages of softening: a glassy-to-leathery transformation as the crystalline chains begin to unfold and assume an amorphous state, and a leathery-to-rubbery transformation as the chains are easily deformed under load.

Two properties are used to define the softening temperature of polymers: heat deflection temperature (HDT) and glass transition temperature (T_g). The HDT is a measure of the ability of the polymer to resist deformation under load at elevated temperature. The definition of the HDT is simply the temperature required to deflect a polymer by a certain amount under heat and load. The stress applied to the polymer is usually 0.46 MPa (66 psi) or 1.8 MPa (264 psi). The higher the HDT then the greater is the temperature needed to deform a polymer when under load. The HDT is also sometimes called the 'deflection temperature under load' or 'heat deflection temperature'. A polymer used in an aircraft part should not exceed an operating temperature of about 80% of the HDT to avoid the likelihood of distortion.

The glass transition temperature T_g is used more often than the HDT to define the softening temperature of polymers. The T_g can be defined as the temperature when an amorphous polymer undergoes the glassy-to-rubbery transformation and there is a significant change in properties.

Below the T_g, the polymer is glassy and hard whereas above the T_g it is rubbery and soft. Many physical properties change suddenly around the T_g, including the viscosity, elastic modulus and strength and, therefore, it is considered an important property for defining the temperature operating limit of polymers. The T_g of various aerospace polymers are given in Table 13.4. It is common practice to set the maximum operating temperature of polymers as the glass transition temperature less a certain temperature interval such as 30 °C (i.e. $T_{max} = T_g - 30$ °C). This ensures that polymers do not suffer excessive softening during service. The T_g of elastomers is below room temperature, and therefore they respond as a rubbery and soft material under normal operating conditions. Elastomers with a low T_g are often used in seals because they are soft and compliant at most operating temperatures for aircraft.

Table 13.4 also gives the average melting temperatures for the thermoplastics, which is usually 1.5 to 2 times higher than the glass transition temperature. Thermosets do not have a melting temperature because the crosslinks stop the chains flowing like a liquid, and instead these polymers decompose at high temperature without melting.

13.10   Polymer additives

Additives are often blended into polymers during processing to improve one or more properties. Additives include chemical pigments to give colour;

Table 13.4 Temperature properties of aerospace polymers

	Glass transition temperature (°C)	Melting temperature (°C)
Thermoplastics		
Polyether ether ketone (PEEK)	140	245
Polyphenylene sulfide (PPS)	75	285
Polyetherimide (PEI)	218	220
Polycarbonate (PC)	150	155
Thermosets		
Polyester	110	–
Vinyl ester	120	–
Epoxy	110–220	–
Phenolic	100–180	–
Bismaleimide	220	–
Polyimide	340	–
Cyanate ester	250–290	–
Elastomers		
Polybutadiene	–90	120
Polychloroprene	–50	80
Polyisoprene	–73	30

filler particles to reduce cost, increase mechanical performance and provide dimensional stability; toughening agents and plasticisers to increase fracture toughness and ductility; stabilisers to increase operating temperature or resistance against ultraviolet radiation; and flame retardants. An important class of additives is fibrous fillers, such as glass, carbon, boron and aramid fibres, used as the reinforcement in fibre–polymer composites, which are described in chapters 14 and 15.

Additives are not always used in polymers for aircraft applications, and when they are used it is often for a specific purpose. Additives may be included in the polymer matrix phase of fibre composites to increase flame resistance and environmental durability. Additives are often used in structural adhesives to increase toughness and durability, particularly against hot–wet conditions. Elastomers used for aircraft tyres contain filler particles called carbon black to increase the tensile strength and wear resistance.

Toughening agents are used in thermosets to increase the fracture toughness, ductility and impact hardness. High-strength thermosets, such as epoxy resins, have low ductility with strain-to-failure values as low as a few percent. Adhesives used in aircraft structural joints require high ductility and fracture toughness and, therefore, toughening agents are used to improve fracture toughness. The most common toughening agent comprises small particles of rubber, usually carboxyl-terminated butadiene nitrile (CTBN) rubber. The rubber particles are only a few micrometres in size and are dispersed through the adhesive to impede crack growth. Fine thermoplastic fillers are also used as toughening agents.

Plasticisers are additives used to increase the toughness of crystalline thermoplastics, in which rubber toughening particles (such as CTBN) are usually ineffective. Plasticisers are low-molecular-weight polymers that increase the spacing between chains of crystalline polymer to make them more flexible and, thereby, tougher.

A safety concern with using polymers and polymer composites in aircraft is fire. Most polymers are flammable and release large amounts of heat, smoke and fumes when they burn. It is often necessary to use flame-retardant additives in polymers to achieve the fire resistant standards specified by safety regulators such as the FAA. Many types of additives are used to improve the flammability resistance of polymers, with the most common containing halogenated or phosphorus compounds. Flame-retardant additives are either discrete particles within the polymer or are chemically incorporated into the polymer chain structure. New and more effective flame retardant additives are being developed, with carbon nanotubes (see chapter 14) offering new possibilities for increased fire resistance combined with improved mechanical performance.

13.11   Polymers for radar-absorbing materials (RAMs)

Radar is the most common technique for the detection and tracking of aircraft. Although radar is an indispensable tool in aviation traffic management, it is a problem in offensive military operations which require aircraft to attack their target and then escape without being detected. Radar involves the transmission of electromagnetic waves into the atmosphere; they are then reflected off the aircraft back to a receiving antenna. Conventional aircraft, such as passenger airliners, are easily detected by radar because of their cylindrical shape together with bumps from the engines and tail plane. Metals used in the aircraft are also strong reflectors of electromagnetic waves, and can be easily detected using radar. Composites are also detected using radar, although usually not as easily as metals.

Radar-absorbing material (RAM) is a specialist class of polymer-based material applied to the surface of stealth military aircraft, such as the F-22 *Raptor* and F-35 *Lightning II* (Fig. 13.19), to reduce the radar cross-section and thereby make them harder to detect by radar. These materials are also applied in stealth versions of tactical unmanned aerial systems, such as the Boeing X-45. RAM is applied over the entire external skin or (more often) to regions of high radar reflection such as surface edges. RAM works on the principle of the aircraft absorbing the electromagnetic wave energy to minimise the intensity of the reflected signal. RAMs are used in combination with other stealth technologies, such as planar design and hidden engines,

(a)

(b)

13.17 Examples of stealth aircraft that use radar-absorbing materials: (a) F-22 *Raptor* and (b) F-35 *Lightning* II. (a) Photograph supplied courtesy of S. Austen. (b) Photograph supplied courtesy of B. Lockett, Air-and-Space.com.

to make military aircraft difficult to detect. It is possible to reduce the radar cross-section of a fighter aircraft to the size of a mid-sized bird through the optimum design and application of stealth technologies.

Information about the composition of RAMs is guarded by the military. Most RAMs consist of ferromagnetic particles embedded in a polymer matrix having a high dielectric constant. One of the most common RAMs is called iron ball paint, which contains tiny metal-coated spheres suspended in an epoxy-based paint. The spheres are coated with ferrite or carbonyl iron. When electromagnetic radiation enters iron ball paint it is absorbed by the ferrite or carbonyl iron molecules which causes them to oscillate. The molecular oscillations then decay with the release of heat, and this is an effective mechanism of damping electromagnetic waves. The small amount of heat generated by the oscillations is conducted into the airframe where it dissipates.

Another type of RAM consists of neoprene sheet containing ferrite or carbon black particles. This material, which was used on early versions of the F-117A *Nighthawk*, works on the same principle as iron ball paint by converting the radar waves to heat. The USAF has introduced radar-absorbent paints made from ferrofluidic and nonmagnetic materials to some of their stealth aircraft. Ferrofluids are colloidal mixtures composed of nano-sized ferromagnetic particles (under 10 nm) suspended in a carrier medium. Ferrofluids are superparamagnetic, which means they are strongly polarised by electromagnetic radiation. When the fluid is subjected to a sufficiently strong electromagnetic field the polarisation causes corrugations to form on the surface. The electromagnetic energy used to form these corrugations weakens or eliminates the energy of the reflected radar signal. RAM cannot absorb radar at all frequencies. The composition and morphology of the material is carefully tailored to absorb radar waves over a specific frequency band.

13.12 Summary

There are three main classes of polymers: thermosets, thermoplastics and elastomers. The polymers most often used in aircraft are thermosets such as epoxy resins and bismaleimides, which are used as the matrix phase in fibre composites and as structural adhesives. The aerospace applications for thermoplastics and elastomers is much less.

The properties of polymers that make them useful aerospace materials are low weight, excellent corrosion resistance and good ductility. Some polymers are also transparent for use in window applications. However, polymers have low stiffness, strength, fatigue life and creep resistance and, therefore, should not be used on their own in structural applications.

Polymers are produced by polymerisation reactions called addition or condensation reactions, which join the monomer units together into long chain-like macromolecules. The chemical and physical structure of the molecular chains has a major influence on the mechanical properties. Thermosets have higher stiffness, strength and creep resistance compared with thermoplastics because the molecular chains are interconnected by crosslinks. These crosslinks resist the sliding of chains under stress and, thereby, increase the mechanical properties.

Epoxy resin is the most used polymer for aerospace, and finds applications in carbon-fibre composites and as a structural adhesive. Epoxy resin has good stiffness, strength and environmental durability and can be cured at moderate temperatures (under 180 °C) with little shrinkage. Other thermosets used in composite materials include bismaleimide and polyimide, which are suited for high-temperature applications. Phenolic resin has excellent flammability resistance and is used in moulded parts and composites for cabin interiors.

Thermoplastics have high toughness, ductility and impact resistance, but the high cost of processing and low creep resistance limits their use in aerospace. Thermoplastics such as PEEK are used as the matrix phase in fibre composites requiring impact toughness. Polycarbonate is used for aircraft windows because of its high transparency, scratch resistance and impact strength.

Elastomers lack the stiffness and strength needed for aerospace applications, although their high elasticity makes them useful as a sealing material. Elastomers have high elasticity owing to their coiled molecular structure which responds under load in a similar way to a spring.

Adhesives used for bonding aircraft components and sandwich composites are usually thermosets, such as epoxy resin. Adhesives must have low shrinkage during curing to avoid the formation of residual tensile stress in the bond. Correct surface preparation is essential to ensure a high strength and durable bond is achieved with a polymer adhesive.

The mechanical properties of polymers are inferior to aerospace structural metals, and their use is restricted to relatively low-temperature applications. Polymers transform from a hard and glassy condition to a soft and rubbery state during heating, and the maximum working temperature is defined by the heat-deflection temperature or glass transition temperature.

Polymers often contain additives for special functions such as colour, toughness or fire.

Radar-absorbing materials (RAMs) are a special class of polymer that convert radar (electromagnetic) energy to some other form of energy (e.g. heat) and thereby improve the stealth of military aircraft. RAM must be used with other stealth technologies such as design adaptations of the aircraft shape to minimise the radar cross-section.

13.13 Terminology

Addition polymerisation: Polymerisation process where monomer molecules add to a growing polymer chain one at a time.

Amorphous: Amorphous polymer is a material in which there is no long-range order in the positions and arrangement of the molecular chains.

Atactic: Polymer chain configuration in which the side groups are positioned randomly on one or the other side of the polymer back-bone.

Block copolymer: Polymer formed when two different types of monomer are linked in the same polymer chain and occur as alternating sequences (blocks) of each type.

Condensation polymerisation: Polymerisation process involving the reaction of monomer molecules that produces a polymer chain and chemical by-products that do not form part of the chain structure.

Copolymer: A polymer whose chemical structure consists of long chains of two chemically different monomers, which repeat at more or less regular intervals in the chain.

Crosslinks: Covalent bonds linking one polymer chain to another chain.

Crystalline: Polymer with a regular order or pattern of molecular arrangement.

Elastomer: A type of polymer with coiled molecular chains that are crosslinked.

Glass transition temperature: Temperature at which a reversible change occurs in a polymer when it is heated to a certain temperature and undergoes a transition from a hard and glassy to a soft and ductile condition.

Graft copolymer: A special type of branched polymer with the main backbone to the chain composed of one type of monomer while the branches are composed of a chemically different monomer.

Heat deflection temperature: Temperature at which a material begins to soften and deflect (usually by 0.010 inches) under load (usually 264 psi).

Homopolymer: Polymer which is formed from only one type of monomer.

Isotactic: Polymer chain configuration in which the side groups are all located on same side of the polymer back-bone.

Molecular weight: A number defining how heavy one molecule (or unit) of a chemical is compared with the lightest element, hydrogen, which has a weight of 1.

Monomer: Chemical compound that can undergo polymerisation, which is a chemical reaction in which two or more molecules combine to form a larger molecule that contains repeating units.

Polymerisation: Chemical reaction in which small molecules combine to form a larger molecule that contains repeating units of the original molecules. It is a process of reacting monomer molecules together in a chemical reaction to form linear chains or a three-dimensional network of polymer chains.

Radar cross-section: A measure of how detectable an object such as an aircraft or helicopter is with radar. A larger radar cross-section indicates that the object is more easily detected.

Stress rupture: The fracture of a material after carrying a sustained load for an extended period of time; it usually involves viscoelastic deformation and occurs more rapidly at elevated temperature.

Syndiotactic: Polymer chain configuration in which the side groups are located at alternate positions along the polymer backbone.

Thermoplastic: Polymer which may be softened by heat and hardened by cooling in a reversible physical process owing to the lack of crosslinks between the chains.

Thermoset: Thermosetting polymer obtained by crosslinking of the chains that make it an infusible and insoluble material.

Van der Waals bond: A weak attractive force between atoms or nonpolar molecules caused by a temporary change in dipole moment arising from a brief shift of orbital electrons to one side of one atom or molecule, creating a similar shift in adjacent atoms or molecules.

Vulcanisation: Chemical reaction whereby the properties of an elastomer are changed by causing it to react with sulfur or another crosslinking agent.

13.14 Further reading and research

Ebnesajjad, S., *Adhesives technology handbook*, William Andrew Inc., 2008.

May, C. A., *Epoxy resins: chemistry and technology*, Marcel Dekker, 1988.

Powell, P. C. and Ingen Housz, A.J., *Engineering with polymers*, Stanley Thornes, 1998.

Tadmor, Z. and Gogos, C. G., *Principles of polymer processing (2nd edition)*, Wiley Interscience, 2006.

Ward, I. M. and Sweeney, J., *An introduction to the mechanical properties of solid polymers*, John Wiley & Sons, 2004.

13.15 Case study: space shuttle *Challenger* accident

The space shuttle *Challenger* (STS-51) exploded just over one minute after take-off on 28 January 1986, killing seven astronauts. After an exhaustive investigation by NASA and other US agencies the cause of the accident was found. The space shuttle is fitted with two solid rocket boosters that generate an extraordinary amount of thrust during take-off that launches the main vehicle into space. Without the boosters the shuttle cannot generate enough thrust to overcome the gravitational pull of Earth. There is a booster rocket attached to each side of the external fuel tank, and each booster is 36 m long and 7.3 m in diameter (Fig. 13.18). The boosters are constructed from hollow metal cylinders, with the joint connecting the cylinders containing two O-rings made with an elastomer. The elastomer is needed to create a tight seal to prevent hot gases escaping from the rocket motor during take-off.

The *Challenger* accident was caused by several factors, with a critical problem being that one of the elastomer O-rings in a booster rocket did not form a tight seal owing to cold weather during take-off. Elastomers shrink and lose elasticity at low temperature and, at take-off, the O-ring was unable to expand sufficiently to form a seal between two cylinders. This caused hot combustion gases (over 5000 °F) inside the rocket motor to rapidly degrade the elastomer O-ring, thus allowing hundreds of tons of propellant to escape and ignite, thereby causing the space shuttle to explode (Fig. 13.19).

13.18 Rocket boosters on the space shuttle. Photograph reproduced with permission from NASA.

13.19 Explosion of the space shuttle *Challenger* (STS-51).

14
Manufacturing of fibre–polymer composite materials

14.1   Introduction

The rapid growth in the use of fibre–polymer composites in aircraft and helicopters since the mid-1990s has led to it now sitting side-by-side with aluminium as the most commonly used structural materials. A critical factor underpinning the growth in composites has been the development of improved, cheaper manufacturing processes. Similarly to the casting and forming processes used to manufacture metal structures, the processes used to produce composite components have a major impact on the cost, quality and material properties. The choice of manufacturing process impinges on the design of the composite component; some processes are suited to flat or slightly curved structures whereas others are better for complex, highly curved parts. The choice of process also determines the mechanical properties of the composite by affecting the volume fraction and orientation of the reinforcing fibres.

The aerospace industry has invested heavily in the development of manufacturing processes capable of producing many types of composite components for aircraft structures and engines. The industry has also developed processes to reduce the manufacturing cost; which typically accounts for over 60–70% of the total production cost (with the remainder being materials, nondestructive inspection and other process-related costs). Figure 14.1 shows the approximate reduction in the cost of manufacturing composite components for civilian and military aircraft since 1980, and costs have fallen considerably owing to advances in automation, rapid processing and other manufacturing technologies.

Composites for aircraft applications are manufactured in two basic material forms: laminate and sandwich composites (Fig. 14.2). Laminates consist of multiple layers of fibre and resin (called plies) which are bonded together into a solid material. The fibres are oriented along the in-plane directions of principal loading to provide high stiffness, strength and fatigue resistance and the polymer matrix binds the fibres into the material. Laminates made of carbon fibre–epoxy resin composite are used in the most heavily-loaded aircraft structures. Sandwich composites consist of thin face skins (usually carbon–epoxy laminate) bonded (often with adhesive film) to a thick,

303

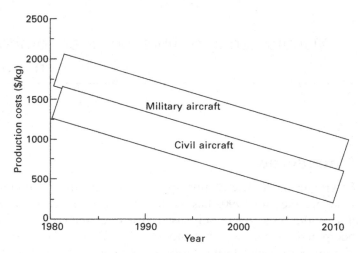

14.1 Approximate reductions in the production costs of aerospace composites owing to improvements in manufacturing technologies.

lightweight core material. Sandwich structures are often used in lightly loaded structures that require high resistance to bending and buckling.

The cost of manufacturing aircraft parts is usually higher with composites than with aluminium. This is because carbon–epoxy is more expensive than aluminium, the tooling costs are higher, and the production processes are often slower and more labour intensive. However, there are numerous benefits in the manufacture of composite components. The number of assemblies, parts and fasteners needed in the construction of composite structures is usually much less than the same structure made using metal. For instance, the fin box of the vertical tailplane of a single aisle aircraft contains over 700 parts and about 40 000 fasteners when made with aluminium, but this falls to only 230 parts and around 10 000 fasteners with composite. As another example, one metal fuselage barrel for a mid-sized airliner requires around 1500 sheets of aluminium held together by about 50 000 rivets. With composite construction, the barrel can be made as a single-piece and the number of fasteners and rivets drops by about 80%.

In this chapter the processes used in the manufacture of composite components for aircraft are studied. Firstly, the processes used to produce carbon, glass and aramid fibres, which are the main types of reinforcement used in aircraft composites, are examined. The processing methods used to produce fibres have a major impact on their mechanical properties and subsequently the structural properties of the composite part. Therefore, the manufacture of these fibres is important to fully understand. Composite manufacturing processes can be roughly divided into two categories depending on how the polymer is combined with the fibres: prepreg-based processes and resin-

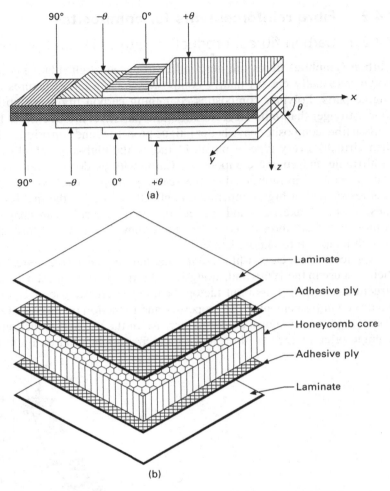

90°   −θ   0°   +θ

90°   −θ   0°   +θ

(a)

Laminate

Adhesive ply

Honeycomb core

Adhesive ply

Laminate

(b)

14.2 Two material forms of composites: (a) laminate and (b) sandwich composite.

infusion processes. Several prepreg-based processes are described: autoclave curing, automated tape lay-up and automated fibre placement. A variety of resin infusion processes are explained, including resin transfer moulding, resin film infusion, vacuum-bag resin infusion, and filament winding. Most manufacturing processes for composites are capable of producing parts to the near-net shape, and a significant amount of machining is not required (unlike many metal products). After manufacturing, most composites only require a small amount of trimming and hole drilling, and the machining processes used to remove material are briefly explained.

14.2   Fibre reinforcements for composites

14.2.1 Carbon fibres: production, structure and properties

Carbon (graphite) fibres are very stiff, strong and light filaments used in polymer (usually epoxy) matrix composites for aircraft structures and jet engine parts. Polymer composites containing carbon fibres are up to five times stronger than mild steel for structural parts, yet are five times lighter. Carbon fibre composites have better fatigue properties and corrosion resistance than virtually every type of metal, including the high-strength alloys used in airframe and engine components. Carbon fibres do not soften or melt and, when used in reinforced carbon–carbon composites, have exceptional heat resistance for high-temperature applications such as thermal insulation tiles, aircraft brake pads and rocket nozzles (as described in chapter 16). Although carbon fibres do not soften/melt, they do ultimately lose strength at high temperature via oxidation.

Applications for carbon-fibre composites have expanded dramatically since their first use in the 1960s and, alongside aluminium, is the most used aircraft structural material. Important factors behind the greater use of composites have been improvement to the properties and reduction in the cost of carbon fibres. Figure 14.3 shows the large increase in the use and reduction in the average price of carbon fibres since their first application in aircraft. It is

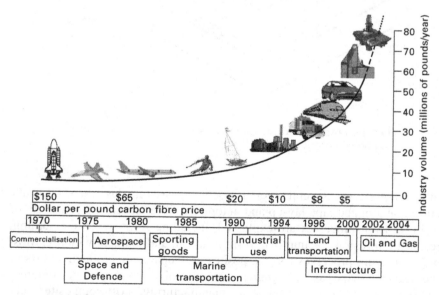

14.3 Changes to the consumption and average price of carbon fibre owing to increasing use in both aerospace and non-aerospace applications.

important to note that large variations exist in the price depending on the fibre properties, and therefore the figure provides an indicative trend.

The properties of carbon fibre that make it an important aerospace material (stiffness, strength and fatigue resistance), are determined by the process methods used in its production. Carbon fibre is extracted via heat treatment from a carbon-rich precursor material, the most common of which are polyacrylonitrile (PAN), pitch and rayon. Carbon fibres produced for aerospace applications are made using PAN, which is an organic polymer. Other precursor materials such as pitch and rayon are rarely used in the production of aircraft-grade carbon fibres because of their lower mechanical performance, but they are used to produce fibres for non-aerospace applications such as sporting equipment, civil infrastructure and marine craft.

Before the production of the carbon fibre, the PAN precursor material is stretched into long, thin filaments. This stretching causes the PAN molecules to align along the filament axis, which subsequently increases the stiffness of the carbon fibre after processing. The greater the stretch applied to the PAN, the higher the preferred orientation of the molecules along the filament, resulting in stiffer carbon fibre. PAN filaments are heat treated in a multistage process while under tension to produce carbon fibres. The process begins by stretching and heating the PAN filaments to 200–300 °C in air. This treatment oxidises and crosslinks the PAN into thermally stable carbon-rich fibres. The fibres are stretched during heating to prevent them contracting during oxidation. The PAN is then pyrolysed at 1500–2000 °C in a furnace having an inert atmosphere (e.g. argon gas), which stops further oxidation of the carbon. This heat treatment is called carbonisation because it removes non-carbon atoms from the PAN molecules (e.g. N, O, H) leaving a carbon-rich fibre with a purity content of 93–95 %. The heat treatment changes the molecular bond structure into graphite. Carbon fibre consists of closely packed layers of graphite sheets in which the carbon atoms are arranged in a two-dimensional hexagonal ring pattern that looks like a sheet of chicken wire. The graphite sheets are stacked parallel to one another in a regular pattern (as represented in Fig. 14.4) and the sheets are aligned along the fibre axis.

The mechanical properties of carbon fibre vary over a large range depending on the temperature of the final heat treatment. There are two general categories of carbon fibre produced depending on the final temperature: high-modulus or high-strength (Table 14.1). The stiffness and strength of carbon fibre is determined by the pyrolysis temperature for the final processing stage, as shown in Fig. 14.5. Peak strength is reached by heating the fibre at 1500–1600 °C whereas the elastic modulus increases with temperature. The production of high-strength carbon fibres involves carbonisation of the PAN filaments at 1500–1600 °C. An additional heat treatment step called graphitisation is performed after carbonisation to produce high modulus fibres, which are

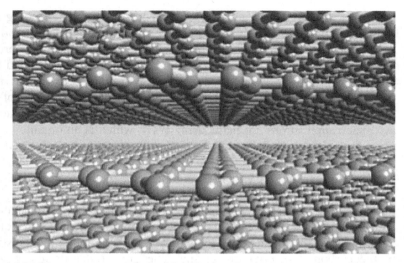

14.4 Sheet structure of graphite. Each sphere represents a carbon atom arranged in a hexagonal pattern in sheets.

Table 14.1 Properties of PAN-based carbon fibres

Property	High-modulus carbon fibres	High-strength carbon fibres
Density (g cm^{-3})	1.9	1.8
Carbon content (%)	+99	95
Tensile modulus (GPa)	350–450	220–300
Tensile strength (MPa)	3500–5500	3500–6200
Elongation-to-failure (%)	0.1–1.0	1.3–2.2

heated in an inert atmosphere at 2000–3000 °C. The higher temperature used for graphitisation increases the preferred orientation of the graphite sheets along the fibre axis and thus raises the elastic modulus.

By appropriate choice of the final process temperature it is possible to control the elastic modulus and strength of carbon fibres for specific structural applications. For instance, carbon fibres used in composite structures which require a high strength-to-weight ratio, such as the wing box, are pyrolysed at 1500–1600 °C to yield high strengths of 3500–6000 MPa. Structures needing a high stiffness-to-weight ratio such as the control surfaces may contain high stiffness fibres heated at 2500–3000 °C which have an elastic modulus of 350–450 GPa. Structures that require both high stiffness and strength such as the fuselage and wings may contain intermediate modulus fibres.

The properties of carbon fibres are determined by the internal arrangement of the graphite sheets, which is controlled by the final temperature. A

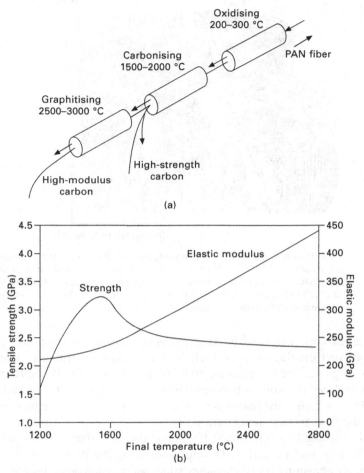

14.5 (a) Production process for carbon fibres; (b) effect of final processing temperature on the tensile modulus and strength of carbon fibre.

single carbon filament is a long rod with a diameter of 7–8 μm. Packed within PAN-based fibres are tiny ribbon-like carbon crystallites which are called turbostratic graphite. In turbostratic graphite, the graphite sheets are haphazardly folded or crumpled together as shown in Fig. 14.6. The intertwined sheets are orientated more or less parallel with the fibre axis, and this crystal alignment makes the fibres very strong along their axis. This arrangement also makes the fibres highly anisotropic; the interatomic forces are much stronger along than between the sheets. For instance, the stiffness along the fibre axis is typically 15 to 30 times higher than across the fibre.

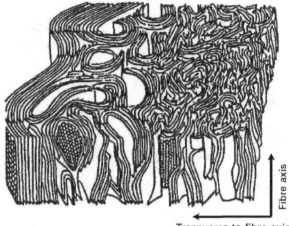

Fibre axis

Transverse to fibre axis

14.6 Three-dimensional representation of the turbostratic graphite structure of PAN-based carbon fibre (from S. C. Bennett and D. J. Johnston, 'Structural heterogeneity in carbon fibers', *Proceedings of the 5th London carbon and graphite conference*, Vol. 1, Society for Chemical Industries, London, 1978, pp. 377–386).

The stiffness of carbon fibre is dependent on the alignment of the crystalline ribbons and the degree of perfection of the graphite structure. The length and straightness of the graphite ribbons determines the fibre stiffness. The order between the ribbons becomes better with increasing temperature (up to 3000 °C), which is the reason for the steady improvement in elastic modulus with the final process temperature.

Fibre strength is determined by the number and size of flaws, which are usually tiny surface cracks. During processing, the fibres are damaged by fine-scale abrasion events when they slide against each other after leaving the furnace and during collimation into bundles. The rubbing action and other damaging events introduces surface cracks that are usually smaller than 100 nm. Although tiny, these cracks have a large impact on fibre strength.

After final heat treatment, carbon fibres are surface treated with various chemicals called the size which serves several functions. A thin film of chemicals is applied to the fibre surface to increase its bond strength to the polymer matrix. Carbon does not adhere strongly to most polymers, including epoxies, and it is necessary to coat the fibre surface with a thin sizing film to promote strong bonding. Size agents are also used to reduce friction damage between fibres. Shortly after the final heat treatment, the individual carbon fibres are collimated into bundles (which are also called tows) for ease of handling.

Section 14.11 at the end of this chapter presents a case study of carbon nanotubes in composites.

14.2.2 Glass fibres: production, structure and properties

Glass fibre composites are used sparingly in aircraft structures owing to their low stiffness. Glass fibre has a low elastic modulus (between 3 to 6 times lower than carbon fibre) which gives glass-reinforced composites a comparatively low stiffness-to-weight ratio. Glass fibre composites are only used in structures where specific stiffness is not a design factor, which is mainly secondary components such as aircraft fairings and helicopter structures such as the cabin shell. Glass fibre composites have low dielectric properties and, therefore, are used when transparency to electromagnetic radiation (i.e. radar waves) is important, such as radomes and aerial covers.

Despite the restricted use of glass-reinforced composites in components where stiffness is a critical property, whenever possible they are used instead of carbon-fibre materials because of their lower cost. Glass fibres are anywhere from 10 to 100 times cheaper than carbon fibres. The greatest use of glass-reinforced composite is inside the aircraft cabin. The majority of cabin fittings, including overhead luggage storage containers and partitions, are fabricated using glass fibre–phenolic resin composite that is inexpensive and lightweight with good flammability resistance.

Glass fibre is a generic name similar to carbon fibre or steel, and as for these materials there are various types having different properties. Glass fibres are based on silica (SiO_2) with additions of oxides of calcium, boron, iron and aluminium. It is the different concentrations of metal oxides that allow different glass types to be produced. There are two types of glass fibres used for aircraft applications: E-glass (short for electrical grade) and S-glass (structural grade). E-glass is cheaper and lower in strength than S-glass and, therefore, is used mostly in composites for aircraft cabin fittings that do not require high structural properties. The higher strength S-glass composite is used in structural components. The composition and engineering properties of E- and S-glass fibres are given in Table 14.2.

Table 14.2 Composition and properties of E-glass and S-glass fibres

	E-glass	S-glass
Composition (%)		
SiO_2	52	64
CaO	17	
$Al_2O_3 + F_2O_3$	14	25
B_2O_3	11	
MgO	4.6	10
$Na_2O + K_2O$	0.8	0.3
Properties		
Density (g cm^{-3})	2.6	2.5
Young's modulus (GPa)	76	86
Tensile strength (MPa)	3500	4600

The internal structure of glass fibre is different from carbon fibre. Glass consists of a silica network structure containing metal oxides, as shown in Fig. 14.7. The network is a three-dimensional structure and, therefore, the fibre properties are isotropic, thus (unlike carbon) the elastic modulus is the same parallel and transverse to the fibre axis.

Glass fibre is manufactured using a viscous drawing process whereby silica and metal oxide powders are initially blended and melted together in a furnace at about 1400–1500 °C. The molten glass flows from the furnace via bushings or spinnerets containing a large number of tiny holes. The glass solidifies into thin continuous fibres as it passes through the holes. The fibre diameter is around 12 μm. Upon leaving the furnace the fibres are cooled using a water spray mist and then coated with a thin layer of size. The size consists of several functional components including chemicals to improve adhesion with the polymer matrix and lubricants to minimise surface abrasion during handling.

14.2.3 Aramid (Kevlar) fibres: production, structure and properties

Synthetic organic fibres are used in polymer composites for specific aerospace applications. Organic fibres are crystalline polymers with their molecular chains aligned along the fibre axis for high strength. Examples of organic fibres are Dyneema® and Spectra®, which are both ultra-high-molecular-weight polyethylene filaments with high-strength properties. Of the many types of

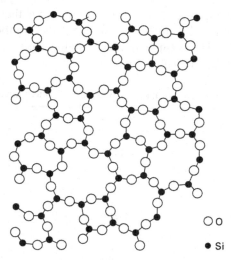

14.7 Internal molecular structure of glass fibre. The metal oxides in glass are not shown.

organic fibres, the most important for aerospace is aramid, whose name is a shortened form of aromatic polyamide (poly-p-phenylene terephthalamide). Aramid is also called Kevlar which is produced by the chemical company Du Pont.

The most common aerospace application for aramid fibre composites is for components that require impact resistance against high-speed projectiles. Aramid fibres absorb a large amount of energy during fracture, thus providing high perforation resistance when hit by a fast projectile. For this reason, aramid fibre composites are used for ballistic protection on military aircraft and helicopters. They are also used for containment rings in jet engines in the event of blade failure. Aramid composites, similarly to glass-reinforced composites, have good dielectric properties, making them suitable for radomes. Aramid composites also have good vibration damping properties, and therefore are used in components such as helicopter engine casings to prevent vibrations from the main rotor blades reaching the cabin. Figure 14.8 shows the vibration damping loss factor for several aerospace materials; aramid–epoxy has ten times the loss decrement of carbon–epoxy and nearly 200 times higher than aluminium.

The process of producing aramid fibres begins by dissolving the polymer in strong acid to produce a liquid chemical blend. The blend is extruded through a spinneret at about 100 °C which causes randomly oriented liquid crystal domains to develop and align in the flow direction. Fibres form during

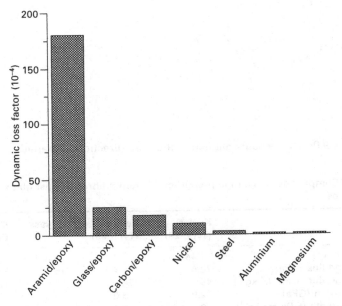

14.8 Vibration loss damping factor of aramid–epoxy composite compared with other aerospace materials.

the extrusion process into highly crystalline, rod-like polymer chains with near perfect molecular orientation in the forming direction. The molecular chains are grouped into distinct domains called fibrils. The fibre essentially consists of bundles of fibrils that are stiff and strong along their axis but weakly bonded together, as shown in Fig. 14.9. As for carbon fibres, aramid fibres are highly anisotropic with their modulus and strength along the fibre axis being much greater than in the transverse direction. The properties of two common types of aramid fibre are given in Table 14.3, and they exceed the specific stiffness and strength of glass fibres. Aramid fibre composites are lightweight with high stiffness and strength in tension. However, these materials have poor compression strength (which is only about 10–20% the tensile strength) owing to low micro-buckling resistance of the aramid fibrils. Therefore, aramid composites should not be used in aircraft components subject to compression loads. Aramid fibres can absorb large amounts of water

14.9 Polymer chains aligned in the fibre direction in aramid.

Table 14.3 Comparison of average properties of aramid fibres against glass and carbon fibres

Property	Kevlar (Type 29)	Kevlar (Type 49)	E-glass	IM Carbon
Density (g cm^{-3})	1.44	1.44	2.6	1.8
Young's modulus (GPa)	70.5	112.4	76	230
Specific modulus (GPa m^3 kg^{-1})	49	78	29	128
Tensile strength (GPa)	2.9	3.0	3.5	4.2
Specific strength (GPa m^3 kg^{-1})	2	2.1	1.3	2.3
Elongation-to-failure (%)	3.6	2.4	4.8	1.9

and are damaged by long-term exposure to ultraviolet radiation. Therefore, the surface of aramid composites must be protected to avoid environmental degradation.

14.3   Production of prepregs and fabrics

Composite aircraft components contain anywhere from tens of thousands to many millions of fibres. For example, an average-sized wing panel made using carbon–epoxy composite comprises of the order of 20 000 million fibres. Fibres are too fine to easily handle and process into composite parts, with carbon fibres being only 1/10 to 1/20 the width of human hair (Fig. 14.10). Therefore, upon leaving the furnace the fibres are bundled together into tows or yarns. Aerospace fibres are collimated into tows that usually contain about 3K (3000), 12 K (12 000) or 24 K (24 000) filaments (Fig. 14.11). These bundles are then used to make prepreg (resin pre-impregnated) material or dry fabric for the production of composite parts.

14.3.1 Prepregs

Major aircraft manufacturers use prepreg tape in the production of composite structural components. The term prepreg is short form for pre-impregnated fibres. Prepreg is a two-part sheet material consisting of fibres (e.g. carbon) and partially cured resin (e.g. epoxy). The most common prepreg used in aircraft is carbon–epoxy, although many other types are used including carbon–bismaleimide, S-glass–epoxy and aramid–epoxy. The benefits of using prepreg include accurate control of the fibre volume content and the ability to achieve high fibre content, thus allowing high-quality composite components to be produced with high mechanical properties.

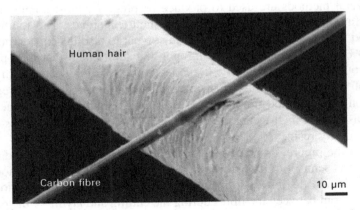

14.10 Comparison of the width of carbon fibre and human hair.

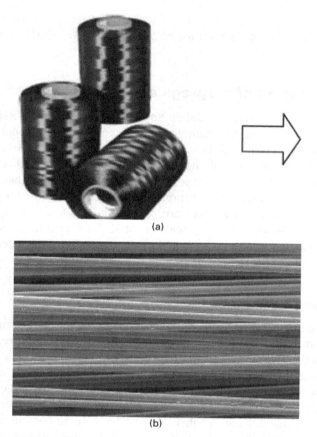

(a)

(b)

14.11 (a) Fibres bundled into a tow. (b) Internal view of a tow showing many individual fibres.

Several methods are used to produce prepreg, the most common being the solution dip, solution spray, and hot-melt techniques. The solution dip method involves dissolving the resin in solvent to a solids concentration of 40–50%. Fibres are then passed through the solution so that they pick up an amount of resin solids. The fibres leave the solution coated with a thin film of resin, and they are then pressed into thin sheets of prepreg. The solution spray method simply involves spraying liquid resin onto the fibres whereas the hot-melt technique involves direct coating of the fibres with a low viscosity resin.

After the prepreg is produced, the resin is partially cured to a condition where it is semisolid; it is too hard to flow like a liquid but soft enough to be pliable and flexible. The resin needs to be hard enough so it does not leak out from between the fibres during the cutting and laying of the prepreg plies. The resin also needs to be soft enough to allow the prepreg to be easily deformed to the shape of the composite component. The resin must be tacky enough (sticky to touch) in order to bond the prepreg layers together during the lay-up of the composite. Epoxy resin used in carbon-fibre prepreg is slightly cured to a semisolid condition, whereby the crosslinking between the polymer chains is about 15–30% complete. Prepreg in this condition is called B-stage cured. The prepreg must be stored at low temperature (about −20 °C) inside a freezer to avoid further curing and crosslinking of the resin matrix which occurs at room temperature.

B-stage cure prepreg is produced as a thin sheet (usually 0.1–0.4 mm thick) with a fibre content of 58–64%. The prepreg is protected on both sides with easily removable separators called backing paper. Backing paper stops the prepreg sheets from sticking to each other before the lay-up of the composite part. Prepreg is used in the manufacture of composite parts by simply cutting to shape and size, removing the backing paper, laying the prepreg sheets in a stack in the preferred fibre orientation, and then consolidating and curing the final composite material using an autoclave as described in 14.5.1.

14.3.2 Dry fabrics

The aerospace industry is increasingly using dry carbon fabric instead of carbon-fibre prepreg to manufacture aircraft structures. There are several advantages gained by using fabric rather than prepreg, including lower material cost, infinite storage life, no need for storage in a freezer, and better formability into complex shapes. The most common fabric is woven fabric produced on weaving looms, and the main styles are plain, twill and satin weaves as shown in Fig. 14.12. Woven fabrics contain fibre tows aligned in the warp direction (which is the weaving direction) and the weft direction, which is perpendicular to the warp. In plain woven fabric, each warp tow alternately crosses over and under each intersecting weft tow. Twill and satin fabrics are woven such that the tows go over and under multiple tows. When observed from the side, weaving causes periodic out-of-plane waviness of the fibres. The waviness results in significant loss in stiffness and strength of the composite because maximum structural performance is achieved when the fibres are absolutely straight and in-plane. Twill and satin weaves have lower degrees of fibre waviness than plain weave, and therefore their composites have higher in-plane mechanical properties. For this reason, twill and satin woven fabrics are preferred over plain woven fabric in the fabrication of aerospace composite components.

Other types of fabric used in aircraft composite parts include non-crimp, braided and knitted fabrics. Non-crimp fabric consists of multiple layers of straight tows oriented at different angles, which are bound together by through-thickness stitches. The tows in non-crimp fabric are not forced to interlace with other tows to produce a weave, and therefore the fibres are

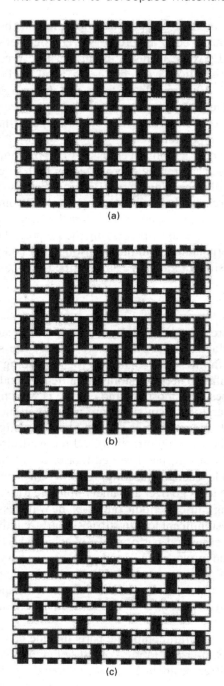

14.12 Common weave architectures: (a) plain, (b) twill and (c) satin.

straight and in-plane for high structural performance. Braided and knitted fabrics are used when high impact collision resistance and high conformed shapes in composite components are required, respectively, although they have low in-plane mechanical properties.

A family of fabrics containing fibre tows aligned in the in-plane direction together with tows running in the through-thickness direction are available for damage tolerant aircraft composite components. The fabrics have a three-dimensional array of fibres, as shown schematically in Fig. 14.13, which allows in-plane and through-thickness loads to be carried by the composite. Composite laminates made using woven fabrics (such as those shown in Fig. 14.13) do not have through-thickness fibres. Such composites have low through-thickness strength and damage resistance. Composites containing a three-dimensional fibre structure have superior damage tolerance because the through-thickness fibres can carry out-of-plane loads. There are various types of three-dimensional fabrics that are produced by orthogonal weaving, stitching, tufting and other specialist techniques. The application for these types of fabrics in aircraft structures is currently limited, although their use is expected to grow.

14.4   Core materials for sandwich composites

Many aircraft structural components are required to carry only light or moderate loads. When a small load is applied on a composite laminate, the structure is designed with a thin skin and few ribs, frames and spars to reduce weight. When the skin becomes too thin and the stiffeners too few then the structure loses its resistance against buckling and, therefore, some additional form of stiffening is required. A common solution is to build the laminate skins as a sandwich by inserting a lightweight filler or core layer. Under bending, the skins carry in-plane tension and compression loads

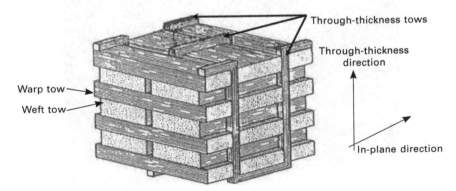

14.13 Three-dimensional fibre tow structure of fabrics.

whereas the core is subjected to shear, as shown in Fig. 14.14. The skin-core construction greatly increases the bending stiffness and thereby makes sandwich structures more resistant to buckling, with only a small increase in weight. Examples of aircraft components made using sandwich composites are control surfaces (e.g. ailerons, flaps) and vertical tailplanes.

Various types of materials are used for the core, with aluminium honeycomb and Nomex being the most common in aircraft sandwich components. Nomex is the tradename for a honeycomb material based on aramid fibres in a phenolic-resin matrix. Aluminium and Nomex core materials have a lightweight cellular honeycomb structure, which provides high shear stiffness (Fig. 14.15). The aerospace industry is increasingly using polymer foams instead of aluminium honeycomb and Nomex because of their superior durability and high-temperature properties. Examples of polymer foams are polyetherimide (PEI) and polymethacrylimide (PMI), and these materials

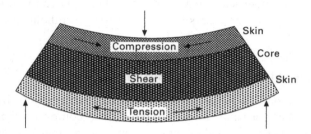

14.14 Basic loading of sandwich composite.

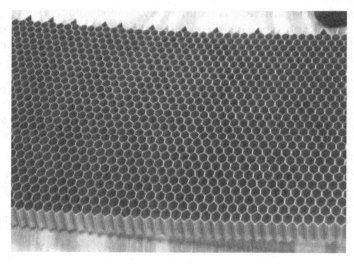

14.15 Honeycomb structure of aluminium.

have a low density cellular structure which has high specific stiffness and strength (Fig. 14.16).

14.5   Composites manufacturing using prepreg

14.5.1 Manual lay-up

Composites have been fabricated using prepreg since the 1970s, and the original fabrication method involved the manual lay-up of the prepreg plies into the orientation and shape of the final component. Manual lay-up involves cutting the prepreg to size, removing the backing paper, and then stacking the prepreg plies by hand onto the tool surface. This process is slow, labour-intensive and inconsistent because of human error in the accurate positioning of the plies. For these reasons, manual hand-lay is rarely used in the construction of large composite components, although it is still used for making small numbers of parts when it is not economically viable to automate the process.

The prepreg is laid-up by hand directly onto the tool, which has the shape of the final component. The prepreg plies are oriented with their fibres aligned in the main loading directions acting on the component when used in service. The most common pattern used in the lay-up of plies for aircraft components is [0/+45/–45/90], which is shown in Fig. 14.17. A composite with this lay-up is called quasi-isotropic because the stiffness and strength properties are roughly equal when loaded along any in-plane direction. The 0° and 90° fibres carry the in-plane tension, compression and bending loads whereas the 45° fibres support shear loads. Another common ply pattern is

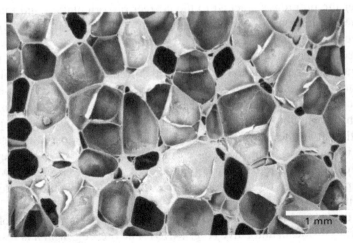

14.16 Cellular microstructure of polymer foams.

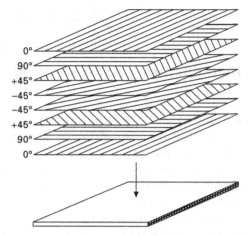

0°
90°
+45°
−45°
−45°
+45°
90°
0°

14.17 Ply orientations for quasi-isotropic cross-ply composite (reproduced with permission from D. R. Askeland, *The science and engineering of materials*, Stanley Thornes (Publishers) Ltd., 1998).

[0/90], known as cross-ply, which is used for composite components subject to in-plane tension or compression loads in service, but not shear loads.

It is important that the orientation of the plies is symmetric around the mid-plane to ensure the material is balanced. That is, the plies in the upper half of the prepreg stack must be arranged as a mirror image of the plies in the lower half. If this does not occur, then the ply pattern is asymmetric and it is possible that the composite may distort or warp after the prepreg is cured. The symmetric lay-up of plies is required for all composite fabrication processes, and not just for prepreg.

After the prepreg is stacked to the correct ply orientation and thickness, it is then compacted by vacuum bagging to remove air from between the ply layers. The prepreg stack is sealed within a plastic bag from which air is removed using a vacuum pump, as shown in Fig. 14.18. The bag is a flexible membrane that conforms to the shape of the prepreg and the underlying tool to ensure good dimensional tolerance. Release film and bleeder layer cloth are placed over the prepreg stack within the vacuum bag. Release film, which is a nonstick flexible sheet containing tiny holes, is used to stop the prepreg from sticking to the bleeder cloth. This cloth is used to absorb excess resin, which is squeezed from the prepreg during consolidation. The prepreg is consolidated and cured within an autoclave, and excess resin flows from the prepreg through the fine holes in the release film into the bleeder layer where it is absorbed.

The final stage in the fabrication of prepreg composite involves consolidation and curing inside an autoclave (Fig. 14.19). The prepreg, which is resting on the tool and enclosed within the vacuum bag, is placed inside an autoclave,

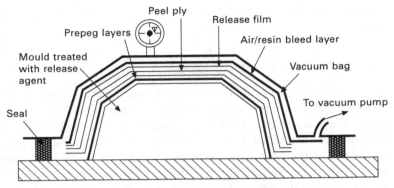

14.18 Schematic of vacuum bagging operation (from azom.com).

14.19 Autoclave used for consolidation and curing of prepreg composite.

which is a closed pressure chamber in which consolidation and cure processes occur under the simultaneous application of pressure and high temperature. An autoclave is a large pressure cooker, in which the prepreg is compacted using pressurised nitrogen and carbon dioxide to over 700 kPa. At the same time, the autoclave is heated to cure the polymer matrix, which for a carbon–epoxy prepreg requires temperatures of 120–180 °C. The combined action of pressure and heat consolidates the composite, removes trapped air, and cures the polymer matrix. The autoclave process produces high-quality composites with high fibre contents (60–65% by volume) which are suitable for primary and secondary components for aircraft and helicopters.

The production of composite components using the autoclave process is practised throughout the aircraft industry. However, the industry is also using other manufacturing processes, as described in the following sections, that avoid the need for an autoclave. Autoclaves are expensive and can only produce components of limited size (less than about 15 m long and 4 m wide).

14.5.2 Automated tape lay-up

Automated tape lay-up (ATL) is an automated process used to lay-up prepreg tape in the fabrication of composite aircraft structures. The process is used in the manufacture of carbon–epoxy prepreg components for both military and commercial aircraft. Examples of military parts made using ATL are the wing skins of the F-22 *Raptor*, tilt rotor wing skins of the V-22 *Osprey*, and skins of the wing and stabiliser of the B-1 *Lancer* and B-2 *Spirit* bombers. ATL is used in the production of empennage parts (e.g. spars, ribs, I-beam stiffeners) for the B777, A340-500/600 and A380 airliners.

ATL is used instead of manual lay-up to reduce the time (by more than 70–85%) and cost spent in the lay-up of prepreg tape on the tool. For instance, the lay-up rate in the production of the primary wing spar to the Airbus A400M is increased from 1 to 1.5 kg h^{-1} of prepreg with manual lay-up to about 18 kg h^{-1} with ATL. The first wing spars to be produced using manual lay-up took about 180 h whereas the lay-up time using the ATL process is only 1.5 h. Other benefits of the ATL process include less material scrap and better repeatability and consistency of the manufactured parts.

The key component of the ATL process is a computer numerically controlled tape-laying head that deposits prepreg onto the tool at a fast rate with great accuracy (Fig. 14.20). The head device is suspended via a multi-axis gantry above the tool surface. A roll of prepreg (75–300 mm wide) in the tape-laying head is deposited on the tool in the desired orientations

14.20 ATL process in the manufacture of aircraft panels.

according to a programmed routine. As the tape is laid down, the head removes the backing paper and applies a compaction force onto the prepreg. The head can also apply moderate heat to the prepreg to improve tackiness and formability. The head is programmed to follow the exact contour of the tool at a speed of about 50 m min^{-1} for the rapid deposition of prepreg (4–45 kg h^{-1}). When the tape-laying head reaches the location where the ply terminates then the tape is automatically cut by blades within the head, the head changes direction, and the tape laying process continues. The head can lay any number of prepreg plies on top of each other in a series of numerically controlled steps with a high degree of accuracy, thereby assuring consistent shape, thickness and quality for the part. After the automated tape lay-up is complete, the prepreg is consolidated and cured. The ATL process is capable of producing composite parts with a flat or slightly curved profile; it is not suited for highly contoured parts which are better made using the automated fibre placement process.

14.5.3 Automated fibre placement

Automated fibre placement (AFP) is used in the automated production of large aircraft structures from prepreg (Fig. 14.21). The process involves the lay-up of individual prepreg tows onto a mandrel using a numerically controlled fibre-placement machine. In the AFP process, prepreg tows or narrow strips of prepreg tape are pulled off holding spools and fed into the fibre placement head. Within the head, the tows, which contain about 12 000 fibres and are approximately 3 mm wide, are collimated into bundles of 12, 24 or 32 tows to produce a narrow band of prepreg material that is deposited onto the mandrel which has the shape of the final component. The placement head is computer controlled via a gantry system suspended above the mandrel. During fibre placement, the mandrel is rotated so the prepreg is wound into the shape of the component. The head moves along the rotating mandrel to steer the fibres so they follow the applied stresses acting on the finished component in service. The head is able to stop, cut the prepreg, change direction and then re-start the lay-down of prepreg tows until the material is built up to the required thickness. Each tow is dispensed from the head at a controlled speed to allow it conform to the mandrel surface, thereby allowing highly curved components to be produced, not possible with ATL. After the lay-up process is complete, the prepreg is cured in the same way as composites manufactured by manual lay-up or ATL.

The AFP process is used by the aerospace industry to fabricate large-circumference and highly contoured structures such as fuselage barrels, ducts, cowls, nozzle cones, spars and pressure tanks. The process is used in the construction of carbon–epoxy fuselage sections to the B787 *Dreamliner*, V-22 *Osprey*, and the *Premier* 1 and Hawker *Horizon* business jets. For the

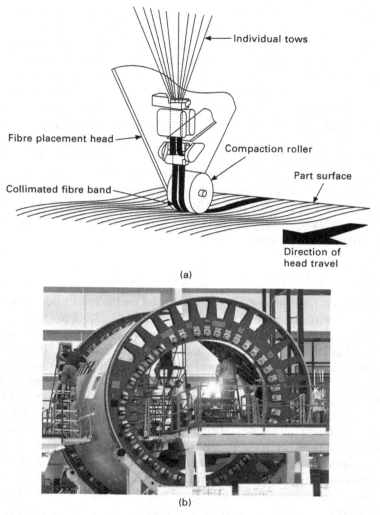

Individual tows

Fibre placement head

Compaction roller

Part surface

Collimated fibre band

Direction of
head travel

(a)

(b)

14.21 (a) Schematic of the automated fibre placement process. (b)
ATL process in the construction of fuselage barrels for the B787
aircraft. Photograph by B. Nettles, Charleston Post and Courier.

Premier 1, the AFP process is used to construct one-piece carbon–epoxy
fuselage barrels measuring 4.5 m long and 2 m at the widest point. AFP is used
to fabricate inlet ducts, side skins and covers to the F-18 E/F *Superhornet*.

14.6   Composites manufacturing by resin infusion

The aerospace industry is increasingly using manufacturing processes that
do not rely on prepreg to produce composite components. Composites

made using prepreg are high-quality materials with excellent mechanical properties owing to their high fibre content. However, prepreg is expensive, must be stored in a freezer, has limited shelf-life (usually 1–2 years), and must be cured in an autoclave which is slow and expensive. The aerospace industry is also keen to use 'out of autoclave' processes because of the faster manufacturing times and lower cost.

The industry is producing composite components using dry fabric which is infused with liquid resin and then immediately consolidated and cured. The types of fabrics used include the plain, twill, satin, knitted and non-crimped materials. These fabrics are infused with resin using one of several manufacturing processes, including resin transfer moulding, vacuum bag resin infusion, resin film infusion, filament winding and pultrusion. Many aerostructure companies also use their own proprietary manufacturing process that is developed in-house. Not all of the many processes are described here; instead just a few are outlined to illustrate the diversity of the processes.

14.6.1 Resin transfer moulding (RTM)

Resin transfer moulding (RTM) is used to fabricate composite components of moderate size (typically under 3 m), such as the fan blades of F135 engines and the spars and ribs for the mid-section fuselage and empennage of the F-35 *Lightning II* fighter. RTM is a closed-mould process that is illustrated in Fig. 14.22. Fabric is placed inside the cavity between two matched moulds with their inner surfaces having the shape of the final component. Fabric plies are stacked to the required orientation and thickness inside the mould, which is then sealed and clamped. Liquid resin is injected into the mould by means of a pump. The resin flows through the open spaces of the fabric until the mould is completely filled. The resin viscosity must be low enough for easy flow through the tiny gaps between the fibres and tows of the fabric. Aerospace-grade resins with low viscosity have been specifically developed for the RTM process. After injection, the mould is heated in order to gel and cure the polymer matrix to form a solid composite part. After curing, the mould is opened and the part removed for edge trimming and final finishing.

The RTM process can produce composites with high fibre volume content (up to 65%), making them suitable for primary aircraft structures that require high stiffness, strength and fatigue performance. However, it can be difficult to completely infuse some types of fabric with resin which leaves dry spots or voids in the cured composite. Furthermore, the fibre architecture can be disturbed by the high flow pressures needed to force the resin through some fabrics. To minimise these problems, a variant of the RTM process called vacuum-assisted resin transfer moulding (VARTM) is used.

The VARTM process is shown in Fig. 14.22b, and is different to conventional

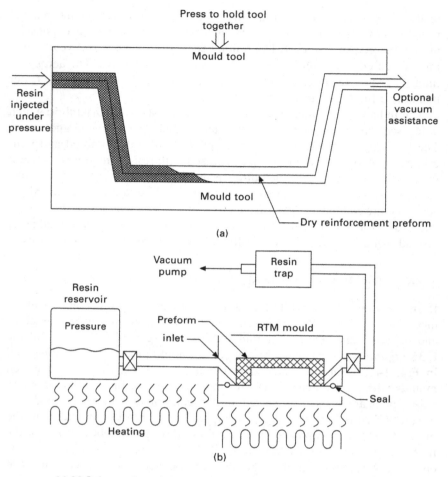

14.22 Schematics of the (a) resin transfer moulding and (b) vacuum
assisted resin transfer moulding processes.

RTM in that a vacuum-pump system is used to evacuate air from the mould
and draw resin through the fabric. After the mould containing the fabric is
closed and sealed, a vacuum pump is used to extract air from the cavity
placing it in a state of low pressure. Rather than resin being injected into
the mould cavity under pressure as in conventional RTM, with VARTM the
resin is drawn into the mould under the pressure differential created by the
vacuum. Resin percolates between the fibres and tows of the fabric until
the mould is filled, at which stage the infusion process stops and the part is
cured at elevated temperature.

14.6.2 Vacuum bag resin infusion (VBRI)

The vacuum bag resin infusion (VBRI) process and similar processes are used to fabricate various types of carbon–epoxy structural components. Figure 14.23 illustrates the basic configuration of the VBRI process, which uses an open mould rather than the two-piece closed mould of the RTM and VARTM processes. This results in VBRI having lower tooling costs, which can be a large capital cost. Fabric plies are stacked on the tool with the top layer being a resin distribution fabric. The fabric stack is enclosed and sealed within a flexible plastic bag that is connected at one point to a liquid resin source and at another point to a vacuum-pump system. Air is removed by the vacuum pump which causes the bag to squeeze the fabric layers to the shape of the mould surface. This part of the process is similar to the vacuum bagging of prepreg before autoclave curing. In the VBRI process, liquid resin flows into the bag under the pressure differential created by the vacuum pump. Resin is drawn through the tightly consolidated fabric as well as along the top resin distribution fabric, from where it seeps downwards into the fabric. After the fabric is completely infused with resin, the material is cured at high temperature within an oven.

14.6.3 Resin film infusion (RFI)

The resin film infusion (RFI) process is suited for making relatively large structures such as stiffened skins and rib-type structures (Fig. 14.24). The process uses an open mould upon which layers of dry fabric and solid resin film are stacked. The film is a B-stage cured resin similar to the cure condition of the resin matrix in prepreg. Film is placed at the bottom, top or between the layers of fabric. The materials are sealed within a vacuum bag and then air is removed using a vacuum pump. The entire assembly is placed into an autoclave and subjected to pressure and heat. The temperature

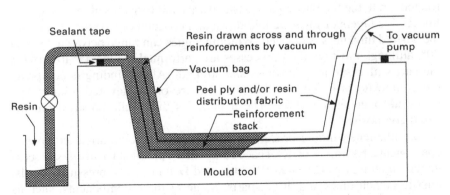

14.23 Schematic of the vacuum bag resin infusion process.

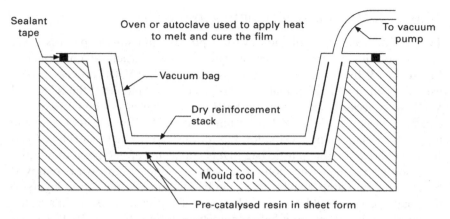

14.24 Schematic of the resin film infusion process.

is increased to reduce the resin viscosity to a level when it is fluid enough to flow into the fabric layers under the applied pressure. Once the infusion is complete the pressure and temperature are raised to consolidate and fully cure the component.

14.6.4 Filament winding

Filament winding is a manufacturing process where cylindrical components are made by winding continuous fibre tows over a rotating or stationary mandrel, as shown in Fig. 14.25. There are two types of filament winding process: wet winding and prepreg winding. Wet winding involves passing continuous tows through a resin bath before reaching a feed head which deposits them onto a cylindrical mandrel. Prepreg winding involves depositing thin strips of prepreg onto the mandrel in a similar manner to the ATP process. The feed head is rotated around the stationary mandrel or, more often, the mandrel is rotated while the feed head passes backwards and forwards along its length. Successive layers of tows are laid down at a constant or varying angle until the desired thickness is reached. Winding angles can range between 25° and 80°, although most winding processes are performed around 45° to provide the part with high hoop stiffness and strength. After winding is complete, the composite is cured at room temperature or at elevated temperature inside an oven or autoclave. The mandrel is then pulled away from the cured composite.

The filament-winding process is used to produce cylindrical composite components. Examples of aerospace components produced using this process include motor cases for Titan IV, Atlas and Delta rockets, pressure vessels, missile launch tubes and drive shafts. Some of these components, such as rocket motor cases, are very large (exceeding 4 m in diameter).

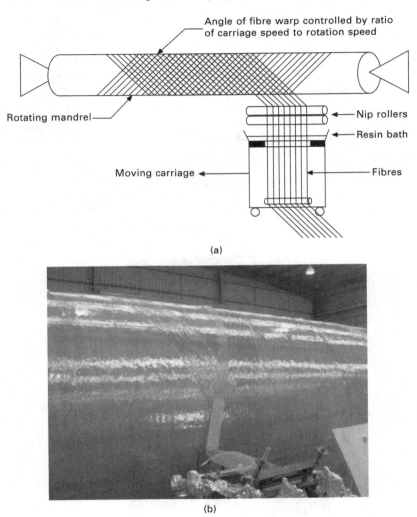

Angle of fibre warp controlled by ratio
of carriage speed to rotation speed

Rotating mandrel

Nip rollers

Resin bath

Moving carriage

Fibres

(a)

(b)

14.25 (a) Schematic and (b) photograph of the filament winding
process.

14.6.5 Pultrusion

Pultrusion is an automated, continuous process used to manufacture composite
components with constant cross-section profiles. Figure 14.26 illustrates the
pultrusion process which is a linear operation starting from the right-hand
side of the diagram. Continuous fibres (tows) are pulled off storage spools
and drawn through a liquid resin bath. The resin-impregnated fibres exit the
bath and are pulled through a series of wipers that remove excess polymer.
After this, the fibre–resin bundles pass through a collimator before entering

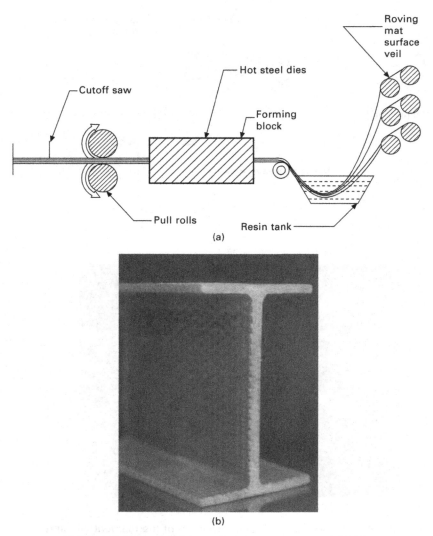

14.26 (a) Schematic of the pultrusion process and (b) I-beam demonstrator for aircraft produced by pultrusion (photograph courtesy of the Cooperative Research Centre for Advanced Composite Structures).

a heated die which has the shape of the final component. As the material passes through the die it is formed to shape while the resin is cured. Heated dies are often about 1 m long, and it is essential that, as the material travels through the die, there is sufficient time to fully cure the resin. When the pull-through speed is too fast the composite exits the die in a partially cured condition and when the speed is too slow then the production rate of the process is too slow. The pulling speed for epoxy-based composites is in the range of 10 to 200 cm min^{-1}. The cured composite leaves the die and is cut by a flying saw to a fixed length. Unlike most other manufacturing processes, pultrusion is a continuous process with the material being pulled through the die by a set of mechanically or hydraulically driven grippers.

There are few examples of pultruded composite components used in aircraft or helicopters. In part, this is because the process produces components having a constant cross-sectional shape with no bends or tapers. Few aerospace components are flat with a constant cross-section. Another problem is that the process is designed for large production runs whereas the production runs for aircraft are usually measured in the hundreds spread over several years. Nevertheless, the pultrusion process has potential for the production of high-strength floor beams and other selected parts for aircraft.

14.7   Machining of composites

The majority of processes used to manufacture composites produce components to the near-net shape. This is one of the advantages of manufacturing with composites rather than metals, which often require extensive milling and machining to remove large amounts of material to produce the final component. Most machining operations for composites simply involve trimming to remove excess material from the edges and hole drilling for fasteners. Trimming can be performed using high-speed saws and routers, although care is required to avoid edge splitting (delamination damage). The preferred method of trimming carbon–epoxy composites is water jet cutting because of high cut accuracy with little edge damage. Water jet cutting is a process involving the use of a high-pressure stream of water containing hard, tiny particles that cut through the material by erosion. Most water jet systems used by the aerospace industry are high-pressure units that use garnet or aluminium oxide particles as the abrasive.

Hole drilling of composites requires the use of specialist drill bits, several of which are shown in Figure 14.27. The aerospace industry often uses flat two-flute and four-flute dagger drills for carbon–epoxy. Drilling must be performed using a sharp bit at the correct force and feed-rate otherwise the material surrounding the hole is damaged. The application of excessive force causes push-down damage involving delamination cracking ahead of the drill bit (Fig. 14.28). Drilling at a high feed rate can also generate high friction temperatures at the hole, thereby overheating the polymer matrix. Aramid fibre composites are particularly difficult to drill without the correct bit because the fibres have a tendency of fuzz and fray. The aramid drill contains a 'C' type cutting edge that grips the fibres on the outside of the hole and keeps them in tension during the cutting process, thereby avoiding fraying.

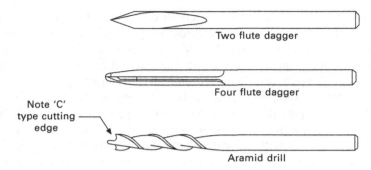

14.27 Drill bits used for carbon–epoxy (two and four flutes daggers) and aramid fibre composites (reproduced from F. C. Campbell, *Manufacturing Technology for Aerospace Structural Materials,* Elsevier Science & Technology, 2006).

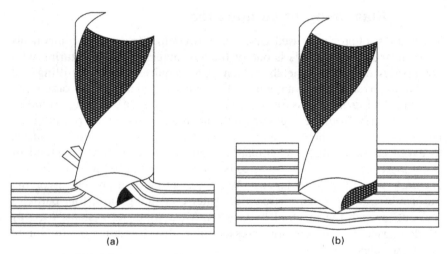

14.28 (a) Peel-up and (b) push down damage to composite materials caused by incorrect drilling.

14.8   Summary

The three types of fibre reinforcement most often used in aircraft composite materials are carbon, glass and aramid. Carbon fibre is used in primary and secondary structures for aircraft, helicopters and spacecraft. Glass fibre is used in the radomes and components when stiffness is not a critical design property, and are used extensively inside cabins for fittings and furnishing. Aramid fibre is often used in structures requiring high vibration damping or impact resistance.

The production of aircraft components using composites instead of metals often results in significantly fewer parts and fasteners because the manufacturing processes for composites have better capability for making integrated parts.

The mechanical properties of carbon fibres are determined by the processing temperature. High stiffness fibres are produced at the highest treatment temperature (2500–3000 °C) whereas high-strength fibres are treated at about 1500 °C. The properties of glass fibres are determined mainly by the addition of metal oxides to the base silica material.

Carbon–epoxy prepreg is commonly used in primary and secondary aircraft structures. Various processing methods are used to produce composite components using prepreg, including manual hand lay-up, automated tape lay-up (for flat and slightly curved sections) and automated tow placement (for highly curved sections). The prepreg is usually consolidated and cured inside an autoclave.

Composites fabricated using prepreg have high stiffness, strength and other structural properties, although they are also expensive owing to the high cost of prepreg material, the need to store prepreg in a freezer, and the need to cure using an autoclave. Aerospace composites are increasingly being fabricated from dry fabric using out-of-autoclave processes. Various types of carbon fabrics are used, including twill and satin weaves as well as non-crimp fabrics. Fabrics are infused with resin using using several processes, including vacuum-assisted resin transfer moulding, vacuum-based resin infusion, and resin film infusion.

Composites are susceptible to delamination cracking and other damage during trimming, drilling and other material removal processes. Specialist trimming methods, such as water jet cutting, and drilling bits are required for composites.

14.9   Terminology

Carbonisation: Process used for the conversion of an organic substance into carbon or a carbon-containing residue through pyrolysis or other method.

Fibril: Very fine filament which forms part of a larger fibre.

Liquid crystal domains: Randomly oriented domains of highly crystalline polymer chains within a liquid polymer melt.

Polyacrylonitrile (PAN): Organic resin used as the chemical precursor for high-quality carbon fibre.

Pitch: Viscous carbon-rich resin usually formed from petroleum that is used in the production of non-aerospace grade carbon fibre.

Rayon: Organic substance manufactured from regenerated cellulose fibre (extracted from plants or produced artifically) that is used in the production of non-aerospace grade carbon fibre.

Size: Thin coating on fibres containing chemical agents which are used for various functions, including increased adhesion with polymer matrix, antistatic properties, binding and lubrication.

Tow: Untwisted bundle of continuous filaments. Tows are designated by the number of fibres they contain, e.g., 12K tow contains about 12 000 fibres.

Turbostratic graphite: Basal sheets of graphite that are haphazardly folded or crumpled together.

Yarn: A continuous strand of twisted fibres used in weaving or knitting.

14.10   Further reading and research

Baker, A., Dutton, S. and Kelly, D. (eds.), *Composite materials for aircraft structures*, American Institute of Aeronautics and Astronautics, Inc., Reston, VA, 2004.

Campbell, F. C., *Manufacturing processes for advanced composites*, Elsevier, Amsterdam, 2004.

Harris, C. E., Starnes, J. H. and Shuart, M. J., 'Design and manufacturing of aerospace composite structures, state-of-the-art assessment', *Journal of Aircraft*, **39** (2002), 545–560.

Sheikh-Admad, J. Y., *Machining of polymer composites*, Springer, 2009

Strong, A. B., *Fundamentals of composite manufacturing*, Society of Manufacturing Engineers, 2008.

14.11   Case study: carbon nanotubes in composites

Carbon nanotubes were discovered in the early 1990s and, since then, have caused great excitement in the scientific community and attracted the interest of the aerospace industry because of their extraordinary properties. Strictly speaking, any tubes with nanometer dimensions are called 'nanotubes', but the term is generally used to refer to carbon nanotubes. Graphene is a tessellation of hexagonal rings of carbon that look like a sheet of chicken wire and a carbon nanotube is basically a graphene sheet rolled up to make a seamless hollow cylinder. Nanotubes can have a hemispherical cap of graphene at each end of the cylinder. Tubes typically have an internal diameter of 5 nm and external diameter of 10 nm. The tubes occur as a single cylinder, known as a single-wall nanotube, or two or more overlaid cylinders, called a multiwall nanotube, as shown in Fig. 14.29.

The mechanical properties of carbon nanotubes are extraordinary compared with conventional engineering materials. The Young's modulus of single-walled nanotubes is 1000 to 1300 GPa, which is much higher than conventional carbon fibres (200–400 GPa) and many times greater than metallic alloys used in aircraft structures. The tensile strength of single-walled nanotubes is close to 200 GPa, which again is much higher than carbon fibres. Carbon nanotubes can be mixed with polymers to create nanocomposites which have good mechanical properties and improved flammability resistance. There is

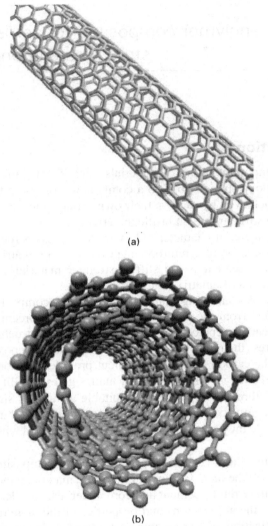

(a)

(b)

14.29 Images of (a) single-wall and (b) multiwall carbon nanotubes.

strong interest in the possible use of carbon nanotubes in nano-mechanical and nano-electronic devices, which may be used in future aircraft. It is also possible that carbon nanotubes may be used to strengthen and toughen polymer composites used in aircraft structures and gas turbine engines. However, the future of polymer nanocomposites is uncertain because several technical challenges exist in making high-performance composites containing carbon nanotubes, and currently their future in aircraft structures is uncertain.

15

Fibre–polymer composites for aerospace structures and engines

15.1   Introduction

Composites are an important group of materials made from a mixture of metals, ceramics and/or polymers to give a combination of properties that cannot otherwise be achieved by each on their own. Composites are usually made by combining a stiff, strong but brittle material with a ductile material to create a two-phase material characterised by high stiffness, strength and ductility. Composites consist of a reinforcement phase and a matrix phase and, usually, the reinforcement is the stiffer, stronger material which is embedded in the more ductile matrix material.

The main purpose of combining materials to create a composite is to gain a synergistic effect from the properties of both the reinforcement and matrix. For example, carbon–epoxy composite used in aircraft structures is a mixture of carbon fibres (the reinforcement phase) embedded in epoxy resin (the matrix phase). The carbon fibre reinforcement provides the composite with high stiffness and strength while the epoxy matrix gives ductility. Used on their own, carbon fibres and epoxy are unsuitable as aircraft structural materials, the fibres being too brittle and the epoxy too weak, but when combined as a composite they create a high-performance material with many excellent properties.

This chapter examines fibre-reinforced polymer matrix composites used in aircraft. In chapter 14, the manufacture of composite materials, including the production of the fibre reinforcement and core materials, was described. This chapter examines the applications and properties of the manufactured composites. The application of fibre–polymer composites in aircraft and jet engines is described, including the benefits and problems with using these materials. The mechanical properties of composites are discussed, including the mechanics-based theories used to calculate the elastic and strength properties. Control of the mechanical properties of composites by the type, volume fraction and orientation of the fibre reinforcement as well as by the selection of the polymer matrix is explained. Other properties of polymer matrix composites are dealt with in following chapters, including their fracture properties (chapters 18 and 19), fatigue properties (chapter 20), creep properties (chapter 22) and methods for their recycling and disposal (chapter 24).

338

15.2   Types of composite materials

Composites are usually classified according to the material used for the matrix: metal matrix, ceramic matrix or polymer matrix. There is also a special type of composite called fibre–metal laminate that does not really fit into any of these three classes. The composites used in aircraft are almost exclusively polymer matrix materials, with the polymer often being a thermoset resin (e.g. epoxy, bismaleimide) and occasionally a high-performance thermoplastic (e.g. PEEK). Metal matrix composites, ceramic matrix composites and fibre–metal laminates are used in much smaller amounts, as described in chapter 16.

The matrix phase of composite material has several important functions, including binding the discrete reinforcement particles into a solid material; transmitting force applied to the composite to the stiff, strong fibres, which carry most of the stress; and protecting the fibres from environmental effects such as moisture and abrasion.

Composites can also be classified based on the shape and length of the reinforcing phase. The three main categories are particle-reinforced, whisker-reinforced and fibre-reinforced composites (Fig. 15.1). The reinforcement, regardless of its shape and size, is always dispersed within the continuous-matrix phase. The main functions of the reinforcement are to provide high stiffness, strength, fatigue resistance, creep performance and other mechanical properties. In some instances, the reinforcement may be used to alter the electrical conductivity, thermal conductivity or other nonmechanical properties of the composite. The stiffening and strengthening attained from the reinforcement is dependent on its shape, and increases in the order: particle, whisker and fibre. For this reason, the composite materials used in aircraft structures contain long, continuous fibres. Composites containing particles or whiskers are rarely used in structural components.

The size of the reinforcement used in composites can range from ultrafine

Particle-reinforced   Whisker-reinforced   Fire-reinforced
composite   composite   composite

15.1 Different types of composites based on the shape of the reinforcement.

particles in the nanometre size range to large particles up to 1 mm or more. The reinforcement used in aircraft composite materials is mostly in the micrometre size range, typically 5–15 μm, in diameter or slightly larger, as shown in Fig. 15.2. The composites used in aircraft structures are reinforced with continuous fibres rather than whiskers or particles. The mechanical properties in the fibre direction increase with the fibre length, as shown in Fig. 15.3. Therefore using continuous fibres that can extend the entire length of the structure maximises the structural efficiency.

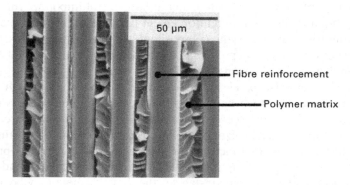

15.2 Microstructure of an aircraft-grade composite material containing continuous-fibre reinforcement having a diameter in the micrometre size range.

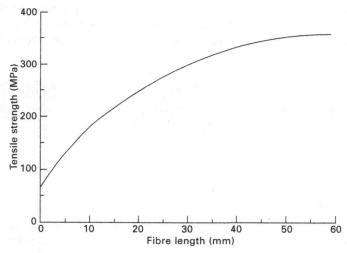

15.3 Effect of fibre length on the tensile strength of the carbon-reinforced polymer composite.

There is growing interest in nanoparticle composites, such as polymers reinforced with carbon nanotubes, for use in aircraft structures due to their exceptional mechanical properties (see chapter 14). In some cases, hybrid composites consisting of two or more types of reinforcing fibres (e.g. carbon and glass or carbon and aramid) are used together to optimise the material properties. For example, aramid fibres used in combination with carbon fibres increase the ballistic resistance of carbon-fibre composite materials. Hybrid composites containing both glass and carbon fibres are used in helicopter rotor blades.

It is worthwhile at this point to distinguish between composites and metal alloys, which are also created by combining two or more materials. The constituent materials in a composite, which are the reinforcement and matrix, are combined but there is no dissolution of one phase into the other phase. In other words, the reinforcement and matrix remain physically discrete phases when combined into the composite. In contrast, in metal alloys the solute material (i.e. alloying elements) dissolves into the solvent (i.e. base metal) and, therefore, the solute does not retain its original physical condition. Only materials which retain the physical properties of the constituents are considered composites.

Fibre–polymer composites alongside aluminium alloys are the most used materials in aircraft structures. The use of composites in civil aircraft, military fighters and helicopters has increased rapidly since the 1990s, and it is now competing head-to-head with aluminium as the material of choice in many airframe structures (as described in chapter 2). The use of composites in gas turbine engines for both civil and military aircraft is also growing. The main reasons for using composites are to reduce weight, increase specific stiffness and strength, extend fatigue life, and minimise problems with corrosion.

Composites used in aircraft structures are in the form of laminates or sandwich materials. Laminates consist of continuous fibres in a polymer matrix, and the most common type used in aircraft is carbon fibre–epoxy. The fibres are stacked in layers, called plies or laminae, often in different orientations to support multidirectional loads. Laminates made of glass fibre–epoxy, carbon fibre–bismaleimide and carbon fibre–thermoplastic may also be used in aircraft, although in smaller amounts than carbon–epoxy. Laminates are used in the most heavily-loaded composite structures and are often supported by ribs, spars and frames. The thickness of the laminate can range from a few millimetres to around 25 mm, depending on the design load.

Sandwich composites are used in lightweight secondary structures requiring high buckling resistance and flexural rigidity, and are often constructed with thin carbon–epoxy face skins covering a lightweight core of polymer foam, Nomex or (in older aircraft types) aluminium honeycomb.

15.3   Aerospace applications of fibre–polymer composites

15.3.1 Composites in military aircraft

The use of composites in aircraft has been led by the military, particularly with fighter aircraft and helicopters. Carbon fibre–epoxy composites have been used in the primary structures of fighter aircraft for many years, including the wings and fuselage, to minimise weight and maximise structural efficiency. The amount of composite varies between different types of fighter aircraft, and the usage has increased greatly since the 1970s (see Fig. 2.12). Composite material used in fighter aircraft ranges from a small amount, as in the F-14 *Tomcat* (4% of the structural mass) and F-16 *Fighting Falcon* (5%), to greater amounts as in the F/A-18 E/F *SuperHornet* (20%), AV8B *Harrier II* (22%), *Dassualt Rafale B/C* (25%) and F-22 *Raptor* (25%). Composites are the most used structural material in late-generation fighters such as *Eurofighter* (40%) and F-35 *Lightning II* (35%). The structural applications of composites, mostly carbon fibre–epoxy and carbon fibre–bismaleimide, in the F-35 are shown in Fig. 15.4.

15.3.2 Composites in passenger aircraft

Composites have been used in single- and twin-aisle passenger aircraft since the late-1960s, with the first applications being in non-safety critical components such as fairings and undercarriage doors. The use of composites in the primary structures of passenger airliners has lagged behind fighter aircraft. The first primary composite structure was the carbon fibre–epoxy horizontal stabilizer for the Boeing 737, which was certified in 1982. The uses of composites in airframe structures developed gradually during the 1990s, and typical applications are shown in Fig. 15.5.

A major expansion in the use of composites occurred with the Airbus 380, which is a 280 tonne airliner (typical operating empty weight) built from 25% composite materials. Carbon-fibre composites are used in many major structural components including the vertical and horizontal tail planes, tail cone, centre wing box, and wing ribs (see Fig. 15.6). The aircraft is also constructed of fibre–metal laminate, which is another type of composite material and is discussed in the next chapter. The Boeing 787 represents the latest in composite applications to large airliners, with 50% of the structural mass consisting of carbon fibre–epoxy composite material. The Airbus A350 is constructed using a similar amount of composite material (about 50% of the structural mass). Table 15.1 gives a summary of the structural components in Airbus and Boeing aircraft made using composite material. Although the use of composites is now widespread in commercial aircraft, it is unusual for the entire airframe to be built with these materials. Apart from a limited

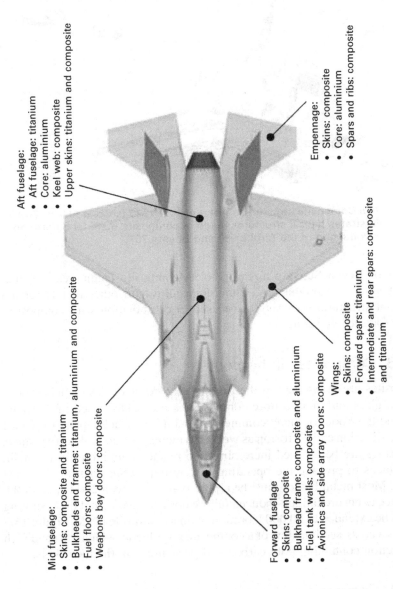

Aft fuselage:
- Aft fuselage: titanium
- Core: aluminium
- Keel web: composite
- Upper skins: titanium and composite

Empennage:
- Skins: composite
- Core: aluminium
- Spars and ribs: composite

Mid fuselage:
- Skins: composite and titanium
- Bulkheads and frames: titanium, aluminium and composite
- Fuel floors: composite
- Weapons bay doors: composite

Forward fuselage
- Skins: composite
- Bulkhead frame: composite and aluminium
- Fuel tank walls: composite
- Avionics and side array doors: composite

Wings:
- Skins: composite
- Forward spars: titanium
- Intermediate and rear spars: composite and titanium

15.4 Application of composites in the F-35 *Lightning II.*

15.5 Composite structures on the Airbus A320 (shaded black). This illustrates typical structures made of composite material before the introduction of the Airbus 380 and Boeing 787.

number of Beech Starship aircraft, no all-composite commercial airliner has yet gone into production. Aluminium alloy and other metals remain important structural materials to be used in combination with composites for the foreseeable future.

15.3.3 Composites in helicopters

Composites are often used in the fuselage and rotor blades of helicopters. Carbon, glass and aramid fibre composites are regularly used in the main body and tail boom of many commercial and military helicopters to reduce weight, vibration and corrosion as well as to increase structural performance. Composites are being used increasingly to replace aluminium in the main rotor blades to prolong the operating life by improving resistance against fatigue. Most metal blades must be replaced after between 2000 and 5000 h of service to ensure fatigue-induced failure does not occur, and the operating life can be extended to 20 000 h or more with a composite blade. Figure 15.7 presents a cross-section view of a composite rotor blade, which is a sandwich construction containing both carbon and glass fibres in the face skins.

15.3.4 Composites in gas turbine engines

Composites are used in gas turbine engine components including the fan blades, front fan case, nacelle, outlet guide vanes, bypass ducts, nose cone

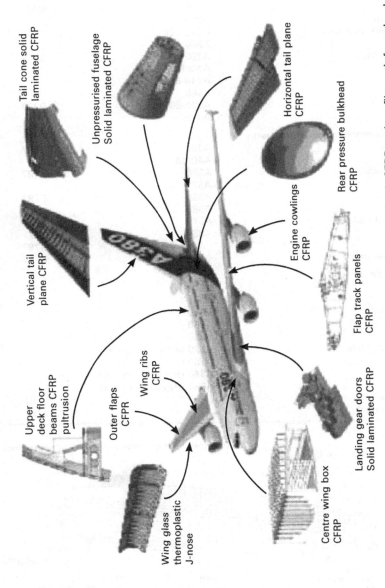

Tail cone solid laminated CFRP

Unpressurised fuselage Solid laminated CFRP

Horizontal tail plane CFRP

Rear pressure bulkhead CFRP

Engine cowlings CFRP

Vertical tail plane CFRP

Flap track panels CFRP

Upper deck floor beams CFRP pultrusion

Wing ribs CFRP

Outer flaps CFPR

Landing gear doors Solid laminated CFRP

Wing glass thermoplastic J-nose

Centre wing box CFRP

15.6 Application of composites in the Airbus 380 (image courtesy of Airbus). CFRP, carbon-fibre reinforced polymer.

Table 15.1 Composite components on Airbus and Boeing airliners

Composite component	Airbus	Boeing
Fuselage and nose components		
Fuselage	A380	B787
Belly fairing	A380	
Rear pressure bulkhead	A340-600	
	A380	
Floor beams	A350	B787
	A400M	
Keel beam	A340-600	
	A380	
Nose cone	A340-600	B787
	A380	
Wing and empennage components		
Wingbox	A380	B787
	A400M	
Wing beams	A350	B787
	A380	
Wing skins	A350	B787
	A380	
Horizontal stabiliser	A340	B737
	A350	B777
		B787
Vertical tailplane	A380	B777
Engine components		
Nacelles	A340	B787
	A380	
Reversers	A340	B787
	A380	
	A350	
Reverser details	A320 (Reverser doors)	
	A380 (Gutter fairing	
	A340 (Reverser doors)	
	A380 (Reverser doors)	
Fan blades		B787
Cone spinners		B787

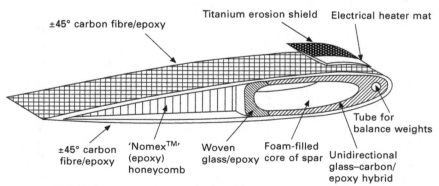

±45° carbon fibre/epoxy · Titanium erosion shield · Electrical heater mat · ±45° carbon fibre/epoxy · 'Nomex™' (epoxy) honeycomb · Woven glass/epoxy · Foam-filled core of spar · Unidirectional glass–carbon/epoxy hybrid · Tube for balance weights

15.7 Helicopter rotor blade constructed of sandwich composite material (image courtesy of Westland Helicopters).

spinner and cowling (Fig. 15.8). The replacement of incumbent metal components with composites provides a saving in the vicinity of 20–30% of the total engine lifetime operation cost. The use of composites is restricted to engine parts required to operate at temperatures below about 150 °C to avoid softening and heat distortion. Carbon-fibre laminates with a high-

(a)   (b)

(c)

15.8 (a) Fan blade and (b) casing made of composite material for the GEnx gas turbine engine. (c) Composite nacelle and cowling of a gas turbine engine.

temperature polymer matrix (such as bismaleimide) is often the composite material of choice for jet engines.

The three major reasons for the use of composites in engines are lower weight, improved structural performance and reduced operating/maintenance cost. The trend in gas turbine engines is towards larger (and consequently heavier) fan sections. The addition of 1 kg to the fan blade assembly needs a compensatory increase in the weight of the fan blade containment case of 1 kg. This 2 kg increase requires a compensatory weight increase of about 0.5 kg to the rotor as well as incremental increases to the weights of the wings and fuselage. Any weight savings gained by using light materials in the fan blade translates to significant savings in the weight of other components. For example, the use of carbon-fibre composite in the two GEnx engines on the Boeing 787 provides an overall weight saving of about 350 kg, which translates directly into fuel burn saving and greater aircraft range. The other benefit of using lightweight composite fan blades is reduced centrifugal force (owing to their light mass) thus increasing the fatigue life.

15.3.5 Composites in spacecraft

Polymer matrix composites have been used for many years in space structures, including truss elements, antennas, and parabolic reflectors. The main truss of the Hubble space telescope, for example, is made of carbon fibre–epoxy composite for lightness, high stiffness and low coefficient of thermal expansion. As another example, the main cargo doors of the space shuttle orbiter are made of sandwich composite material and the arm of the remote manipulator system is made of carbon-fibre composite (Fig. 15.9).

15.4   Advantages and disadvantages of using fibre–polymer composites

15.4.1 Advantages of using fibre–polymer composites

There are many benefits to be gained from using carbon–epoxy and other types of fibre–polymer composite materials instead of aluminium alloy in aircraft. The main advantages are summarised here and a comparison between carbon–epoxy composite and an aircraft-grade aluminium alloy is given in Table 15.2.

Weight

The lower density and superior mechanical properties of carbon-fibre composite compared with aluminium results in significant weight saving. With careful design, it is possible to a achieve weight savings of 10–20% using carbon

(a)

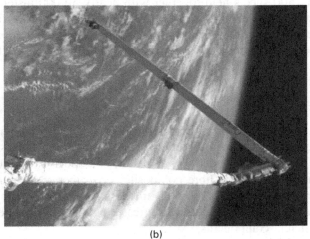

(b)

15.9 (a) Cargo bay doors and (b) remote arm of the space shuttle orbiter are made of composite materials.

Table 15.2 Comparison of properties of carbon–epoxy laminate (quasi-isotropic) and aluminium alloy (2024 Al)

Property	Carbon–epoxy	Aluminium alloy
Density (g cm^{-3})	1.6	2.7
Structural efficiency		
Specific modulus, E/ρ (MPa m^3 kg^{-1})	71	26
Specific strength, σ/ρ (MPa m^3 kg^{-1})	0.4	0.2
Fatigue resistance (endurance stress limit after one million load cycles)	70–80% of the ultimate strength	20–35% of the ultimate strength
Corrosion resistance	Excellent	Fair
Thermal conductivity (W m^{-1} K^{-1})	20	250
Coefficient of thermal expansion (K^{-1})	2×10^{-6}	22×10^{-6}

fibre–epoxy composite. For instance, the Boeing 787 is about 20% lighter than an equivalent aircraft made entirely of aluminium which, together with other factors, translates to a fuel saving of about 20% and reduced seat-mile costs by around 10%.

Integrated manufacture

Composite manufacturing processes are better suited than metal forming processes to produce one-piece integrated structures with fewer parts. Typically, assembly accounts for about 50% of the production cost of the airframe, and the possibility of producing integrated structures offers the opportunity to reduce the labour, part count, and number of fasteners. For example, one mid-sized aircraft fuselage barrel made using aluminium requires about 1500 sheets held together with nearly 50 000 fasteners. An equivalent barrel made of composite material only requires about 3000 fasteners because of the fewer number of parts. The F-35 *Lightening II* has just one-half the number of parts of the F-18 *SuperHornet*, one of the aircraft it may replace.

Structural efficiency

The mechanical properties of composites can be tailored by aligning the fibre reinforcement in the load direction, thereby providing high stiffness and strength where it is needed. As a result, the specific stiffness and strength properties of carbon-fibre composites are superior to aluminium alloy. Furthermore, the choice of high-stiffness or high-strength carbon fibres provides the aircraft engineer with greater flexibility in the design of structurally efficient components.

Fatigue resistance

Composites have high fatigue resistance under cyclic stress loading, thereby reducing the maintenance cost and extending the operating life. The superior fatigue resistance of carbon-fibre composite compared with aluminium is a major reason for its use in aircraft structures, helicopter rotor blades, and fan blades for gas turbine engines.

Corrosion resistance

Carbon-fibre composites are immune to corrosion, which is a major problem with aluminium. The high cost of inspecting and repairing metal structures damaged by corrosion is minimised with the use of composites. The corrosion resistance of composite material allows for a higher humidity level inside the cabin, which creates a more comfortable environment for crew and passengers. The humidity for an aluminium fuselage must be kept at a very low level (around 5%) to avoid corrosion, and this can be increased in a composite fuselage (15–20%) for greater comfort.

Radar absorption properties

Using specialist designs, materials and manufacturing processes, composites can be made with high radar absorption properties. When radar absorbing composite material is used in stealth military aircraft they are difficult to detect using radar.

Heat insulation

The thermal conductivity of composites is much lower than metals, which makes them a good heat insulator. This property is used in fighter aircraft to reduce heat conduction from the engines through the fuselage, thereby making the aircraft harder to detect using infrared devices.

Low coefficient of thermal expansion

Carbon-fibre composites have a very low coefficient of thermal expansion, which means they experience little or no expansion or contraction when heated or cooled. This property is utilised in structures that need high dimensional stability with changes in temperature. For example, the truss supporting the antenna and deep-space observation devices in the Hubble space telescope is made of composite material. When the temperature of the telescope changes the truss does not change shape, thereby providing a stable platform for the antenna and devices.

15.4.2 Disadvantages of using fibre–polymer composites

Composites, like any material, are not without their drawbacks and limitations. The major issues are discussed in this section.

Cost

The cost of producing aircraft components from composite material is often more expensive than using aluminium. The higher cost of composite is caused by several factors, including the high cost of carbon fibres, the labour-intensive nature of many of the manufacturing processes, and high tooling costs.

Slow manufacture

The production of aircraft components using composites can be slower than with aluminium owing to the long time needed to lay-up the fabric or prepreg ply layers and the curing time of the polymer matrix. The use of automated lay-up processes and multilayer non-crimp fabrics reduce production time, although most manufacturing processes are not well suited to the rapid production of a large number of thermoset matrix parts.

Anisotropic properties

Designing aircraft components made of composite materials is more challenging than for metal alloys owing to their anisotropic properties. Careful design and manufacture is required to ensure the fibres are aligned in the load direction, otherwise the composite may have inferior mechanical performance.

Low through-thickness mechanical properties

The stiffness, strength, damage tolerance and other mechanical properties of composites are low in the through-thickness direction owing to the absence of reinforcing fibres. Figure 15.10 shows typical values for the in-plane and through-thickness mechanical properties of carbon fibre–epoxy composite. The through-thickness property values are just a small fraction of the in-plane values, and the application of through-thickness loads on to composites must be avoided.

Impact damage resistance

Composites are susceptible to delamination cracking when impacted at low energies because of their low through-thickness strength and fracture

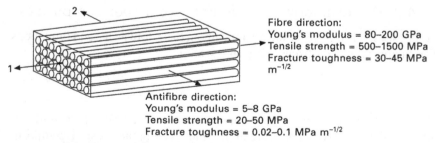

Fibre direction:
Young's modulus = 80–200 GPa
Tensile strength = 500–1500 MPa
Fracture toughness = 30–45 MPa m$^{-1/2}$

Antifibre direction:
Young's modulus = 5–8 GPa
Tensile strength = 20–50 MPa
Fracture toughness = 0.02–0.1 MPa m$^{-1/2}$

15.10 Elastic modulus, strength and fracture toughness properties of a carbon–epoxy composite in the fibre and antifibre (through-thickness) directions. The properties vary over a range of values depending on the fibre type, fibre volume content, and matrix material.

toughness. Impact events such as bird strike, large hail stone impacts, and tools which are accidentally dropped during maintenance can damage composites more severely than occurs with a metal alloy. A related problem is impact damage, which can reduce the compressive strength and other mechanical properties of the composite.

Damage tolerance

The growth of damage (e.g. delamination cracks) in composite materials is difficult to control and predict. A large amount of damage growth can occur rapidly with little or no warning. For this reason, primary composite aircraft structures must be designed according to the so-called 'no growth' damage tolerance philosophy, which means that pre-existing damage must not grow over a specified period of time of aircraft service (usually two or more inspection intervals). As a result, composite structures must be over-designed to ensure adequate damage tolerance, thus increasing their weight and cost.

Notch sensitivity

The reduction in the failure strength of composites owing to notches (e.g. bolt holes, windows) can be greater than the loss suffered by metal alloys. Composite materials with highly anisotropic properties (i.e. a large percentage of fibres aligned in one direction) experience a high concentration of stress near notches when under load, resulting in a large knock-down in strength.

Temperature operating limit

Composites soften and distort at lower temperatures than aluminium owing to the glassy-to-rubbery transformation of the polymer matrix. The maximum operating temperature of epoxy matrix composites is typically in the range 100–150 °C, which limits their use for high-temperature applications.

Flammability

Composite materials are flammable, and burn, produce smoke and release heat when exposed to a high-temperature fire, as may occur in a post-crash accident.

Low electrical conductivity

Composite materials are much poorer conductors of electricity than the metals used on aircraft. An electrically conductive material (such as copper mesh) must be incorporated into composite materials used on the external surface of aircraft to dissipate electricity in the event of lightning strike.

15.5   Mechanics of continuous-fibre composites

15.5.1 Principles of composite mechanics

Understanding the mechanical behaviour of continuous-fibre–polymer composites is different to understanding the mechanical properties of metals and their alloys, which has been a major focus of this book. The properties of metals are controlled by vacancies, dislocations, grain boundaries, precipitates, solute (alloying and impurity) elements, and other microstructural features. These features have no role to play in the mechanical properties of composite materials. Instead, understanding the mechanical properties is based on the key concept of load sharing between the fibre reinforcement (which is stiff and strong) and the polymer matrix (which is compliant and weak). The properties of composites are determined by the fibre properties, matrix properties, interfacial properties between the fibre and matrix, and how effectively the load is shared between the fibres and matrix.

When an external load is applied to a composite, a certain proportion of that load is carried by the fibre reinforcement and the remainder by the polymer matrix. At the simplest level, the applied load is shared between the fibres and matrix based on their relative volume fractions. This is expressed mathematically as:

$$P = P_f V_f + P_m V_m \qquad [15.1]$$

where P is the total applied load, $P_f V_f$ is the proportion of load on the fibres, and $P_m V_m$ is the proportion of the load on the matrix. P_f and P_m are the loads carried by the fibres and matrix, and V_f and V_m are the volume fractions of the fibres and matrix, respectively.

Provided the response of the composite to the applied load is elastic, then the distribution of load-sharing between the fibres and matrix is independent of the stress level. In other words, the relative distribution of load between the fibres and matrix remains the same for any stress up to the elastic stress limit of the composite, which quite often is the ultimate failure stress. Because the fibre reinforcement is much stiffer and stronger than the polymer matrix, it is important that as high a proportion of the applied load as possible is borne by the fibres. For this reason, the fibre content of aircraft composites is as high as practically possible. The fibre volume content is typically in the range of 55–65%, the maximum values that can be achieved using the manufacturing processes described in chapter 14. The relative proportions of the load borne by the fibres and the matrix is dependent not only on their relative volume fractions, but also on the shape and orientation of the fibres together with the elastic properties of the fibres and matrix.

The concept of load-sharing between the fibres and matrix is the basis for the theoretical understanding of the mechanical properties of composites as well as the practical aspects in the design and manufacture of composites with maximum mechanical performance. The concept of load-sharing is applied to the micromechanics of composites, which is the mathematical analysis of the mechanical properties based on the properties and interactions of the fibres and matrix. Analytical and finite element models are used to simulate the microstructure of composites, and from this the mechanical properties (such as elastic modulus or strength) are calculated in terms of the properties and volume fractions of the fibres and matrix. Figure 15.11 illustrates the basic premise of micromechanics, where the properties of the fibre reinforcement are combined with the properties of the polymer matrix to calculate the 'average' properties of the composite material based on volume-averagis of the two constituents.

The averaging approach to micromechanical analysis assumes that the fibres are arranged in a regular pattern. In actual composite materials, the fibres are randomly distributed throughout the matrix phase, with regions of higher than average volume content of fibres and other regions with lower than average fibre content. Micromechanical analysis assumes that the fibres are packed in a regular pattern. For example, Fig. 15.12 shows two regular fibre packing patterns: the square array and the hexagonal array. Either array can be viewed as a repetition of a single unit cell of the composite which contains the fibre and matrix. The unit cell represents the basic building block of the composite material, and is the simplest geometry for analysing using the volume averaging approach to micromechanics.

The simplest approach to micromechanical analysis of composite materials involves treating the fibre and resin separately within a unit cell, and assuming that there is no interaction between the two constituents. The property averaging concept is called 'rule-of-mixtures', and it is applied to derive simple-to-use equations for mechanical and other properties including:

- density,
- in-plane and transverse elastic properties,
- in-plane and transverse tensile strength,
- in-plane compressive strength, and
- in-plane and transverse thermal conductivity.

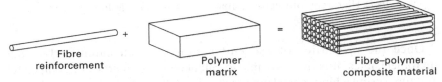

15.11 Representation of the averaging approach used in micromechanics.

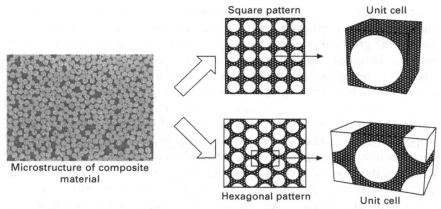

15.12 Unit cell model for the micromechanical analysis of composites.

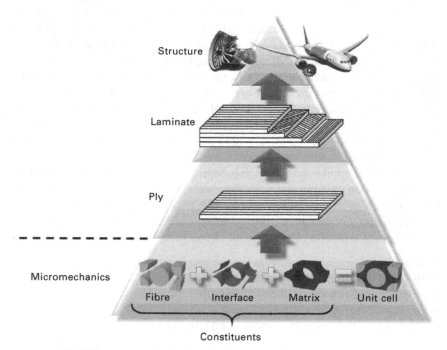

15.13 Hierarchy of micromechanics-based analysis for composite structures.

Once the properties of the unit cell have been calculated using micromechanics, it is then possible to calculate the properties of composite structures using a hierarchical approach to modelling as shown in Fig. 15.13. Micromechanical modelling of the unit cell is the foundation for higher levels of modelling which follow the sequence of single-ply modelling, laminate (or ply-by-ply) modelling, and finally structural modelling. Single-ply modelling calculates the properties of one ply layer within a composite, in which the fibres and matrix are treated separately (as depicted in Fig. 15.11). All of the fibres within the single ply are assumed to be aligned in the same direction (i.e. unidirectional fibre laminate). Taking the properties calculated for a single ply, the next level of modelling involves analysing multiple-ply layers with different fibre angles to calculate the properties of a multidirectional composite. Laminate modelling treats each ply layer separately and the fibres and matrix within each ply are treated as a continuum. Structural (also called macroscopic) modelling analyses the composite as a single orthotropic material with the geometric features of the final component. Structural modelling is used to predict the stiffness, strength and other properties of the final composite structure, and this is performed using finite element and other numerical methods.

15.5.2 Elastic properties of composite materials

Longitudinal Young's modulus of composites

One of the main reasons for using continuous-fibre composites in aircraft structures is their high specific stiffness compared with many metal alloys. The elastic modulus properties of carbon-fibre composites are superior to aluminium, which is an important reason for the increasing use of composites and the corresponding decline in the application of aluminium in aircraft structures.

The elastic modulus of a unidirectional composite reinforced with straight, continuous fibres when loaded in the fibre direction can be calculated using rule-of-mixtures modelling. The unidirectional composite shown in Fig. 15.10 is loaded in tension along the fibre direction (which is also called the 1– or longitudinal direction). In this load state, the fibre and matrix are assumed to act in parallel and both constituents experience the same elastic strain (ε_1). This iso-strain condition is true even though the elastic moduli of the fibres and matrix are different. The iso-strain condition is expressed as:

$$\varepsilon_1 = \varepsilon_f = \frac{\sigma_f}{E_f} = \varepsilon_m = \frac{\sigma_m}{E_m} \qquad [15.2]$$

where ε_f and ε_m are the strain values of the fibre and matrix; σ_f and σ_m are the stresses carried by the fibre and matrix, and E_f and E_m are the Young's modulus of the fibres and matrix, respectively.

Based on the iso-strain condition, the longitudinal Young's modulus can be calculated using:

$$E_1 = E_f V_f + E_m V_m \qquad [15.3]$$

where $V_f + V_m = 1$ when no voids are present.

Because the elastic modulus of the matrix is much lower than the fibre reinforcement, the fibres contribute between 95 and 99% of the in-plane stiffness of a unidirectional composite containing carbon at the typical volume contents used in aerospace materials ($V_f \sim 0.55–0.65$). The type of polymer matrix (e.g. epoxy, bismaleimide, PEEK) does not have much effect, and the matrix is chosen for reasons other than longitudinal stiffness, such as cost, maximum operating temperature or durability.

Rule-of-mixtures analysis is remarkably accurate in the calculation of the longitudinal Young's modulus of unidirectional composites. For instance, Fig. 15.14 compares the measured and calculated Young's modulus values for a unidirectional composite over a range of fibre contents, and the excellent agreement demonstrates that the in-plane stiffness can be accurately predicted using this simple analysis. Rule-of-mixtures modelling can be used to determine the longitudinal modulus for various types of composite materials to identify which provides the highest stiffness, as shown in Fig. 15.15. Rule-of-mixtures is only accurate over the fibre volume content between about 0.2 and 0.7, which is within the range used in aerospace composites.

Transverse Young's modulus of composites

Composite materials must be designed to ensure that the external load is applied parallel to the fibres. The load should never be applied in the antifibre direction because of the low transverse stiffness and strength of the composite.

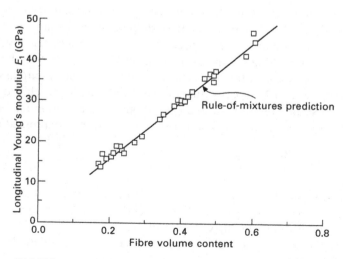

15.14 Measured (data points) and calculated (line) Young's modulus for unidirectional fibreglass composite with increasing fibre content.

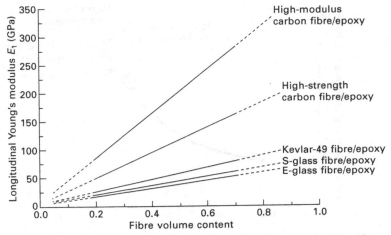

15.15 Effect of fibre type and fibre volume content on the Young's modulus of different types of unidirectional composite materials. The solid line indicates the approximate range of fibre contents over which rule-of-mixtures analysis is valid.

The simplest model to calculate the transverse modulus assumes that the fibres and matrix act in series under an external load, and is expressed as:

$$E_2 = \left[\frac{V_f}{E_f} + \left(\frac{V_m}{E_m} \right) \right]^{-1} \qquad [15.4]$$

Figure 15.16 shows the effects of fibre type and fibre volume content on the transverse Young's modulus of different types of unidirectional composites. The fibre reinforcement has a much smaller effect on the transverse Young's modulus compared with the longitudinal modulus, with virtually identical stiffness properties for composites containing fibres with vastly different modulus values.

The transverse modulus is not always accurately calculated using equation [15.4] because the strain is not uniform when a composite is loaded in the antifibre direction. Owing to the large difference between the elastic modulus of the fibres and matrix, the strain is distributed unevenly in the matrix under a transverse load. Figure 15.17 shows the heterogeneous strain field in a composite loaded in transverse tension, where the darkest fringe lines (located above and below the fibres) indicate regions where the matrix is highly strained. The heterogeneous strain affects the accuracy of the micromechanical analysis given in equation [15.4], with the model underpredicting the transverse modulus.

Other models have been developed to account for the heterogeneous strain

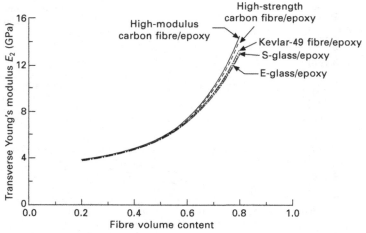

15.16 Effect of fibre type and fibre volume content on the transverse modulus of different types of unidirectional composite materials.

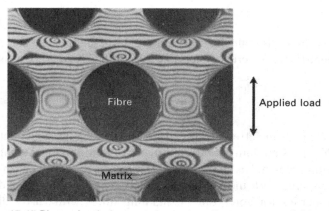

15.17 Photoelastic image of a composite-type material loaded normal to the fibre direction showing the heterogeneous strain distribution indicated by the fringe lines (from A. Puck, Zur Beanspruchen und Verformung von GFK-Mehrschichtenverbund-Bauelmenten, *Kunststoffe*, **57** (1967), 965–673.

distribution through the polymer matrix. One such model is the Halpin–Tsai model, which states that the transverse modulus is calculated using:

$$E_2 = \frac{E_m(1 + \xi \eta V_f)}{(1 - \eta V_f)}$$ [15.5]

where the interaction ratio is:

$$\eta = \dfrac{\left(\dfrac{E_{2f}}{E_m} - 1\right) \cdot}{\left(\dfrac{E_{2f}}{E_m} + \xi\right)}$$

and ξ is an adjustable factor.

Other elastic properties as well as the thermal properties of composites can be calculated using rule-of-mixtures analysis. Table 15.3 gives equations to predict the shear modulus, Poisson's ratio, thermal conductivity, and specific heat capacity of unidirectional composites.

Effect of fibre orientation on Young's modulus of composites

The Young's modulus of a unidirectional composite is much higher when loaded in the longitudinal direction than in the transverse direction owing to the high stiffness provided by the fibres. For example, the longitudinal modulus for a high modulus carbon fibre–epoxy composite is more than 200 times higher than the transverse modulus. The longitudinal (or 0°) and transverse (90°) directions are the two extremes for loading of a unidirectional composite: parallel and perpendicular to the fibre direction. Between these

Table 15.3 Rule-of-mixtures equations for unidirectional composites

Property	Equation
In-plane shear modulus	$G_{12} = \left[\dfrac{V_f}{G_f} + \dfrac{V_m}{G_m}\right]^{-1}$
	G_f: shear modulus of fibre
	G_m: shear modulus of matrix
In-plane Poisson's ratio	$v_{12} = v_f V_f + v_m V_m$
	v_f: Poisson's ratio of fibre
	v_m: Poisson's ratio of matrix
Transverse Poisson's ratio	$v_{21} = (v_f V_f + v_m V_m)\dfrac{E_2}{E_1}$
In-plane thermal conductivity	$k_1 = k_f V_f + k_m V_m$
	k_f: thermal conductivity of fibre
	k_m: thermal conductivity of matrix
Transverse thermal conductivity	$k_2 = k_m \dfrac{k_f(1 + V_f) + k_m V_m}{k_f V_m + k_m (1 + V_f)}$
In-plane specific heat capacity	$C = \dfrac{1}{\rho}[C_f \rho_f V_f + C_m \rho_m V_m]$
	C_f: specific heat capacity of fibre
	C_m: specific heat capacity of matrix

two extremes the Young's modulus changes continuously with the fibre angle. For instance, Fig. 15.18 shows the effect of fibre angle on the elastic modulus properties of a unidirectional carbon–epoxy composite. The Young's modulus decreases with increasing fibre angle, particularly from 0° to 30° when the stiffness falls sharply by nearly 80%. The high sensitivity of the Young's modulus to fibre angle occurs for any type of unidirectional composite material, and their fibres must be closely aligned to the load direction to achieve high stiffness.

The Young's modulus of a unidirectional composite loaded at any fibre angle (ϕ) between 0° and 90° can be calculated using:

$$E(\phi) = \left[\frac{\cos^4\phi}{E_1} + \frac{\sin^4\phi}{E_2} + \left(\frac{1}{G_{12}} - \frac{2v_{12}}{E_1} \right) \sin^2\phi \cos^2\phi \right]^{-1} \quad [15.6]$$

The other elastic properties of unidirectional composites are also dependent on the fibre angle. Figure 15.17 also shows the variation in the shear modulus of carbon/epoxy with fibre angle, and this property is highest at 45° and lowest when the load is applied in the fibre (0°) and anti-fibre directions (90°). The change in the shear modulus with fibre angle can be determined using:

$$G(\phi) = 2\left(\frac{2}{E_1} + \frac{2}{E_2} + \frac{4v_{12}}{E_1} - \frac{1}{G_{12}} \right) \sin^2\phi \cos^2\phi + \frac{1}{G_{12}}(\sin^4\phi \cos^4\phi)$$

$$[15.7]$$

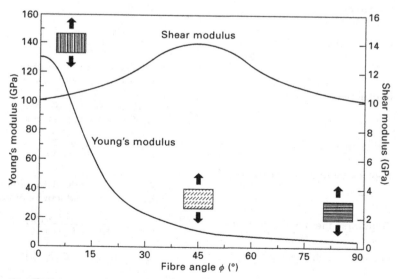

15.18 Effect of load angle on the Young's modulus (E) and shear modulus (G) of a unidirectional carbon–epoxy composite.

The highly orthotropic nature of the elastic properties of unidirectional composites means they should not be used in aircraft and other engineering structural applications. The loads applied to aircraft structures are rarely unidirectional. The majority of aircraft structures are subject to multidirectional loads which cannot be effectively carried using a composite material in which all the fibres are aligned in the same direction.

The composites used in aircraft structures are designed with the fibres aligned in two or more directions to support multidirectional loads. The two most common fibre patterns are cross-ply [0/90] where 50% of the fibres are aligned in the 0° direction and 50% in the 90° direction and quasi-isotropic [0/±45/90] where 25% of the fibres are aligned in each of the 0°, +45°, −45° and 90° directions. The quasi-isotropic fibre pattern is shown in Fig. 14.18. The 0° and 90° fibres in the quasi-isotropic pattern is used to carry in-plane tension, compression and bending loads whereas the +45° and −45° fibres carry in-plane shear loads. The cross-ply pattern is used when shear loads are absent. The effect of load angle on the Young's modulus of quasi-isotropic and cross-ply composites is shown in Fig. 15.19, and the variation in stiffness with angle is much less extreme than the unidirectional condition. The Young's modulus of a quasi-isotropic composite is relatively insensitive to the angle, and its in-plane elastic behaviour is approaching that of a fully isotropic material. The elastic properties of composite laminates with multidirectional fibre patterns can be calculated using laminate analysis.

Other fibre patterns are used occasionally in composite aircraft structures, although the most common is the quasi-isotropic pattern followed by the

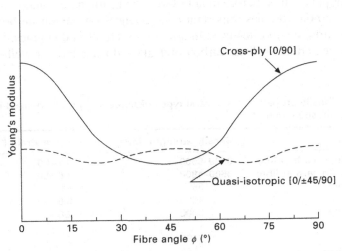

15.19 Effect of load angle on the Young's modulus of quasi-isotropic and cross-ply composites.

cross-ply pattern. An important consideration in the fibre arrangement of multidirectional composites is that the fibre pattern is symmetric about the mid-thickness point; that is, the fibre pattern in the upper half of the composite is a mirror image of the fibre pattern in the lower half. As mentioned in the previous chapter, symmetric fibre arrangement is essential to avoid warping and distortion of the composite component.

15.5.3 Strength properties of composite materials

Tensile strength properties of composites

The longitudinal tensile strength of composite materials is determined mostly by the strength and volume content of the fibre reinforcement. The breaking strength of the fibres is much greater than the strength of the polymer matrix, and therefore the fibres determine the ultimate strength of the composite. Table 15.4 lists the tensile strength and failure strain values for several types of fibres and polymers used in aircraft composites. The fibre strengths are typically 50 to 100 times higher than the matrix and, consequently, the strength of the matrix has little influence on the in-plane tensile strength of composite materials.

The fibres used in aircraft composites have high stiffness and strength, but they have little or no ductility and as a result fail at low strain. Figure 15.20 shows tensile stress–strain curves for various types of fibres used in aircraft composite materials. Carbon fibres are brittle materials that sustain no plastic deformation before breaking at low strain (less than 0.5–2.0%). Similar behaviour occurs with glass fibres whereas aramid fibres experience a small amount of plastic flow before final failure. The brittle nature and low failure strain of carbon fibre has important consequences on the tensile behaviour of carbon-fibre composites used in aircraft structures and engines.

Brittle materials such as carbon and glass do not have a well-defined

Table 15.4 Tensile properties for several types of fibres and polymers used in aircraft composite materials

Material	Tensile strength (MPa)	Failure strain (%)
High-modulus carbon fibre	3500–5500	0.1–1.0
High-strength carbon fibre	3500–4800	1.5–2.0
S-glass fibre	3500	4.5
E-glass fibre	4600	5.0
Kevlar-49 fibre	3000	2.8
Epoxy	90	3
Bismaleimide	50	3
Polyetheretherketone (PEEK)	96	50

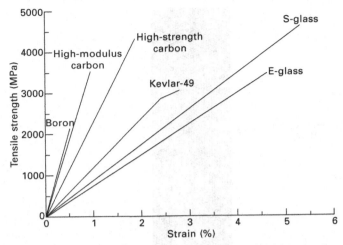

15.20 Tensile stress–strain curves for fibres used in aerospace composite materials.

tensile strength, unlike metals or polymers. For example, the aircraft-grade 7075-T76 aluminium alloy has a fixed tensile yield strength close to 470 MPa. Likewise, epoxy resin has a constant strength of about 100 MPa when fully cured. The strength of carbon fibre, however, varies over a very wide range (1400 to 4800 MPa or more) despite being produced, like metals and polymers, under controlled conditions which should ensure a consistent strength value.

The strength of brittle fibres is extremely variable because the failure stress is highly sensitive to the presence of flaws and defects. Tiny cracks and voids are present in fibres, and these have a major influence on fibre strength. For example, a surface crack in carbon fibre as small as 0.3–0.4 μm reduces the breaking strength by more than 50%. Cracks are accidentally created at the fibre surface when the individual filaments are collimated into bundles after being produced. The fibres slide and rub against each other during winding and handling which introduces tiny scratches on the surface. Even when fibres are wound and handled with great care it is difficult to avoid surface damage. Lubricants within the size coating are applied to fibres to minimise sliding friction, but this does not completely eliminate surface damage. The small cracks and other surface damage caused to the fibres are not all the same size, but vary randomly along the fibre length as illustrated is Fig. 15.21. Fracture of the fibre always occurs at the largest flaw, which determines the tensile strength, and the other (smaller) flaws have no influence on fibre strength. Because the maximum flaw size can vary from the fibre to fibre, despite being manufactured and handled under the same conditions, the strength is different from fibre to fibre.

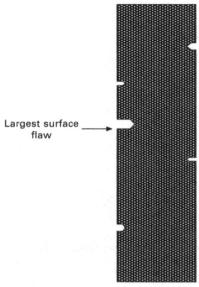

15.21 Schematic of surface flaws of different lengths that occur at
random locations along the fibre. Failure occurs at the largest flaw.

The tensile strength of brittle fibres (σ_f) such as carbon is related to the
largest crack length (c) according to:

$$c = \frac{\pi}{4}\left(\frac{K_c}{\sigma_f}\right)^2$$ [15.8]

where K_c is the fracture toughness of the fibre material. (The concept of
fracture toughness and the fracture properties of composites and other aerospace
materials are described in chapters 18 and 19). The fracture toughness of
carbon is very low (about 1 MPa m$^{1/2}$), which means its strength is very
sensitive to crack size. Figure 15.22 shows the dependence of carbon fibre
strength on the largest crack size, and the strength decreases rapidly with
very small increases in the crack length. Owing to this high sensitivity of
fibre strength to crack size, and the random nature of the maximum crack size
between fibres, the tensile strength varies over a wide range. Figure 15.23
shows the strength distribution plot for carbon fibre and, unlike metals and
polymers that have a single strength value, the fibre strength varies over a
wide range.

Longitudinal tensile strength of unidirectional composites

The tensile strength of unidirectional composite materials such as carbon
fibre–epoxy is much higher than many aerospace alloys, including the

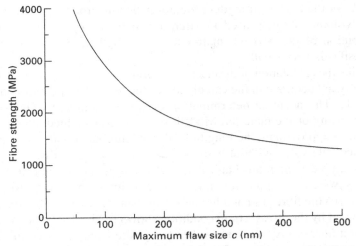

15.22 Effect of maximum surface flaw size on the tensile strength of carbon fibre.

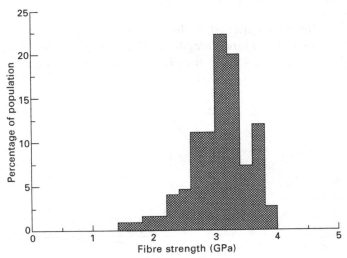

15.23 Strength distribution plot for carbon fibre.

aluminium alloys used in aircraft structures. For example, Table 15.2 shows that the specific tensile strength of carbon fibre–epoxy is about twice that of 2024 aluminium alloy. Owing to the failure stress of the fibres used in aerospace composites being much higher than the polymer matrix, the longitudinal tensile strength is strongly influenced by the tensile strength and volume fraction of the fibres. For example, the fibres typically account for

95–98% of the tensile strength of a unidirectional carbon–epoxy composite at the volume contents used in aircraft structures ($V_f \sim 0.55$–0.65). The contribution of the polymer matrix to the strengthening of unidirectional composites is very small.

Unlike the calculation of the elastic properties, micromechanical modelling is usually not accurate in the calculation of the tensile strength of composite materials. The accurate determination of the strength properties requires tensile testing of the material. Modelling based on weighted averaging can be used to approximate the longitudinal strength, but should never be used as an accurate value without being validated by tensile testing.

The simplest micromechanical model for longitudinal tensile strength assumes two cases: (i) the fibre failure strain is lower than the matrix failure strain or (ii) the fibres fail at a higher strain than the matrix. The first case is represented by the stress–strain condition shown in Fig. 15.24, and it applies to carbon fibre–epoxy and most other composite materials used in aircraft.

When the failure strain of the fibres is lower than the matrix, then the longitudinal tensile strength (X_T) of a unidirectional composite is given by:

$$X_T = \sigma_f^u V_f + \sigma_m^* V_m \qquad [15.9]$$

where σ_m^* is the matrix strength at the fibre failure strain.

As mentioned, the failure strength of brittle fibres is determined by the size of the largest flaw, which varies from fibre to fibre. As a result, fibres

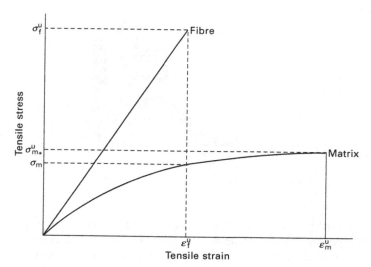

15.24 Tensile stress–strain curves for the fibre reinforcement and polymer matrix where the ultimate failure strain of the fibre (ε_f^u) is lower than the matrix (ε_m^u).

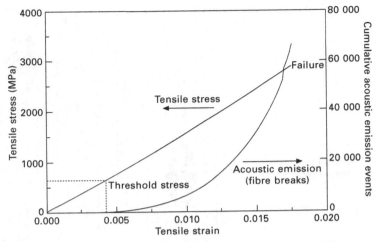

15.25 Effect of increasing tensile stress on the number of fibres broken (as measured by the cumulative number of acoustic emission events) within a unidirectional carbon–epoxy composite.

break over a range of stress values when a composite is loaded in tension. Those fibres with the lowest strength (because they contain the largest flaw) are the first to fail within the material. As the tension stress applied to the composite increases the fibres containing smaller flaws break until all the fibres are broken, and this event defines the ultimate tensile strength of the composite material. For example, Fig. 15.25 shows the effect of increasing tensile stress on the number of broken fibres within a unidirectional carbon–epoxy composite. The fibres begin to break at a threshold stress level, above which the total number of broken fibres rises at an increasing rate until the final failure stress. The design load limit for composite materials used in aircraft structures is well below this threshold limit to ensure that no fibres are broken during routine flight operations.

An important process in the tensile failure of fibres is shear lag, which is illustrated in Fig. 15.26. When a fibre breaks, its capacity to carry the applied tensile stress does not immediately drop to zero. The broken fibre can continue to carry load by transferring the stress across the break by shear lag. This is a process whereby tensile stress is transferred between the two ends of a broken fibre by shear flow of the surrounding polymer matrix. The efficacy of the shear lag process to transfer stress across a broken fibre decreases with increasing tensile strain because the two fibre ends become further apart. Therefore, the tensile load capacity of broken fibres decreases with increasing strain, and this has the result of more stress being exerted on the unbroken fibres. These fibres then become overstressed which causes them to break. As more fibres break the number of remaining intact fibres

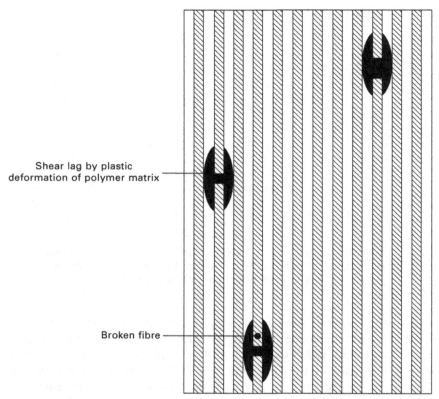

Shear lag by plastic
deformation of polymer matrix

Broken fibre

15.26 Schematic of the shear lag process whereby the applied
tensile stress is carried across the broken ends of fibres by shear
deformation of the matrix.

becomes fewer until eventually the composite fails. Without this shear lag
process, the longitudinal tensile strength of composite materials would be
lower than it actually is.

Transverse tensile strength of composites

The transverse tensile strength of composite materials is much lower than
their longitudinal strength owing to the absence of fibres aligned in the
transverse direction. Figure 15.27 compares the longitudinal and transverse
tensile strengths of a carbon–epoxy composite over a range of fibre contents,
and the failure strength in the transverse direction is just a small fraction
(typically less than 5–10%) of the longitudinal strength. This is one reason
for the poor impact damage resistance of composite structures, with impact
events such as bird strike causing significant damage owing to the low

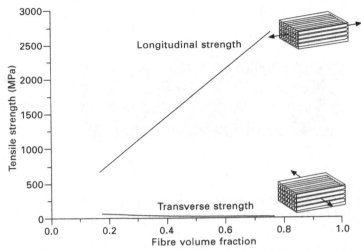

15.27 Effect of fibre content on the longitudinal and transverse tensile strengths of a unidirectional carbon–epoxy composite.

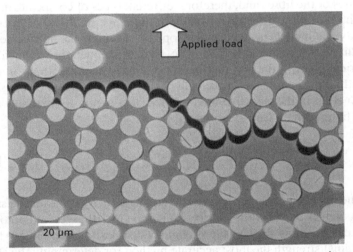

15.28 Cracking between the fibres and matrix in a composite material under transverse tensile loading.

transverse strength. (The impact damage resistance of composites is described in chapter 19).

The transverse strength is controlled by the failure strength of the interface between the fibres and matrix. Failure under transverse tensile loading usually occurs by cracking along the fibre–matrix interface, as shown in Fig. 15.28. The use of sizing compounds on the fibre surface to promote strong adhesion bonding with the polymer matrix is one of the common ways to

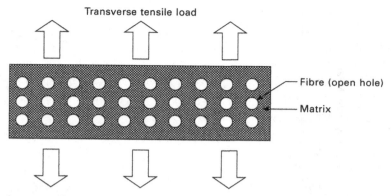

Transverse tensile load

Fibre (open hole)

Matrix

15.29 Representation of a composite material with a square pattern
of fibres subjected to transverse tensile loading.

increase the transverse tensile strength. Like the transverse Young's modulus,
the transverse tensile strength is not influenced significantly by the type or
properties of the fibres and, therefore, different types of composite materials
have similar failure strengths (usually in the range of 30–80 MPa).

Analytical models have been developed to calculate the transverse tensile
strength of composite materials. The simplest model treats the fibres as
cylindrical holes in the polymer matrix that have no strength, as illustrated
in Fig. 15.29. The transverse strength (Y_T) is then calculated by considering
the reduction in the effective load-bearing area owing to the holes created
by the fibres, which can be expressed as:

$$Y_T = \sigma_m \left[1 - 2 \left(\frac{V_f}{\pi} \right)^{1/2} \right]$$

[15.10]

when the fibres are arranged in a square grid pattern. σ_m is the tensile strength
of the polymer used for the matrix. Equation [15.10] gives an approximate
estimate of the transverse tensile strength of composite materials, but like the
longitudinal strength the most accurate method for determining the strength
is by tensile testing.

Effect of fibre orientation on tensile strength

The tensile strength of a unidirectional composite is dependent on the fibre
angle in a similar way to the Young's modulus. Figure 15.30 shows the
effect of fibre angle on the tensile strengths of unidirectional carbon–epoxy
and glass–epoxy materials. The strength drops rapidly with increasing angle
owing to the loss in load-bearing capacity of the fibres, which are only

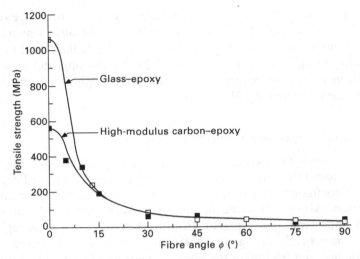

15.30 Effect of fibre angle on the tensile strength of unidirectional carbon–epoxy and glass–epoxy composites.

highly effective when aligned close to the load direction (within 3–5°). This behaviour, along with the highly orthotropic modulus properties, is the reason why unidirectional composites should not be used in aircraft structures.

The effect of fibre angle on the tensile strength of a unidirectional composite can be calculated in several ways, although the easiest method is to treat the failure process using three arbitrary stress states. Failure is assumed to occur in one of three modes:

- Mode I: failure occurs by tensile rupture of fibres at small angles (typically $\phi < 5°$).
- Mode II: failure occurs by shear of the matrix or fibre/matrix interface between small and intermediate angles (5° < ϕ < 45–60°).
- Mode III: failure occurs by transverse tensile fracture at high angles (ϕ > 45–60°).

Failure of a unidirectional composite under off-axis tensile loading can be estimated using the maximum stress criterion. This criterion assumes that the stresses for the three mode conditions are related to the fibre angle as follows:

- Mode I: axial tensile stress: $\sigma = X_T/\cos^2 \phi$
- Mode II: shear stress: $\sigma = 2S_{12}/\sin 2\phi$
- Mode III: transverse tensile stress: $\sigma = Y_T/\sin^2 \phi$

where X_T, S_{12} and X_Y represent the failure strengths of the composite in axial tension ($\phi = 0°$), in-plane shear ($\phi = 45°$) and transverse tension ($\phi =$

90°), respectively. These three equations are solved over the range of fibre angles between 0° and 90°, as shown in Fig. 15.31. The curve which has the lowest strength value over the range of fibre angles is then used to define the failure strength. It is shown in Fig. 15.31 that this approach provides a good estimation of the reduction to the tensile strength of a unidirectional composite with increasing fibre angle.

Longitudinal compressive strength of composites

The failure strength of composite materials under compression loading is usually different than under tension owing to the different failure behaviour. The most common failure modes of composites under in-plane compression are microbuckling or kinking of the load-bearing fibres. Microbuckling involves the lateral (out-of-plane) buckling of fibres over a small region, as shown schematically in Fig. 15.32. Kinking involves the localised out-of-plane rotation and fracture of the fibres, as shown in Fig. 15.33, and occurs more often in carbon fibre–epoxy composites than microbuckling.

Kinking occurs from a local buckling instability, which develops in the load-bearing fibres. The instability arises from a defect (e.g. void, resin-rich region) or free edge which does not provide sufficient lateral restraint to the fibre under compression loading. The instability allows the fibre to rotate

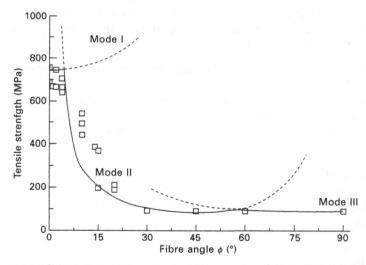

15.31 Calculation of the effect of fibre angle on the tensile strength of a unidirectional carbon–epoxy composite using the maximum stress criterion. The data points are experimental strength values. The full curves indicate the range of fibre angles when the equations for the maximum stress criterion are valid.

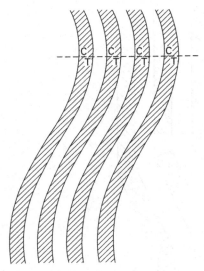

15.32 Schematic of microbuckling of a unidirectional composite under compression loading. 'C' represents compression and 'T' represents tension loading on the fibre surface during microbuckling.

in a very small region towards the defect or free edge, with the amount of fibre rotation increasing with the compressive strain. As the small region of fibres is rotated, the surrounding polymer matrix is plastically deformed by shear stress generated by the rotation process. Eventually, the fibres rotate over a sufficiently large angle that they break over a small length (usually 100–200 μm long), which is called a kink band. Once the first fibres fail by kinking, the other fibres are overloaded and they also then fail by kinking. A kink band propagates rapidly through the composite material causing compressive failure. Although the kink band is short in length, once it has propagated through the load-bearing section of the composite the material can no longer support an applied compression load.

Brittle fibre composites, such as carbon or glass-reinforced materials, fracture along a well-defined kink plane when loaded to the compressive failure stress. The compressive strength of a unidirectional composite that fails by kinking is calculated using:

$$X_C = \frac{\tau_y}{\gamma_y + \phi} \qquad\qquad [15.11]$$

where τ_y and γ_y are the shear yield stress and shear yield strain, respectively, of the polymer matrix, and ϕ is the fibre misalignment angle. This equation shows that the kinking stress is dependent on the fibre angle, and maximum

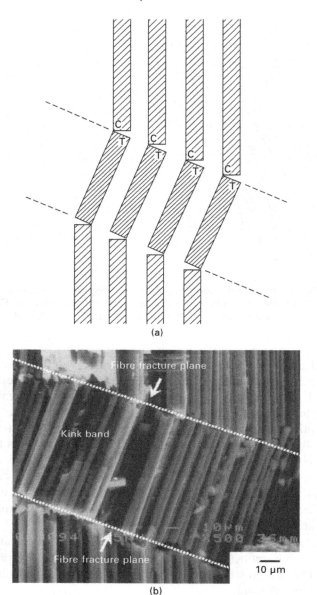

15.33 (a) Schematic of kinking and (b) a kink band within a composite material.

compressive strength is attained when the fibres are aligned with the direction of compression loading. The kinking stress decreases rapidly with increasing fibre angle in a manner similar to the tensile strength (Fig. 15.30). The

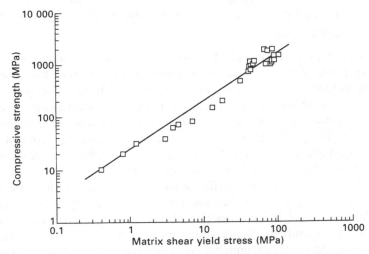

15.34 Effect of matrix shear yield strength on the compressive
strength of unidirectional composite materials (adapted from
P. M. Jelf and N. A. Fleck, 'Compressive failure mechanisms in
unidirectional composites', *Journal of Composite Materials,* **26**
(1992), 2706–2726).

kinking stress is also dependent on the shear properties of the polymer matrix,
and one of the most effective ways of increasing the compressive strength
of composites is to have a polymer matrix with high shear resistance. For
example, Fig. 15.34 shows the large improvement to the compressive kinking
stress that can be attained by increasing the matrix shear strength.

Strength properties of multidirectional composites

As mentioned, unidirectional composites are not used in aircraft structures
owing to their highly orthotropic properties. These materials are only
structurally efficient when the load is applied parallel with the fibre, which
is difficult to ensure for aircraft structures. Instead, quasi-isotropic and cross-
ply composite materials are commonly used, and the effect of load angle on
their tensile and compressive strength properties are similar to that shown in
Fig. 15.19 for the Young's modulus. The strength properties are reasonably
constant at the different load angles in the quasi-isotropic composite owing
to the presence of an equal proportion of 0°, +45°, –45° and 90° fibres. The
tensile and compressive strengths of the cross-ply composite are highest in
the 0° and 90° directions and lowest at 45°.

15.5.4 Effect of microstructural defects on mechanical properties

The mechanical properties of composite materials can be adversely affected by defects which are inadvertently created using manufacture. Manufacturing defects include voids, dry spots, resin-rich regions, irregular fibre distribution, micro-cracks, shrinkage, and misaligned fibres. Controlled manufacturing conditions and stringent quality control checks are applied to ensure that the composites produced for aircraft structures have a defect content which is below an acceptable level. Composites having high defect content must be repaired (if economically feasible) or scrapped.

Of the many types of defects, the most common are voids which are created by air being trapped within the material during ply lay-up or by gas created as a by-product of the cure reaction of the polymer matrix. The processes used to manufacture aerospace composites, such as autoclave curing, resin transfer moulding and vacuum bag resin infusion, result in low void contents (typically under 1% by volume). In addition, the epoxy and other thermoset resins used as the matrix material are formulated to yield little or no gaseous volatiles during curing and, therefore, void formation is minimised. Despite these measures, voids can still occur, thus reducing the mechanical properties. Voids are more detrimental to the mechanical properties influenced by the matrix, the so-called matrix-dominated properties, than the fibre-dominated properties. Matrix-dominated properties include interlaminar shear strength, impact resistance and delamination fracture toughness. Fibre-dominated properties include the tensile modulus and strength, and these are not affected significantly by voids unless present at a high volume content.

Figure 15.35 shows the effect of volume content of voids in an aircraft-grade carbon–epoxy composite on the interlaminar shear strength (matrix-dominated property) and tensile strength (fibre-dominated property). The interlaminar shear strength decreases rapidly with increasing void content above about 1%, and this behaviour is typical for matrix-dominated properties, whereas the tensile strength is largely unaffected. For this reason, the volume content of voids (and other defect types) in aircraft composite materials must be kept to very low levels (under 0.5–1.0%) to avoid any significant loss in mechanical performance.

15.6   Sandwich composites

Sandwich composite materials consist of thin face skins bonded to a thick low-density core material. By the correct choice of materials for the skins and core, sandwich composites provide higher ratios of stiffness-to-weight and strength-to-weight than many metals and fibre–polymer laminates. In most cases, the main reason sandwich construction is used is to reduce weight,

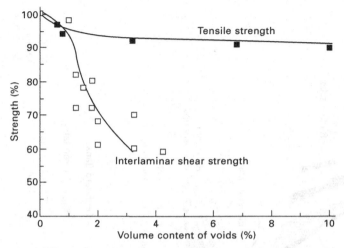

15.35 Effect of void content on the percentage tensile and interlaminar shear strengths of a carbon fibre–epoxy composite.

but it also provides other advantages including improved acoustic damping (noise reduction), thermal insulation, and impact absorption.

Sandwich composites are used almost exclusively in secondary aircraft structures such as the control surfaces (e.g. ailerons, flaps, spoilers, slats), engine nacelles, radomes, floor panels, and the vertical tailplane (Fig. 15.36). Rarely, if ever, are sandwich materials used in primary structures of conventional civil or military aircraft. An exception is the sandwich fuselage of the Beech Starship, which is an all composite aircraft (Fig. 15.37). Sandwich composites are also used in the fuselage and wings of sailplanes and other ultralight aircraft.

Sandwich construction is used most often for aircraft structures subjected to bending loads. The separation of the face skins, which are stiff and strong, using the lightweight core offers the possibility of making the resulting stresses act far from the neutral axis of the structure. Moving the skins away from the neutral axis greatly increases the bending stiffness and thereby makes sandwich structures resistant to buckling. The skins carry most of the in-plane tension, compression and bending loads applied to the sandwich structure. The function of the core is to carry shear stresses resulting from transverse loads, to support the face skins, and to maintain a separation distance between the skins. To fulfil these functions the core material must have sufficient shear stiffness. Table 15.5 compares the structural efficiency of a monolithic material (in this case a 0.8 mm sheet of aluminium) against two sandwich composite materials with different thickness. When loaded in bending, the stiffness and strength increases with the thickness of the sandwich

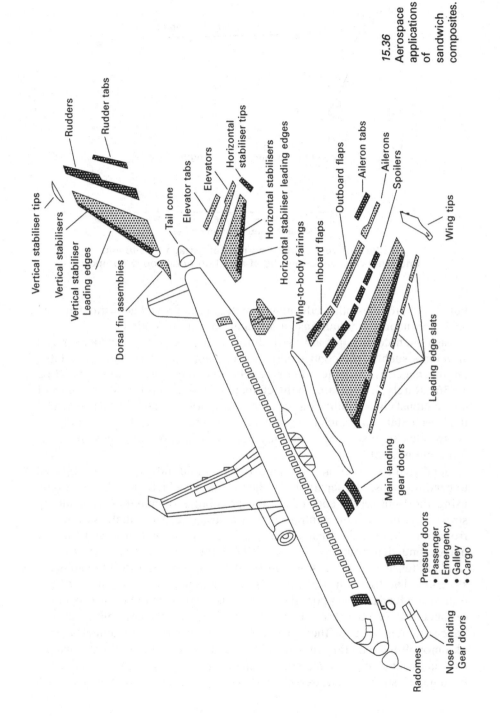

Vertical stabiliser tips

Vertical stabilisers

Vertical stabiliser Leading edges

Dorsal fin assemblies

Rudders

Rudder tabs

Tail cone

Elevator tabs

Elevators

Horizontal stabiliser tips

Horizontal stabilisers

Horizontal stabiliser leading edges

Wing-to-body fairings

Inboard flaps

Outboard flaps

Aileron tabs

Ailerons

Spoilers

Wing tips

Leading edge slats

Main landing gear doors

Pressure doors
• Passenger
• Emergency
• Galley
• Cargo

Radomes

Nose landing Gear doors

15.36 Aerospace applications of sandwich composites.

15.37 Beech Starship has a sandwich composite fuselage construction.

Table 15.5 Structural efficiency of sandwich composite panels. Note that the overall material thickness doubles but the skin material thickness remains the same for cases B and C

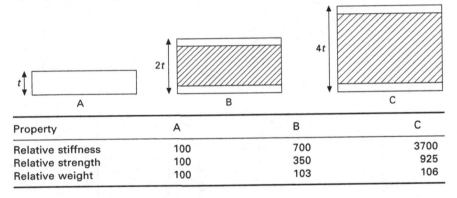

Property	A	B	C
Relative stiffness	100	700	3700
Relative strength	100	350	925
Relative weight	100	103	106

material. At the highest thickness, the strength is increased over 9 times and the stiffness by 37 times, but with only a 6% increase in weight.

The face skins are made of a thin sheet of fibre–polymer laminate such as carbon–epoxy, although in earlier sandwich constructions aluminium alloy was used. There are four main groups of low-density core materials: honeycomb cores, corrugated cores, foam cores, and monolithic cores of homogeneous material. The first three types are used in aircraft sandwich composites, whereas the monolithic core materials (of which the most common is balsa wood) are rarely used.

Honeycomb core consists of very thin sheets attached in such a way that connected cells are formed. The cell structure closely resembles the honeycomb found in a beehive. The two most common honeycomb configurations are shown in Fig. 15.38: hexagonal and rectangular. Various types of material are used for the honeycomb, with aluminium alloy (usually 5052-H39, 5056-H39 or 2024-T3) and Nomex being the most common in aircraft sandwich components. Nomex is the tradename for a honeycomb material based on aramid fibre paper in a phenolic resin matrix. Aluminium and Nomex core materials have a lightweight cellular honeycomb structure which provides high shear stiffness.

The aerospace industry is increasingly using polymer foams instead of aluminium honeycomb and Nomex because of their superior durability and

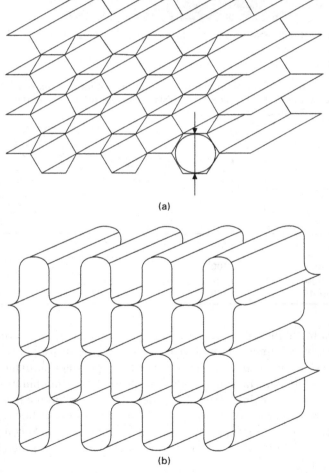

(a)

(b)

15.38 Honeycomb core types: (a) hexagonal core and (b) rectangular core.

resistance to water absorption. The foams used are often closed cell, which means the cell voids are completely surrounded by a thin membrane of the solid polymer. Some types of polymer foams have an open cell structure where the cells are interconnected, although these are used rarely in aircraft in preference to closed-cell foams. Examples of closed cell foams are polyetherimide (PEI) and polymethacrylimide (PMI), and these materials have a low-density cellular structure which has high specific stiffness and strength (Fig. 15.39). New types of foam metals are emerging, such as aluminium foams, although these are not currently used in aircraft sandwich materials.

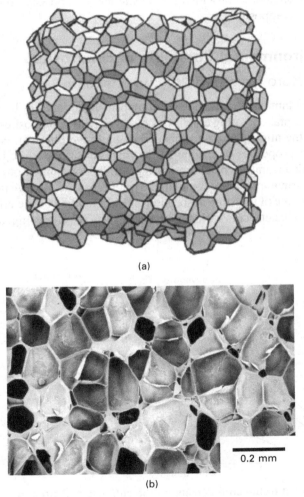

(a)

0.2 mm

(b)

15.39 (a) Schematic and (b) photograph of the closed cell structure of polymer foams.

The mechanical properties of the face skins and core must be balanced to achieve high structural efficiency. The core must have sufficient shear and compressive strength to ensure effective support of the skins. If the core is too weak in shear, the skins bend independently and the bending stiffness of the sandwich material is low. The stiffness and strength properties of core materials increase linearly with their density, as shown in Fig. 15.40, and therefore increased mechanical performance comes with a weight penalty. The core materials used in aircraft structures commonly have density values in the range of 75 to 250 kg m^{-3}, which provides a good balance between mechanical performance and light weight. The stiffness and strength of the skins must be sufficient to provide high tensile and compressive properties to the sandwich composite.

15.7   Environmental durability of composites

15.7.1 Moisture absorption of composites

The properties of laminates and sandwich composite materials may be affected by the environmental operating conditions of the aircraft. Composites absorb moisture from the atmosphere and this can degrade the physical, chemical and mechanical properties over time so that measures have to be taken to minimise any deterioration by the environment. The combination of high humidity and warm temperature found in the tropics has the synergistic effect of increasing the rate of moisture absorption and accelerating the deterioration of the material. This effect is called hygrothermal (hot/wet) ageing, and it is

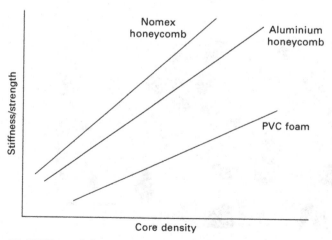

15.40 Effect of the core density on the stiffness and strength of core materials. The PVC foam is shown for comparison, but is not used in aircraft sandwich composite materials.

the most common environmental condition for degrading the properties of composite materials. Aircraft components may also be affected by various fluids used on aircraft, including fuels, fuel additives and hydraulic fluids, as well as by ultraviolet radiation. The long-term durability of composite materials in the aviation environment is essential to maintain structural integrity and safety whilst also minimising maintenance costs and aircraft downtime for inspection and repair.

The absorption of moisture occurs mostly via the diffusion of water molecules through the polymer matrix. Moisture is also absorbed along the fibre–matrix interfacial region, which in some materials can provide a pathway for the rapid ingress of water. Carbon and glass fibres do not absorb moisture and therefore do not assist in the diffusion process. Organic-based fibres, on the other hand, can absorb significant amounts of moisture; for example, aramid fibres can absorb up to 6% of their own weight in water. Because most of the moisture is absorbed by the polymer matrix, the type of resin used in a composite has a large effect on environmental durability. Moisture absorption and its effect on properties can vary by an order of magnitude between different types of polymers. For example, some thermoplastics such as PEEK absorb less than 1% of their own weight in water whereas some thermosets (such as phenolics) can absorb more than 10%. Epoxy resins absorb between about 1% and 4% in water (depending on the type of epoxy), and the epoxies used in external aircraft structures exposed directly to the atmosphere are often selected for their low moisture absorption.

15.7.2 Fickian diffusion of moisture in composites

Exposure of composite materials to humid air and water (rain) allows moisture to diffuse through the outer plies of the composite towards the centre. Water molecules diffuse through the matrix via the 'free space' between the polymer chains under a concentration gradient where the moisture content is highest at the surface and decreases towards the centre. After a period of time at constant humidity, an even distribution of moisture occurs through the material and this is the saturation limit of the polymer matrix.

The diffusion of moisture into aircraft composite materials can usually be described by Fickian diffusion behaviour. Fickian diffusion is characterised by a progressive increase in weight of the material owing to the uptake of water until an asymptotic value is reached at full saturation, as shown in Fig. 15.41. The rate of moisture absorption increases with the temperature, but is not strongly influenced by the relative humidity of the environment. In other words, the moisture uptake by a composite material occurs faster in warm than in cool conditions, but is not affected significantly by whether it is dry or moist air. The saturation limit, defined by the maximum, steady-state weight gain of the composite, increases with the humidity level of the

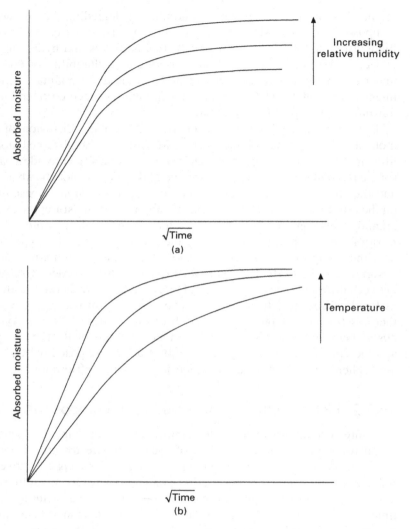

15.41 Weight gain owing to moisture absorption against time $(\sqrt{t})$ for composite materials exhibiting Fickian diffusion behaviour: (a) effect of humidity level; and (b) effect of temperature.

atmosphere. The absorption of moisture by composites which exhibit Fickian diffusion behaviour is often reversible, and when the material is exposed to a dry environment the water can diffuse out into the atmosphere. Therefore, composite structures used in aircraft that operate in different environmental conditions may undergo cyclic absorption and loss of moisture.

The absorption of moisture into a composite material can be calculated using Fickian diffusion kinetics, which allows the time to saturation to be determined. Fickian behaviour assumes that the moisture concentration gradient through-the-thickness of a material can be expressed as a function of time:

$$\frac{\partial c}{\partial t} = \frac{d}{dx}\left(D\frac{\partial c}{\partial x}\right)$$

[15.12]

where c is the moisture concentration, t is time, x is the distance below the material surface, and D is the diffusion coefficient of moisture through the material. The diffusion coefficient is calculated using an Arrhenius equation:

$$D = D_o \exp\left(-\frac{E}{RT}\right)$$

[15.13]

where D_o is the diffusion coefficient at a known temperature (T), E is the activation energy for diffusion, and R is the universal gas constant. The diffusion coefficient is related exponentially to temperature, and therefore the rate of moisture absorption increases rapidly with temperature. Figure 15.42 shows the relationship between the diffusion coefficient for water in a carbon fibre–epoxy composite with temperature. An increase in temperature of 10° C typically doubles the diffusion rate.

The weight gain of a composite with increasing time is calculated using:

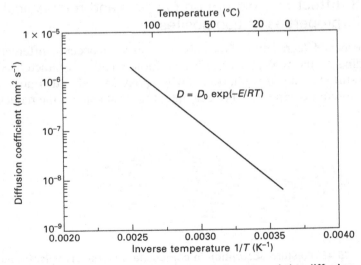

15.42 Relationship between the temperature and the diffusion coefficient for a carbon fibre–epoxy composite.

$$M_t = \left\{ 1 - \frac{8}{\pi^2} \sum_{n=0}^{\infty} \frac{1}{(2n+1)^2} \exp[-D(2n+1)^2\pi^2 t/h^2] \right\} M_\infty \quad [15.14]$$

where t is the time, M_∞ is the mass change at saturation, and h is the thickness of the composite material. This expression can be used to predict the weight gain with time for any fixed environment condition (constant humidity and temperature) where the diffusion coefficient is known.

The diffusion coefficient for anisotropic materials such as a unidirectional composite depends on the fibre orientation, as shown in Fig. 15.43. The diffusion coefficient when moisture ingress is perpendicular to the fibres, which is the case for most composite materials exposed to a humid environment, is calculated using:

$$D_\perp = [1 - 2(V_f/\pi)^{1/2}]D_m \quad [15.15]$$

where D_m is the diffusion coefficient for the matrix phase at the known temperature. When moisture ingress occurs parallel to the fibres then:

$$D_{||} = (1 - V_f)D_m \quad [15.16]$$

where it is assumed that the fibres are arranged in a regular square pattern. Diffusivity of water in the fibre direction in a carbon–epoxy composite can be up to 10 times higher than that perpendicular to the fibres. However, most composite structures are designed with the fibres parallel to the surface, and therefore only moisture ingress perpendicular to the fibres is considered.

15.7.3 Effect of moisture on physical and mechanical properties of composites

Environmental degradation of the polymer matrix can occur in different ways, depending on the type of resin. Plasticisation, which is characterised by a loss in stiffness and strength, occurs with epoxy and other thermoset resins used in aircraft composites. For example, Fig. 15.44 shows the reduction in

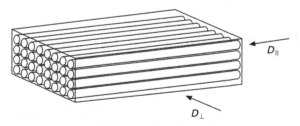

15.43 Moisture absorption in composite materials is anisotropic with the diffusion rate parallel to the fibres ($D_{||}$) being higher than normal to the fibres ($D_\perp$).

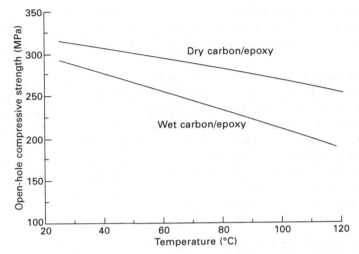

15.44 Effect of moisture absorption on the open-hole compressive strength of carbon fibre–epoxy composite.

the open-hole compressive strength with increasing temperature for a carbon fibre–epoxy composite in a dry or saturated condition. The reduction in strength becomes more extreme in the saturated composite with increasing temperature owing to plasticisation of the matrix phase. Matrix-dominated properties are more sensitive to plasticisation than fibre-dominated properties. As a general rule, for a carbon fibre–epoxy composite cured at 180 °C, moisture reduces the matrix-dominated mechanical properties by up to 10% for subsonic aircraft and 25% for supersonic aircraft (owing to their higher skin temperatures). The absorption of water can also reduce the glass transition temperature and consequently the maximum operating temperature of the composite. Reductions in the glass transition temperature of 30 °C or more can be experienced, as discussed in chapter 13. Plasticisation is a reversible process, and the properties recover when moisture is driven out of the composite by drying.

More severe forms of degradation of the matrix include chemical changes by hydrolysis or chain scission reactions between the polymer and water. The polymer chains can be broken down by chemical reactions with the water. These changes are irreversible, and epoxies and other aerospace resins are formulated to avoid this type of damage. Swelling and volumetric stresses can occur when a large amount of moisture is absorbed, which causes warping and cracking of the composite.

Moisture absorption is a problem for sandwich composites with an open-cell core material, such as Nomex or aluminium honeycomb. Moisture diffuses through the laminate face skin and forms water droplets within the core, as shown in Fig. 15.45. Often after long-term exposure the water builds up at the bottom of the cells where it can soften Nomex and corrode aluminium honeycomb. This problem can be avoided by the use of closed cell polymer foam.

Water trapped within composites can seriously affect the properties when it freezes. Water that collects within voids and cracks in composites and within the cells of core materials freezes when exposed to cold conditions at high altitude. Water expands when it freezes and has a higher bulk modulus than epoxy resin. Therefore, when water freezes it exerts tensile pressure on the surrounding composite material that can cause permanent damage, such as delamination cracking. The water freezes and melts each time the aircraft ascends or descends from high altitude, which causes cyclic stressing of the composite in a process called freeze/thaw.

15.7.4 Effect of ultraviolet radiation on composites

The surface of composite materials can be degraded by ultraviolet (UV) radiation which breaks down the polymer matrix and, if present, organic fibres such as aramid. UV radiation destroys the chemical bonds in epoxy resins and many other types of polymers, and the degraded material is then removed from the surface by wind and rain. UV damage is often confined to the topmost layers of the material, and the bulk properties are unlikely to be affected owing to the slow rate of degradation. It is possible to dope polymers with UV-absorbing compounds to minimise the damage or, alternatively the composite surface can be protected using UV absorbent paint.

15.8   Summary

Composites consist of a reinforcement phase and a matrix phase. The reinforcement can be particles, whiskers or continuous fibres. Aerospace

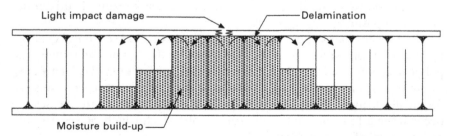

Light impact damage ———————   ——— Delamination

Moisture build-up ——

15.45 Moisture absorption into open-cell cores within sandwich composites owing to skin damage (reproduced from G. Kress, 'Design Criteria', in: *ASM Handbook Volume 21: Composites*, ASM International, Materials Park, OH, 2001).

composites are reinforced with continuous fibres which provide higher stiffness and strength than particles or whiskers. The matrix phase can be metallic, ceramic or polymeric, although the matrix for most aerospace composites is a thermoset resin (e.g. epoxy, bismaleimide) or high-performance thermoplastic (e.g. PEEK). The functions of the polymer matrix are to bind the reinforcement fibres into a solid material; transmit force applied to the composite on to the fibres; and to protect the fibres from environmental damage.

There are many advantages as well as several problems with using carbon-fibre composites rather than aluminium in aircraft. The advantages include reduced weight, capability to manufacture integrated structures from fewer parts, higher structural efficiency (e.g. stiffness/weight and strength/weight), better resistance against fatigue and corrosion, radar absorption properties, good thermal insulation, and lower coefficient of thermal expansion. The disadvantages of composites include higher cost, slower manufacturing processes, anisotropic properties making design more difficult, low through-thickness mechanical properties and impact damage resistance, higher sensitivity to geometric stress raisers such as notches, lower temperature operating limit, and lower electrical conductivity.

The mechanical properties of composite aircraft structures are determined using a hierarchical approach to modelling that begins with micromechanical analysis of the unit cell followed by ply analysis then laminate analysis and, finally, structural analysis.

The elastic modulus and strength properties of unidirectional composites are anisotropic, with the mechanical properties decreasing rapidly with fibre angle from the longitudinal ($0°$) to the transverse ($90°$) direction. The longitudinal tensile properties are controlled mostly by the properties and volume content of the fibre reinforcement. The transverse properties are more dependent on the matrix properties. The anisotropic properties of composites is minimised by using cross-ply [0/90] or, in particular, quasi-isotropic [0/±45/90] fibre patterns.

The carbon and glass fibres used in composite materials are brittle and fail at low strain. The fibre strength is determined by the largest flaw, which varies from fibre to fibre. As a result, the tensile strength of fibres varies over a wide range, and does not have a single strength value.

Failure of composites under compression loading occurs by microbuckling or, more often, kinking. Kinking involves the localised out-of-plane rotation and fracture of load-bearing fibres caused by a buckling instability. The compressive strength is determined by the applied stress required to cause kinking, which increases with the shear properties of the polymer matrix and the closer the fibres are aligned to the load direction.

Microstructural defects such as voids, microcracks, resin-rich regions and wavy fibres can reduce the Young's modulus and, in particular, the strength properties of composite materials. Matrix-dominated properties (e.g.

interlaminar shear strength, impact resistance) are affected more by defects than fibre-dominated properties (e.g. tensile strength).

Sandwich composites are used in aircraft structures that require high bending stiffness and buckling resistance. The core material must have sufficiently high shear stiffness to transfer load to the face skins. Different types of core materials are used, including honeycombs and closed-cell polymer foams.

Composites can absorb moisture and other fluids (e.g. fuel, hydraulic oil) which can adversely affect their physical and mechanical properties. The most common environmental durability issue for aerospace composites is moisture absorption from the atmosphere, particularly under hot/wet conditions. Thermoset resins generally absorb more moisture than high-performance thermoplastics, and epoxies can gain 1–4% of their original weight owing to water absorption. The composite materials used in aircraft structures often exhibit Fickian behaviour which is characterised by a steady uptake of water until the saturation limit is reached. The absorption rate increases rapidly with temperature whereas the maximum amount of moisture absorbed at the saturation limit is determined by the relative humidity of the environment. The absorption of moisture by epoxy matrix composites causes plasticisation, which is reversible when the material is dried. The absorption of moisture reduces the glass transition temperature and mechanical properties (particularly the matrix-dominated properties). Composites are also susceptible to environmental degradation by freeze/thaw and UV radiation.

15.9   Terminology

Fibre-dominated property: Mechanical property that is strongly influenced by the properties of the fibre reinforcement. Examples are tensile modulus and strength.

Fibre-reinforced composite: Composite material reinforced with long, continuous fibres. The length-to-width ratio of the particles is many thousands-to-one.

Freeze/thaw: Process that involves freezing (expansion) and thawing (contraction) of water trapped within a composite material resulting in damage.

Hybrid composite: Composite reinforced with two or more types of fibres.

Kink band: The region over which the load-bearing fibres are rotated during kinking, and the length of the kink band is defined by the distance between the fracture planes of the fibres.

Kinking: Compressive failure of composite materials which involves rotation and fracture of load-bearing fibres over a short distance.

Matrix-dominated property: Mechanical property which is strongly influenced by the properties of the matrix phase. Examples are interlaminar shear strength, transverse tensile strength, interlaminar fracture toughness and impact damage resistance.

Microbuckling: Compressive failure of composite materials which involves buckling of the load-bearing fibres.

Particle-reinforced composite: Composite material reinforced with particles (spherical or irregular shape). The length-to-width ratio of the particles is close to unity.

Reinforcement: Constituent material of a composite that forms the discrete (discontinuous) phase. Reinforcement phase often used to increase structural properties such as stiffness and strength of the composite.

Shear lag: Strengthening process by which broken fibres transfer an applied tensile stress across their ends by shear deformation of the surrounding polymer matrix.

Whisker-reinforced composite: Composite material reinforced with short fibres (known as whiskers). The length-to-width ratio of whiskers is several hundred-to-one.

15.10 Further reading and research

Marks, N.J., 'Polymeric based composite materials' in *High performance materials in aerospace*, edited H.M. Flower, Chapman and Hall, London, 1995, pp. 203–224.

Baker, A., Dutton, S. and Kelly, D. (eds.), *Composite materials for aircraft structures*, American Institute of Aeronautics and Astronautics, Inc., Reston, VA, 2004.

Chawla, K.K., *Composite Materials: Science and Technology*, Springer–Verlag, New York, 1998.

Middleton, D.H. (ed.), *Composite materials in aircraft structures*, Longmans, UK, 1990.

Niu, M.C.Y., *Composite airframe structures*, Comilit Press, Hong Kong, 1992.

Kelly, A. and Zweben, C. (eds.), *Comprehensive composite materials*, Elsevier, Amsterdam, 200.

16

Metal matrix, fibre–metal and ceramic matrix composites for aerospace applications

16.1 Metal matrix composites

16.1.1 Introduction to metal matrix composites

Metal matrix composites (MMCs) are lightweight structural materials used in a small number of aircraft, helicopters and spacecraft. MMC materials consist of hard reinforcing particles embedded within a metal matrix phase. The matrix of MMCs is usually a low density metal alloy (e.g. aluminium, magnesium or titanium). The metal alloys used in aircraft structures, such as 2024 Al, 7075 Al and Ti–6Al–4V, are popular matrix materials for many MMCs. Nickel superalloys may be used as the matrix phase in MMCs for high-temperature applications.

The metal matrix phase is strengthened using ceramic or metal oxide in the form of continuous fibres, whiskers or particles. Boron (or borsic, a SiC-coated boron), carbon and silicon carbide (SiC) are often used as continuous fibre reinforcement, and these are distributed through the matrix phase. Silicon carbide, alumina (Al_2O_3) and boron carbide (B_4C) are popular particle reinforcements. The maximum volume content of reinforcement in MMCs is usually below 30%, which is lower than the fibre content of aerospace carbon–epoxy composites (55–65% by volume). Reinforcement contents above about 30% are not often used because of the difficulty in processing, forming and machining of the MMC owing to high hardness and low ductility.

16.1.2 Properties of metal matrix composites

MMCs offer a number of advantages compared with their base metal, including higher elastic modulus and strength, lower coefficient of thermal expansion, and superior elevated temperature properties such as improved creep resistance and rupture strength. Some MMCs also have better fatigue performance and wear resistance than the base metal.

Lower density is an attractive property of MMCs made using a metal matrix having a higher specific gravity than the ceramic reinforcement. The density of most ceramic materials is moderately low (generally under 3 g cm^{-3}) and, when used in combination with a denser metal, there is an overall reduction

394

in weight. Titanium, steel and nickel matrix composites, for example, have lower densities than their base metal which translates into a weight saving. However, aluminium and magnesium alloys, which have a lower or similar density to the ceramic reinforcement, may incur a weight penalty. Figure 16.1 shows the percentage density change of several aerospace alloys with increasing volume content of silicon carbide reinforcement. The weight saving is an incentive for heavy metals such as nickel-based superalloys, provided this is done without degrading important structural properties such as toughness and creep resistance.

MMCs are characterised by high stiffness, strength and (in most materials) fatigue resistance. The improvement in these properties is controlled by the stiffness, strength, volume content and shape of the reinforcement. This control allows the properties of MMCs to be tailored to an application requiring a combination of high properties. The properties of MMC reinforced with continuous fibres are anisotropic, with their mechanical properties such as stiffness and strength being highest in the fibre direction. The modulus is isotropic in MMCs containing whiskers that are randomly aligned or particles that are evenly dispersed through the metal matrix phase. Figure 16.2 shows the effect of increasing volume content of continuous fibres of alumina on the elastic modulus and yield strength of an aluminium–lithium alloy in the parallel and transverse (anti-fibre) directions. The longitudinal and transverse properties increase linearly with the fibre content, and it is possible to increase the stiffness and strength by 50–100% compared with the base metal. The greatest improvement in longitudinal stiffness and strength is achieved using

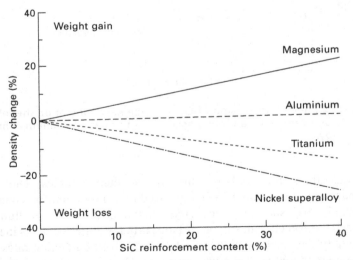

16.1 Percentage change in density of aerospace alloys owing to SiC reinforcement.

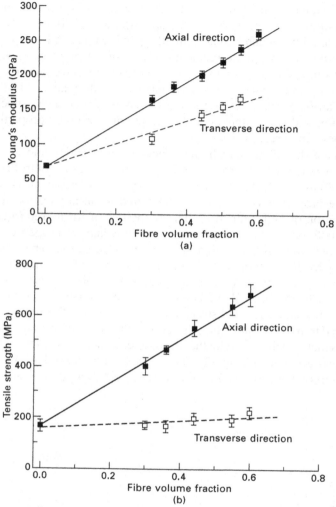

16.2 Effect of increasing Al_2O_3 reinforcement content on the (a)
Young's modulus and (b) tensile strength of an aluminium–lithium
alloy.

continuous fibres, followed by whiskers and then particles. The fatigue
performance of metals can also be improved by ceramic reinforcement. For
example, Fig. 16.3 shows a large increase in the fatigue life of an aluminium
alloy when reinforced with silicon carbide particles. Improvements in fatigue
are generally the result of higher modulus and work-hardening rates of the
composite compared with the base material. However, there are occasions
when the fatigue life is degraded by ceramic reinforcement.

   MMCs are characterised by low fracture toughness and ductility, which
are problems when used in damage-tolerant structures. The toughness and
ductility of MMCs depends on several factors: composition and microstructure
of the matrix alloy; type, size and orientation of the reinforcement; and the
processing conditions. Figure 16.4 shows reductions to the fracture toughness
and ductility of aluminium when the reinforcement content is increased to
30–40%, which are typical concentrations used for high-strength applications.

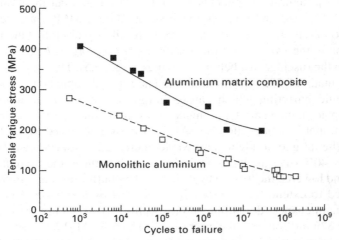

16.3 Fatigue life graphs for an aluminium alloy with and without
ceramic particle reinforcement.

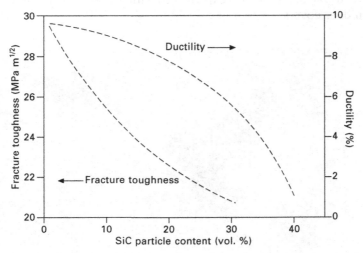

16.4 Typical effect of increasing reinforcement content on the fracture
toughness and ductility of metal matrix composites.

Reductions in toughness of over 30% and ductility to under 2–3% are common with many MMC materials.

16.1.3 Aerospace applications of metal matrix composites

Aerospace applications of MMCs are few, and their use is not expected to grow significantly in the foreseeable future owing to problems with high manufacturing costs and low toughness. The current applications are confined to structural components where the design and certification issues are straightforward and there is low risk of failure. MMCs are currently not used on civil airliners and only rarely in military aircraft. One notable application is the two ventral fins on the F-16 *Fighting Falcon*, which are located on the fuselage just behind the wings (Fig. 16.5). The original ventral fins were made of 2024-T4 aluminium. The fins are subjected to turbulent aerodynamic buffeting which causes fatigue cracking in the aluminium alloy. Replacement ventral fins made using a ceramic particle-reinforced aluminium matrix composite (6092Al–17.5% SiC) are fitted to the F-16 to alleviate the fatigue problem. This MMC increased the specific stiffness of the fins by 40% over the baseline design, which reduced the tip deflections by 50% and lowered the torsion loads induced by buffeting. The use of MMC is expected to extend the service life of the fins by about 400%; with the reduced maintenance, downtime and inspections costs saving the USAF an estimated $26 million over the aircraft life. The 6092Al–17.5% SiC particle composite is also used in fuel access door covers on F-16 aircraft. Similar to the ventral fins, the higher stiffness, strength and fatigue life of the MMC eliminated cracking problems experienced with the original aluminium alloy used in door covers.

16.5 Al–SiC composites are used in the ventral fins (circled) and fuel access doors of the F-16 *Fighting Falcon*.

MMCs are used in the main rotor blade sleeve of the Eurocopter EC120 and N4 helicopters. The sleeve is a critical rotating component because failure results in total loss of the main rotor blade. Sleeve materials require an infinite fatigue life under the operating stresses of the rotor blades, together with high specific stiffness and good fracture toughness. Titanium alloy is normally used for the sleeve, but to reduce cost and weight while maintaining high fatigue performance, strength and toughness, the sleeves for the EC120 and N4 are fabricated using a particle-reinforced aluminium composite (2009Al–15% SiC).

The first successful application of MMCs reinforced using continuous fibres was in the space shuttle orbiter. Struts used to stiffen the mid-fuselage (payload) section of the orbiter are made using aluminium alloy reinforced with 60% boron fibres. This composite is also used in the landing gear drag line of the orbiter. The continuous boron fibres are aligned along the axis of the tubular struts and drag line to provide high longitudinal specific stiffness. About 300 MMC struts are used as frame and rib truss members to form the load-bearing skeleton of the orbiter cargo bay, which provides a 45% weight saving over conventional aluminium construction.

Another space application of continuous-fibre MMCs is the high gain antenna boom for the Hubble space telescope (Fig. 16.6). The boom, which is 3.6 m long, is a lightweight structure requiring high axial stiffness and low coefficient of thermal expansion to maintain the position of the antenna during space manoeuvres. The boom also provides a wave guide function and therefore needs good electrical conductivity to transmit signals between the spacecraft and antenna dish. The boom is made using 6061 aluminium reinforced with continuous carbon fibres. The material provides a 30% weight

16.6 MMC boom for the Hubble space telescope.

saving compared with previous designs based on monolithic aluminium or carbon–epoxy composite.

There are many potential applications for MMCs in aircraft gas turbine engines and scram jet engines owing to their low weight, high-temperature stability and excellent creep resistance. Titanium matrix composites may replace heavier nickel-based superalloys in high-pressure turbine blades and compressor discs for jet engines, although much more development work is needed before this is achieved.

The engine and structural applications for MMCs is limited owing to a number of technological problems that are difficult to resolve. MMCs are expensive materials to process into finished components because of high costs in manufacturing, shaping and machining. MMCs are difficult to plastically form using conventional plastic forming processes such as rolling or extrusion owing to their low ductility and high hardening rate. Relatively low levels of plastic forming can cause micro-cracking in the forged MMC component. MMCs are also difficult to machine by milling, routing, drilling and other material removal processes because of their high hardness, which causes rapid tool wear.

MMCs also have poor ductility and low toughness, which is a major concern for the aerospace applications where damage tolerance is a key design consideration for many structural and engine components. Lastly, aerospace engineers are not familiar with MMCs and the large database of technical information required for aircraft certification is lacking. Until the technical issues are resolved, the technology reaches maturity, and aircraft designers become familiar and confident with MMCs, it is likely that the applications for these materials will remain limited despite their high specific stiffness and strength.

16.2   Fibre–metal laminates

16.2.1 Introduction to fibre–metal laminates

Fibre–metal laminates (FMLs) are lightweight structural materials consisting of alternating thin layers of metal and fibre–polymer composite (Fig. 16.7). FML is made using thin sheets (0.2–0.4 mm) of lightweight metal, such as aluminium or titanium, bonded to thin layers of prepreg composite, with the outer surfaces being metal. The most common FML used in aerospace structures is GLARE® (glass-reinforced fibre–metal laminate). GLARE® consists of thin sheets of aluminium alloy bonded to thin layers of high strength glass–epoxy prepreg composite. GLARE® is used in the upper fuselage and leading edges of the vertical fin and horizontal stabilisers of the Airbus 380 aircraft. GLARE® is also used in the cargo doors of the C-17 Globemaster III. Another FML is ARALL® (aramid aluminium laminate)

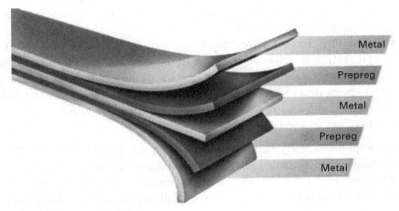

16.7 Fibre–metal laminate.

that is made of alternating sheets of aluminium alloy and aramid–epoxy. ARALL® has exceptional impact resistance and damage tolerance, but its use in aerospace structures is limited because of moisture ingress problems into the aramid–epoxy layers, which reduces material integrity.

16.2.2 Properties of fibre–metal laminates

FMLs have physical and mechanical properties that make them suitable for aerospace structural applications. These properties include:

- Low weight. FMLs are lighter than the equivalent monolithic metal owing to the lower density layers of prepreg composite. For example, GLARE® has a lower density (~2.0 g cm^{-3}) than monolithic aluminium alloy (2.7 g cm^{-3}), which results in a significant weight saving in the A380.
- Mechanical properties. FMLs have higher tensile strength, damage tolerance, impact strength and fatigue resistance than the monolithic metal. These are key reasons for the use of GLARE® in the A380 fuselage. Under flight-loading conditions, the rate of fatigue crack growth in GLARE® is 10–100 times slower than monolithic aluminium alloy because the cracks are stopped or deflected at the metal–composite interfaces.
- Tailored properties. It is possible to tailor the structural properties of FMLs by adjusting the number, type and alignment of the prepreg layers to suit local stresses and shapes through the aircraft.
- Corrosion resistance. Through-the-thickness corrosion is prevented in FML owing to the barrier role played by the composite layers. This limits corrosion damage to the surface metal layer and the internal metal sheets are protected by the composite. Therefore, the incidence of corrosion damage such as pitting and stress corrosion cracking, which are major

problems for monolithic aluminium aircraft structures (see chapter 21), is reduced with GLARE®.

- Fire. FML has high fire resistance owing to its low thermal conductivity. GLARE® has lower through-thickness thermal conductivity than monolithic aluminium which results in slower heat flow through the fuselage structure in the event of post-crash fire. FML has better resistance to burn-through in the event of fire and can potentially substitute for titanium in firewalls.

Despite the many benefits derived from using FMLs such as GLARE®, the future of these materials in aircraft structures is uncertain because of their high cost. FMLs are typically 7–10 times more expensive than the monolithic metal, and the high cost is a major impediment to their use in aircraft.

16.3   Ceramic matrix composites

16.3.1 Introduction to ceramic matrix composites

Ceramics are used mainly in aerospace for their outstanding thermal stability; which includes high melting temperature, low thermal conductivity, and high modulus, compressive strength and creep resistant properties at high temperature. There are many types of ceramics, with the most important for aerospace being polycrystalline materials and glass–ceramics. A characteristic of ceramic materials is high strength under compression loading, but low strength and toughness under tension. The compressive strength of most polycrystalline ceramics is in the range 4000–10 000 MPa, whereas the tensile strength is only 50–200 MPa. The fracture toughness of most ceramics is 10–50 times lower than high-strength aluminium alloy. Ceramics have low tolerance to tiny voids and cracks that occur during fabrication or in service, and these defects cause brittle fracture under low tensile loads. For this reason, the strength, ductility and toughness properties are too low for ceramics to be used in aerospace structures.

Ceramic matrix composites (CMCs) are used to increase the tensile strength and toughness of conventional ceramic material while retaining such properties as lightness, high stiffness, corrosion resistance, wear resistance and thermal stability. CMCs consist of a ceramic matrix phase reinforced with ceramic fibres or whiskers. The reinforcing fibres and whiskers greatly increase the breaking strength and toughness of ceramics, as shown in Table 16.1. Improvements in the strength and toughness of over 100% are gained when ceramics are reinforced with continuous fibres. Although the mechanical properties are improved by the reinforcement, the strength and toughness are still too low for their use in primary aerospace structures. However, the properties are sufficient for the use of CMCs in lightly loaded aerospace components requiring thermal stability.

Table 16.1 Effect of SiC reinforcement fibres on the bending strength and fracture toughness of selected ceramic materials

Material	Flexural strength (MPa)	Fracture toughness (MPa m$^{-1/2}$)
SiC	500	3
SiC/SiC	760	18
Al_2O_3	550	4
Al_2O_3/SiC	790	8
Si_3N_4	470	3
Si_3N_4/SiC	790	41
Glass	60	1
Glass/SiC	830	14

CMCs also have high resistance against thermal shock, which is the resistance to cracking and failure when rapidly heated and cooled (often in the presence of high pressure). Rapid cooling of the surface of a hot material is accompanied by surface tensile stresses. The surface contracts more than the interior, which is still relatively hot. As a result, the surface 'pushes' the interior into compression and is itself 'pulled' into tension. Surface tensile stress creates the potential for brittle fracture. The high thermal shock resistance of CMCs is the result of their high strength at high temperature combined with their low coefficient of thermal expansion. The excellent thermal shock resistance and thermal stability allows CMCs to be used at temperatures many hundreds of degrees higher than the melting point of metal alloys. Current metals-based technology can produce alloys stable to about 1000 °C, and nickel superalloys can be used at slightly higher temperatures when insulated with a thermal barrier coating and when cooling systems are included. CMCs can survive for long periods of time at temperatures well above the melting point of superalloys. CMCs retain their stiffness, strength and toughness properties to temperatures close to their melting temperature, which can exceed 3000 °C.

The CMC materials most often used in aerospace are silicon carbide–silicon carbide (SiC–SiC) composite, glass–ceramic composite, and carbon–carbon composite. All three types are used in aerospace thermal protection systems. SiC–SiC composites are only used in a few niche applications, such as convergent–divergent engine nozzles to fighter aircraft where the temperatures reach ~1400 °C. Several types of glass–ceramic composites are used in heat shields for re-entry spacecraft such as the space shuttle orbiter. The most important ceramic matrix composite is carbon–carbon, which is used in heat shields, rocket engines and aircraft brake pads.

16.3.2 Properties of carbon–carbon composite

Carbon–carbon composites consist of carbon fibres embedded in a carbon matrix. Continuous carbon fibres (rather than short whiskers) are used as the reinforcement to maximise strength and toughness. Similarly to polymer matrix laminates, the tailored arrangement of carbon fibres in selected orientations within a continuous carbon matrix is used to achieved the desired properties. The fibre architectures include unidirectional, bidirectional, three-dimensional orthogonal weaves, or multidirectional weaves and braids.

Carbon–carbon composites are expensive to fabricate and costly to machine into the final product shape. A major drawback of these materials is their high cost which makes them prohibitively expensive for many aerospace applications where otherwise they are well suited. The manufacturing process involves impregnating fibrous carbon fabric with an organic resin such as phenolic. The material is pyrolysed (i.e. heated in the absence of oxygen) at high temperature to convert the resin matrix to carbon. The material, which is soft and porous, is impregnated with more resin and pyrolysed several more times until it forms a compact, stiff and strong composite.

Carbon–carbon composites have many thermal and mechanical properties required by aerospace materials that must operate at high temperature. These properties include stable mechanical properties to temperatures approaching 3000 °C; high ratios of stiffness-to-weight and strength-to-weight; low thermal expansion; and good resistance to thermal shock, corrosion and creep. The material properties of carbon–carbon composites, as for fibre–polymer composites, are anisotropic. The stiffness, strength and thermal conductivity is highest along the fibre axis and lowest normal to the fibre direction. The properties also depend on the fibre fraction, type of carbon fibre, fibre architecture and processing cycle.

Table 16.2 shows the improvement in properties achieved by reinforcing polycrystalline carbon with continuous carbon fibres. The carbon fibres improve the properties of monolithic carbon by ten times or more, with no increase in weight. A unique feature of carbon–carbon composite is that its strength can increase with temperature. Figure 16.8 shows the strength properties of several carbon–carbon composites and other aerospace materials over a wide temperature range. The strength of advanced types of carbon–carbon increase with the temperature up to at least 2000 °C, which is an obvious benefit when used at high temperature. The improvement in strength is caused by

Table 16.2 Mechanical properties of monolithic carbon (polycrystalline) and reinforced carbon–carbon composite

Property	Polycrystalline (monolithic) carbon	Reinforced carbon–carbon composite
Elastic modulus (GPa)	10–15	40–100
Tensile strength (MPa)	40–60	200–350
Compressive strength (MPa)	110–200	150–200
Fracture toughness (MPa m$^{-1/2}$)	0.07–0.09	5–10

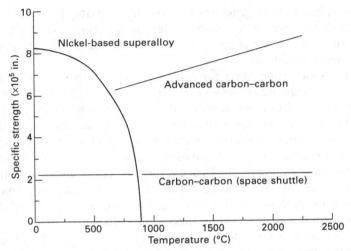

16.8 Dependence of specific strength on temperature for carbon–carbon composites and nickel superalloy (adapted from D. R. Askeland, *The science and engineering of materials,* Stanley Thornes (Publishers) Ltd., 1996).

the closing of microcracks in the interface between the carbon matrix and the fibre matrix when the temperature is raised.

Carbon–carbon composite suffers severe high-temperature oxidation which causes rapid degradation and erosion. Carbon becomes susceptible to oxidation when heated above 350 °C and the oxidation rate rises rapidly with temperature. As a result, there is carbon–carbon breakdown in the presence of air. It is necessary to protect carbon–carbon composites using an oxidation-resistant surface coating. The carbon–carbon tiles on the space shuttle orbiter, for example, are coated with silicon carbide to provide oxidation protection to about 1200 °C.

16.3.3 Aerospace applications of carbon–carbon composites

Heat shields of re-entry space vehicles such as the space shuttle orbiter are made using carbon–carbon composite. This material is used in the nose cone and leading edges of the orbiter where the temperature reaches 1600 °C during re-entry. These materials are also used in the nose cones of intercontinental ballistic missiles. Carbon–carbon is also used in rocket engines and nozzles for its high thermal stability and thermal shock resistance. During blast-off, the thrusters generate temperatures ranging from 1500–3500 °C at heating rates of 1000–5000 °C s⁻¹ and produce combustion pressures approaching

300 atmospheres. This generates extreme thermal shock, a rapid rise in temperature and pressure, which shatters most materials. The excellent thermal shock resistance of carbon–carbon composite ensures their survival in these harsh conditions.

Another space application for carbon–carbon composites is in thermal doublers, which are used to remove heat from a spacecraft (usually generated by internal electronics equipment) and then radiate that heat into space. These composites have the thermal properties and low weight required for thermal doublers on satellites. It is also possible to use carbon–carbon composite in aircraft heat exchangers, which are used to cool hot gases and liquids such as hydraulic and engine fluids. Heat exchangers require materials with high thermal conductivity, corrosion resistance, stiffness, strength, low permeability and high temperature stability; carbon–carbon composite is one of the few materials with this unique combination of properties.

Carbon–carbon composite is used in aircraft brake pads because of its low weight, high wear resistance and high-temperature properties. The friction between aircraft brake discs can generate average temperatures of around 1500 °C, with transient hot-spot temperatures of up to 3000 °C. The disc material must therefore have high temperature strength, thermal shock resistance and high thermal conductivity to rapidly remove heat from the contact zone between the disc surfaces. Carbon–carbon is one of the few materials with this combination of properties required for brakes discs. The weight of aircraft brakes is also significant. A large civil airliner usually has eight sets of brakes with a combined weight of over 1000 kg when made using heat-resistant steel. An equivalent set of brakes made of carbon–carbon weigh less than 700 kg. Carbon–carbon brakes are used in most fighter aircraft and increasingly in civil airliners. The Boeing 767 and 777 airliners and several Airbus aircraft are equipped with carbon–carbon brake discs.

Although not currently used, CMCs such as carbon–carbon have potential applications in jet engines because of their ability to operate uncooled at temperatures beyond the reach of metals. Cycle efficiency improvements, from reducing cooling air to turbine aerofoils and seals, lead to significant fuel consumption benefits. For example, it has been estimated that the replacement of metal seals by advanced CMCs technology would yield a saving of $250 000 on the Boeing 777 over 15 years.

Section 16.7 at the end of the chapter presents a case study of the use of ceramic matrix composites in the space shuttle orbiter.

16.4   Summary

Metal matrix composites offer several improved properties compared with the monolithic base metal, including higher stiffness, strength, wear resistance and, in some materials, fatigue performance and reduced weight. However, there are also drawbacks with MMCs, such as higher manufacturing and machining costs together with lower toughness and ductility.

The aerospace applications of MMCs are limited because of their high cost and low toughness. MMCs have been used in military fighter aircraft and helicopter rotor blades to reduce fatigue problems, fuselage struts in the space shuttle orbiter, and selected components on the Hubble space telescope.

Fibre–metal laminates are lightweight structural materials consisting of thin sheets of metal and fibre–polymer composite. The most popular fibre–metal laminate used is GLARE® that consists of alternating layers of high-strength aluminium alloy and glass fibre–epoxy. The future of fibre–metal laminates is uncertain because of their high manufacturing cost.

GLARE® is used in the A380 fuselage because of its lower weight, higher static and impact strengths, better damage tolerance, longer fatigue life and superior corrosion resistance compared with monolithic aluminium alloy.

Ceramics are intrinsically brittle and lack high tensile strength and toughness. Ceramics used in aerospace are reinforced with ceramic fibres or whiskers for greater toughness and strength.

Ceramic matrix composites are used for high-temperature applications. The materials are characterised by low weight, high stiffness and compressive strength, and excellent thermal stability and thermal shock resistance.

Carbon–carbon composite is the most common aerospace ceramic material, and is used in heat shields for re-entry spacecraft, liners and nozzles for rocket engines, and brake pads for aircraft.

16.5   Terminology

ARALL®. Fibre–metal laminate consisting of several layers of aluminium alloy interspersed with layers of aramid fibre prepreg.

Ceramic matrix composite. A composite material with at least two constituent parts, one being a ceramic which forms the matrix phase.

Fibre–metal laminate. A class of metallic material consisting of a laminate of thin metal layers bonded with layers of fibre–polymer composite material.

GLARE®. Fibre–metal laminate consisting of several layers of aluminium alloy interspersed with layers of fibreglass prepreg.

Metal matrix composite. A composite material with at least two constituent parts, one being a metal which forms the matrix phase.

Thermal shock. Cracking of materials (usually brittle solids such as ceramics) caused by a sudden drop or rise in temperature. Thermal shock occurs when a sharp thermal gradient causes different parts of an object to expand by different amounts, thus generating a stress gradient. At some point, the thermal stress overcomes the strength of the material, causing a crack to form.

16.6   Further reading and research

Chawla, K. K., *Ceramic matrix composites (2nd edition)*, Kluwer Academic Publishers, Norwell, MA, 2003.

Chawla, K. K., *Composite materials: science and technology*, Springer–Verlag, New York, 1998.

Clyne, T. W. and Withers, P. J., *An introduction to metal matrix composites*, Cambridge University Press, Cambridge, 1993.

Dogan, C. P., 'Properties and performance of ceramic–matrix and carbon–carbon composites', in *ASM handbook volume 21: composites*, D. B. Miracle and S. L. Donaldson, ASM International, Material Park, OH, 2001, pp. 859–868.

Kearns, K. M., 'Applications of carbon–carbon composites', in *ASM handbook volume 21: composites*, D. B. Miracle and S. L. Donaldson, ASM International, Material Park, OH, 2001, pp. 1067–1070.

Miracle, D. B., 'Aeronautical applications of metal matrix composites', in *ASM handbook volume 21: composites*, D. B. Miracle and S. L. Donaldson, ASM International, Material Park, OH, 2001, pp. 1043–1049.

Savage, G., *Carbon-carbon composites*, Chapman and Hall, London, 1993.

Volelesang, L. B. and Vlot, A., 'Development of fibre metal laminates for advanced aerospace structures', *Journal of Materials Processing Technology*, **103** (2000), 1–5.

Vlot, A. and Gunnick, J. W. (eds), *Fibre metal laminates: an introduction*, Kluwer Academic Publishers, Dordrecht, The Netherlands, 2001.

Wu, G. and Yang, J.-M., 'The mechanical behavior of GLARE laminates for aircraft structures', *Journal of Materials*, **57** (2005), 72–79.

16.7   Case study: ceramic matrix composites in the space shuttle orbiter

The space shuttle orbiter is the world's first, and so far only, reusable spacecraft. (The Russian built Buran was developed in the 1980s as a reuseable vehicle similar to the space shuttle which performed just one unmanned space mission). The construction of the orbiter is similar to a conventional airliner; the body is built mostly of high-strength aluminium alloys and the payload doors are fibre–polymer sandwich composite material. Although speciality materials are used in highly stressed components of the mainframe, such as titanium and metal matrix composite, most of the structure is built with the same aluminium alloys used in civil and military aircraft. Chapter 3 provides more information on the structural materials used in the orbiter.

The orbiter is covered with ceramic tiles that protect the vehicle, payload and crew from the extreme heat generated during re-entry into the Earth's atmosphere. Figure 16.9 shows the temperature profile over the orbiter surface during re-entry; the nose and leading edges are heated to 1000–1400 °C whereas the maximum temperatures of other sections are 200–1000 °C. The properties of aluminium and polymer matrix composite demand that the orbiter's structure is kept below 150–200 °C, and therefore the heat insulation tiles are essential.

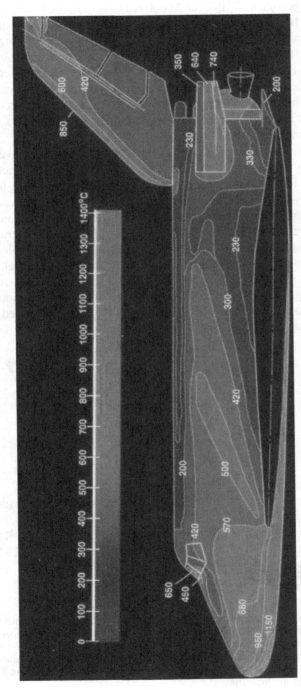

16.9 Temperature surface profile of the space shuttle orbiter during re-entry.

The orbiter is covered with over 25 000 reusable ceramic matrix composite tiles (Fig. 16.10). All the tiles are brittle and can crack when stressed or impacted, as tragically proven when the *Columbia* broke up during re-entry on flight STS-107 (February 2003). Chapter 18 describes the impact fracture of the carbon–carbon tiles that caused the *Columbia* incident. Because the aluminium structure of the orbiter expands and contracts owing to temperature changes over the course of a flight mission, the tiles are not mounted directly onto the skin. Instead, compliant adhesive felt pads are used to bond the tiles to the aluminium.

The forward nose cap and leading edges of the wings are covered with carbon–carbon composite tiles. These tiles are coated with black silicon carbide for oxidation resistance, with the black colour helping to radiate heat during re-entry. Carbon–carbon tiles are used in the hottest regions where the temperature exceeds about 800 °C. The cooler regions, where the temperature is 200–800 °C are covered mostly with white ceramic tiles that reflect solar radiation to keep the space shuttle cool. The tiles consist of porous high-purity silica ceramic covered with borosilicate glass. Two other types of silica-based tiles are used on the orbiter: fibrous refractory composite insulation (FRCI) or toughened unipiece fibrous insulation (TUFI). FRCI is used in a few selected regions whereas the TUFI is applied over the extreme back of the orbiter near the engines.

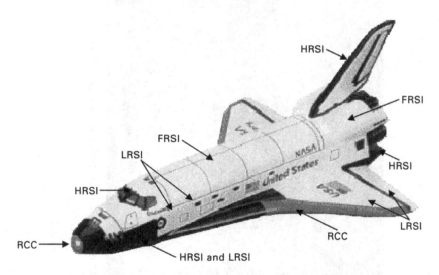

16.10 Heat insulation tiles on the space shuttle orbiter where the black regions are carbon–carbon composite and white regions are mostly high-purity silica ceramic. RCC is reinforced carbon–carbon composite; HRSI is high-temperature reusable surface insulation; LRSI is low-temperature reusable surface insulation; FRSI is fibrous reusable surface insulation.

17
Wood in small aircraft construction

17.1   Introduction

The first structural material used in powered aircraft was wood. The mainframe of *Kitty Hawk* flown by the Wright brothers in 1903 was made from timber covered with fabric. Wood was the material of choice in early aircraft because of its good strength-to-weight ratio. Also, wood can be easily crafted and shaped into spars and beams for the fuselage, wings and other structures. Wood was the most used structural material in aircraft until the introduction of high-strength aluminium and steel in the 1920s. The use of wood in large modern aircraft is now nonexistent, although it remains a useful material in the construction of small aircraft; notably guilders, ultralights, aerobatic aircraft and certain types of piston-driven aircraft (Fig. 17.1). Wood is used in spars, ribs, longerons and stringers of the mainframe of several types of small aircraft.

In this chapter, we examine the properties of wood that are important to its use in small aircraft. The structure, composition and properties of wood are studied. Important factors concerning the structural performance and durability of wood are also examined; including mechanical anisotropy, moisture absorption, and its effect on mechanical performance and durability in the aviation environment.

17.1 Application of wood in aircraft: glider made of spruce and ply (photograph supplied courtesy of C. White).

411

© Woodhead Publishing Limited, 2012

17.2   Advantages and disadvantages of wood

The properties of wood that appeal to the designers and builders of small aircraft are lightness, stiffness, strength and toughness. Table 17.1 gives the properties of two woods used in aircraft, Sitka spruce and Douglas fir, and of several aerospace grade metal alloys and carbon fibre–epoxy composite. The density of wood is very low (typically under 0.5 g cm^{-3}) compared with the other types of aerospace materials, although the Young's modulus, tensile strength and fracture toughness are also lower. Because of the low density, the specific mechanical properties of wood are similar to 2024 aluminium alloy and most magnesium alloys. However, the specific properties of wood are inferior to the other types of aerospace materials.

There are many properties of wood that make it inferior to metals and composites as an aircraft structural material. As mentioned, the stiffness, strength and toughness of wood are lower. Other drawbacks of wood are:

- Lack of uniformity in properties such as density and strength. The properties of timber pieces taken from the same species of tree may differ by 100% or more. It is therefore necessary to apply high design safety factors to wooden structures, which diminishes the potential weight savings otherwise gained by using such a low density material.
- Mechanical properties of wood are highly anisotropic. Large differences occur in the stiffness, strength and toughness in the longitudinal and transverse directions of cut timber.
- Defects in wood, such as knots and pitch pockets, reduce the strength properties.
- Wood is hygroscopic; it shrinks, swells and undergoes changes to the mechanical properties depending on the atmospheric humidity.
- Wood is prone to attack from insects, fungus and other microorganisms when used without effective chemical treatment and surface protection.

17.3   Hardwoods and softwoods

Timber is classified as hardwood or softwood depending on whether it is harvested from a deciduous or evergreen tree. Hardwoods are deciduous trees such as oak, elm and birch, whereas softwoods are evergreens such as spruce, pine and fir. As their name implies, hardwoods are generally harder and stronger than softwoods. For example, the bending strength of most hardwoods is within the range 45–80 MPa whereas softwoods are usually within 30–60 MPa. Hardwoods are generally heavier than softwoods, with their density in the range 0.5–0.8 g cm^{-3} whereas the density of most softwoods is within 0.30–0.6 g cm^{-3}. Certain softwoods have a good combination of density and strength to be suitable for aircraft construction, the most common are Sitka spruce and Douglas fir.

Table 17.1 Comparison of mechanical properties of Sitka spruce and Douglas fir with other aerospace structural materials

Materials	Specific gravity* (g cm^{-3})	Young's modulus* (GPa)	Specific modulus* x10^6 (N m kg^{-1})	Yield strength* (MPa)	Specific strength* x10^3 (N m kg^{-1})	Fracture toughness[†] (MPa m$^{0.5}$)	Specific fracture toughness[†] x10^3 (N m$^{1.5}$ kg^{-1})
Sitka spruce*	0.4	9	22.5	55	138	4	10.0
Douglas fir*	0.5	12	24.0	70	140	6	12.4
Ti alloy (Ti-6Al-4V)	4.6	108	23.5	1003	217	100	21.7
Al alloy (2024 Al-T6)	2.7	70	25.9	385	142	37	13.7
Al alloy (7075 Al-T76)	2.7	70	25.9	470	174	29	10.7
Mg alloy (WE46-T6)	1.7	45	26.5	200	117	20	11.8
Carbon/epoxy composite[‡]	1.7	50	29.4	760	450	38	22.4

*Wood properties for 15% moisture content measured parallel to grain. [†]Wood properties for 15% moisture content measured transverse to grain. [‡][0/±45/90] carbon/epoxy; fibre volume content = 60%.

17.4   Structure and composition of wood

The properties of wood that make it a useful aircraft material (i.e. lightness, strength and toughness), are controlled by its structure and composition. For this reason, it is useful to understand the structure and composition of wood at several levels: the macro level (above ~1 cm in size), microstructural level (~0.1 to 1 cm) and cellular level (below ~0.1 cm).

17.4.1 Macrostructure of wood

The macrostructure of wood consists of several distinct zones that occur across a tree trunk, as shown in Fig. 17.2. The outer layer, or bark, provides some protection for the inside of a tree against birds, insects, fungi and other organisms. Just beneath the bark is a zone called the cambium that contains new growing cells for both the bark and inner region of the tree. During the warmer months of each year, the cambium grows new wood cells on its inner surface and new bark cells on its outer surface. The sapwood is a region adjoining the cambium that contains cells where nutrients are stored and sap (containing water and minerals) is transported along the trunk and into the

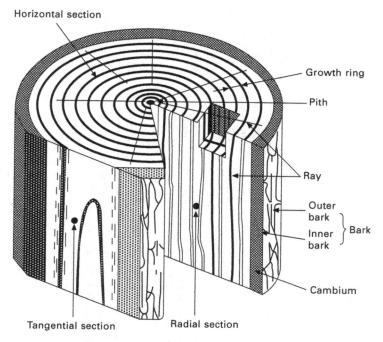

17.2 Cross-section of a tree showing the different regions across the trunk.

branches and twigs. The inner core of the tree is called the heartwood, and is composed entirely of dead cells. The heartwood is denser and stronger than the cambium and sapwood, and provides most of the mechanical support for a tree. Only wood taken from the heartwood of a mature, fully grown tree is used for aircraft construction. Even then, not all the heartwood is suitable because of knots and other defects. Usually less than 5% of timber taken from the heartwood is suitable for aircraft construction. This is another problem with using wood: the large amount of waste material generated for a small amount of quality timber.

Wood is a highly anisotropic material. Three axes are used to describe the directionality of cut timber, as shown in Fig. 17.3. The axes are mutually perpendicular to each other. The longitudinal axis (L) is parallel with the axis of the trunk, and is also known as the fibre or grain direction. The radial axis (R) is across the diameter of the trunk and the tangential axis (T) is normal to the radial direction. The mechanical properties are different along each axis, and therefore it is important that timber used in aircraft construction is cut in the direction that maximises structural performance. For example, stiffness and strength are highest when timber is loaded parallel with the longitudinal axis and lowest in the radial direction for reasons that are explained in 17.5.2. Wood is toughest when cracking occurs in the radial direction and is more susceptible to splitting in the other directions. Care is needed when constructing structures to ensure the wood grain is aligned with the major loads acting on the aircraft.

17.4.2 Microstructure of wood

The microstructure of heartwood consists of long and thin hollow cells that are extended in the longitudinal direction. Figure 17.4 shows the microstructure of

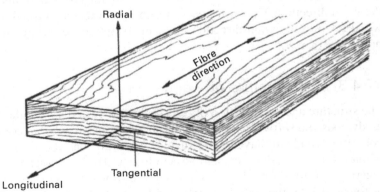

17.3 Axes used to specify directions in cut timber (reproduced with permission from the US Department of Agriculture, Handbook No. 72).

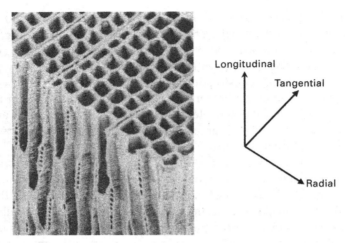

17.4 Wood microstucture showing the cellular network (photograph supplied courtesy of US Forest Products Laboratory, Madison, Wisconsin).

Douglas fir viewed under a microscope at the magnification of 300 times. The microstructure consists of long, hollow cells squeezed together like drinking straws. These cells are called fibres or fibrils because of their very high aspect ratio (length-to-width) of 100 or more. Fibre cells are less than 0.01 mm wide and roughly 1 mm long in hardwoods and 5 mm long in softwoods. The fibres provide wood with most of its stiffness, strength and toughness.

The microstructure of wood is often considered in the materials engineering context as a unidirectional fibre–polymer composite. Wood consists of long fibre cells aligned in the same direction, which is analogous to the fibres in a composite, which are bonded together in an organic matrix. Most of the wood fibres are roughly parallel with the longitudinal axis, and this is called the grain direction. The term grain describes both the cell direction and the wood texture; the grain direction has a major influence on the mechanical properties of wood.

17.4.3 Cell structure of wood

The structure and composition of cells in the heartwood are similar for most hardwoods and softwoods. Figure 17.5 shows a typical cell that consists of a hollow core and multilayer wall structure. The wall consists of several layers of microfibrils, which are crystalline cellulose $(C_6H_{10}O_5)_n$ molecules clumped together into long strands. Each molecular chain consists of long segments of linear, crystalline cellulose separated by shorter segments of amorphous cellulose. Cellulose is a linear molecule with no cumbersome sidegroups, and so it crystallises easily into strands of great stiffness and strength. The

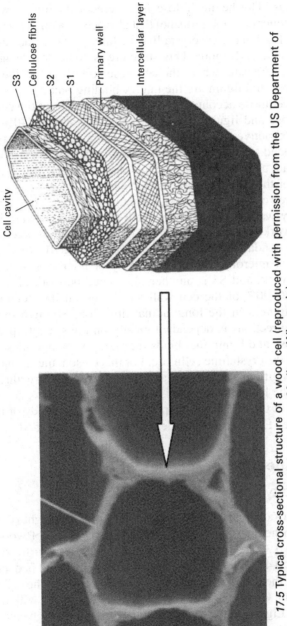

17.5 Typical cross-sectional structure of a wood cell (reproduced with permission from the US Department of Agriculture, Forest Products Laboratory, Madison, Wisconsin).

Cell cavity

S3
Cellulose fibrils
S2
S1
Primary wall
Intercellular layer

crystalline cellulose strands are encased in a layer of hemicellulose. This is a semicrystalline polymer of glucose that acts as a binding agent to the cellulose strands. The hemicellulose is covered with lignin, which is an amorphous polymer (phenol–propane) that also acts as a binder. A microfibril then consists of a bundle of crystalline cellulose chains that are encased with hemicellulose and lignin. This structure is somewhat anagolous to a fibre–polymer composite, where the cellulose strands act as the fibres and the hemicellulose and lignin are the matrix binding phase.

Crystalline cellulose accounts for 40–50% by weight of the cell wall and the hemicellulose and lignin together account for another 40%. The wall also contains extractives, which are organic chemicals and oils that are not chemically bonded into the microfibril structure. Extractives give colour to the wood and act as preservatives against the environment and insects. As much as 10% of the wood may be extractives.

Cell walls are composed of two substructures: the outer primary wall and inner secondary wall. In the primary wall, the microfibrils are in a loose, disordered network. They are not aligned in any particular direction, but are random. The secondary wall consists of three layers: S1 which is a criss-cross network of microfibrils, S2 is an array of microfibrils in a parallel, spiral-type network, and S3 is another disordered network of microfibrils. S2 makes up 70–90% of the cell wall and is responsible for most of the stiffness and strength in the longitudinal direction. Strength in the radial and tangential directions is dependent mostly on the strength properties of the hemicellulose and lignin that binds the wall layers, and these are much weaker than for the crystalline cellulose. For this reason, the strength is much higher in the longitudinal direction, and aircraft wooden structures must be designed with the loads acting in this direction.

Section 17.9 at the end of the chapter presents a case study of the *Spruce Goose*, a Hughes H-4 Hercules made almost entirely of wood.

17.5   Engineering properties of wood

17.5.1 Density

The density of woods vary over a wide range from ultralight (e.g. balsa at ~0.1 g cm^{-3}) to heavy (e.g. oak at ~0.7 g cm^{-3}). Most softwoods have a density between 0.30 and 0.50 g cm^{-3}, and are lighter than hardwoods that are mostly between 0.45 and 0.80 g cm^{-3}. The density of wood is dependent on two parameters: the cell density and the moisture content. The cell density is determined by the volume of the inner cell cavity compared with the volume of the cell wall. Lightweight woods such as balsa have a high cavity-to-wall volume ratio whereas heavy woods such as oak have a smaller ratio.

The density is also controlled by the moisture content. Water is present in the cell cavities and cell walls. Green wood (freshly cut timber) can contain up to 50% water and is not suitable for aircraft because it is too heavy, soft and prone to attack from fungi and parasites. Wood must be dried by seasoning, which takes anywhere between 2 and 10 years at room temperature. More often wood is seasoned by kiln drying at elevated temperature, which takes a few days. When wood dries, the density is reduced because moisture is driven out by evaporation from the cell cavities and, to a lesser extent, cell walls. The drying of wood increases the longitudinal stiffness and strength because the cellulose fibrils in the cell walls pack more closely together into space that was occupied by water molecules.

The target moisture content when drying wood for use in aircraft is about 12%, at which point the density, mechanical properties, durability and dimensional stability have reached a near optimum level. However, the actual moisture content may be anywhere between 10 and 15% depending on the temperature and humidity of the environment.

The density of seasoned timber is usually classified (at 12% moisture content) as being:

- Very light, under 0.3 g cm^{-3}
- Light, 0.3 to 0.45 g cm^{-3}
- Medium, 0.45 to 0.65 g cm^{-3}
- Heavy, 0.65 to 0.80 g cm^{-3}
- Very heavy, 0.80 g cm^{-3} and above.

The requirement for aircraft materials to have low density combined with high strength means that most woods are at the lower band of the medium density classification. This includes the softwoods used in aircraft, such as Sitka spruce (average density of 0.4 g cm^{-3}) and Douglas fir (0.5 g cm^{-3}). The exception is wood used in aircraft propellers which is often heavier because of the requirement for higher stiffness and strength than the airframe.

A single density value for a particular type of wood is often quoted in research studies and aircraft design manuals, although there is much scatter in the density for the same species of tree. For example, Fig. 17.6 shows the variation in density for Sitka spruce used in aircraft. The density varies because it is influenced by the maturity, environment and growth conditions of the tree as well as the part of the trunk from which the timber is cut. For this reason, the density of individual timber pieces should be measured before they are built into an aircraft to ensure the total structural weight is within the design limit.

17.5.2 Mechanical properties

The mechanical properties of wood are highly anisotropic, and therefore care is needed when constructing timber aircraft structures. The longitudinal

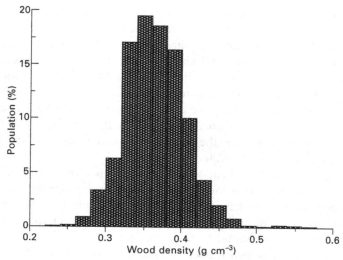

17.6 Histogram plot of density values for Sitka spruce.

modulus and yield strength are much higher because the applied load acts parallel to the microfibrils in the cell wall. The crystalline cellulose chains in the microfibrils are stiff and strong and therefore able to carry a relatively high load. The cell walls collapse by microbuckling and kinking in much the same way as fibre–polymer composites fail in compression. The transverse properties are determined mainly by the hemicellulose and lignin that binds the microfibrils, which are much weaker than crystalline cellulose. The properties are also lower because of the voids and lack of continuity to the cell structure in the transverse direction.

Table 17.2 gives elastic modulus, compressive strength and fracture toughness for a variety of woods measured in the longitudinal and radial directions. The longitudinal modulus is typically 10–20 times higher than the transverse modulus. The longitudinal strength properties are 5–10 times higher than in the radial direction. The fracture toughness is much higher when crack growth occurs in the radial direction. The fracture toughness is anywhere between 5 and 25 times higher across the grain than along it. This is because of two important toughening mechanisms. Firstly, when a crack grows across the grains it is forced to spread, at right angles to the break, along the fibril interfaces, leading to a large fracture surface. After cracking along the interfaces, the fibrils are then pulled out, which requires a high force and strain. Toughness in the grain direction is lower because the crack grows through the brittle lignin phase at cell interfaces without significant branching or fibril pull-out. It is for this reason that wood is easier to split along the trunk than cut across it using an axe.

Table 17.2 Engineering properties of woods

Wood	Density (g cm^{-3})	Young's modulus (GPa)*		Compression strength (MPa)[†]		Fracture toughness (MPa m$^{0.5}$)[‡]	
		Long.	Radial	Long.	Radial	Long.	Radial
Balsa	0.20	4	0.2	12	0.5	0.05	1.2
Mahogany	0.53	13.5	0.8	34	12.1	0.25	6.3
Douglas fir	0.55	16.4	1.1	39	9.0	0.34	6.2
Pine	0.55	16.3	0.8	36	7.4	0.35	6.1
Birch	0.62	16.3	0.9	39	8.3	0.56	–
Ash	0.67	15.8	1.1	38	11.0	0.61	9.0
Oak	0.69	16.6	1.0	39	15.0	0.51	4.0
Beech	0.75	16.7	1.5	39	12.5	0.95	8.9

*Dynamic moduli; moduli in static tests are about two-thirds. Data from M.F. Ashby & D.R.H. Jones, *Engineering Materials 2*, Elsevier, Oxford 2006. [†]Data from L.J. Markwardt, 'Aircraft woods: Their properties, selection and characteristics', *U.S. Forest Products Laboratory Report*, No. 354. [‡]Data from M.F. Ashby & D.R.H. Jones, *Engineering materials 2*, Elsevier, Oxford 2006.

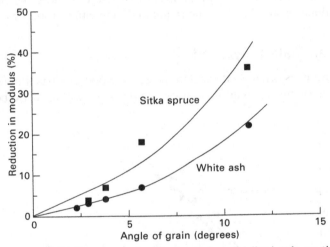

17.7 Effect of grain angle relative to longitudinal axis on the percentage reduction in elastic modulus of Sitka spruce and white ash.

Timber used in aircraft must be cut with their grains parallel to the surface to ensure maximum mechanical performance. Large reductions in the properties are experienced when the grain angle is more than a few degrees from the load direction. Figure 17.7 shows the effect of grain angle on the percentage reduction in the elastic modulus of Sitka spruce and white ash. The modulus decreases rapidly with increasing grain angle, and similar losses are experienced with strength. For this reason, the grains along the longitudinal axis of timber used in aircraft must be reasonably straight. It is recommended that the maximum grain slope should not deviate from the longitudinal axis by more than 1:16, which is equivalent to a grain angle of 3°. The strength properties of wood are seriously degraded by defects which disrupt the grain structure, such as knots and pitch pockets. Only high quality timber that is free of defects should be used in aircraft construction. It is for this reason that only a small amount (usually under 5%) of the timber taken from a tree trunk is suitable for use in aircraft.

In addition to grain direction, many of the key engineering properties considered in the selection of wood for airframes are dependent on the density. Properties such as elastic modulus, strength, fracture toughness and impact resistance increase with the density. For example, Fig. 17.8 shows that the compressive strength of woods measured in the longitudinal and radial directions increase with density. The longitudinal modulus E_L and strength σ_L increase rectilinearly with their density; i.e. $E_L \propto \rho$ and $\sigma_L \propto \rho$. However, the transverse modulus increases as a cubic function of density $(E_T \propto \rho^3)$ and the transverse strength rises as a squared function of density $(\sigma_T \propto \rho^2)$. Therefore, using denser wood in an aircraft structure is more beneficial to the longitudinal properties than the radial and tangential properties.

17.5.3 Laminated plywood

Wood in aircraft is often used as a laminated plywood construction to avoid the problems of mechanical anisotropy. Plywood is produced by bonding

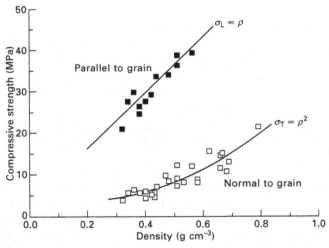

17.8 Comparison of the compressive strengths of woods with different densities measured in the longitudinal and transverse directions.

thin sheets (0.5–1.0 mm thick) of timber together so that the grains of each sheet run perpendicular to the adjacent sheets. In this way the material is nearly isotropic along the plane of a timber structure. Plywood is also more resistant to splitting, cracking and warping than regular wood. Aircraft plywood has traditionally been made from spruce, birch, mahogany or gabbon. High-strength and durable adhesives must be used for bonding the timber sheets; with epoxy, casein and resorcinol formaldehyde being used for aircraft plywood construction.

17.5.4 Environmental problems with wood

The mechanical properties of wood are susceptible to the operating environment of the aircraft. Firstly and most importantly, wood is hygroscopic causing the moisture content to change with the humidity of the atmosphere. This means the density and mechanical properties change with the humidity level. In dry conditions, the water content of seasoned timber can drop to as low as 8%. In hot and wet tropical environments, the moisture content can reach 18–20%, causing density to rise and mechanical performance to fall. Timber also swells and shrinks with changes in moisture content, which can lead to loosening of bolted connections, particularly propeller flange bolts, and warping of the airframe under extreme conditions. Excessive moisture absorption can reduce the mechanical properties, in some cases by more than 20%. Figure 17.9 shows the effect of water content on the longitudinal bending modulus and strength of Sitka spruce. The properties drop rapidly

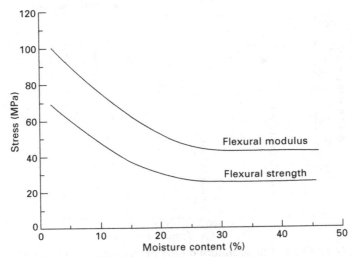

17.9 Effect of moisture content on the modulus and strength of Sitka spruce.

with rising moisture content until it rises above 25% where they remain low and constant. Similar behaviour is observed for the longitudinal tensile and compressive properties. Therefore every effort should be used to seal and protect timber aircraft against water and humidity.

Another problem is that the strength of wood is inversely proportional to the temperature. Figure 17.10 shows the percentage reduction in the compressive strength of balsa with increasing temperature. Although wooden aircraft do not experience extremely high temperatures, they are heated above 50 °C when operating in hot environments. Surface temperatures as high as 70 °C have been measured on aircraft based at airports in hot, arid regions. Softening of the hemicellulose in the microfibrils begins at about 50 °C, weakening wooden structures. The aircraft industry uses a rule-of-thumb for estimating softening caused by hot weather: the tensile stiffness and strength decrease by about 1% for every 1 °C increase in the wood temperature.

17.6   Summary

The stiffness, strength and other mechanical properties of wood are inferior to metallic and composite materials. Therefore, wood is only suitable for use in aircraft structures required to carry relatively low loads.

The mechanical properties of wood are highly anisotropic, with the stiffness and strength being highest in the grain direction and lowest across

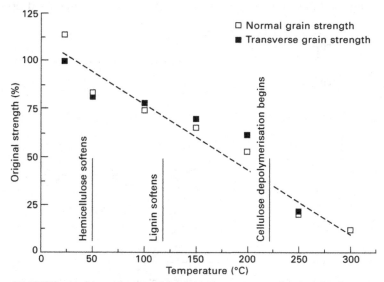

17.10 Effect of temperature on the percentage reduction in the compressive strength of balsa in the longitudinal and transverse directions.

the grain. This must be considered in the design of wooden aircraft structures by ensuring that major loads are applied along the grain. It is recommended that the grain angle is within 3° of the load direction to ensure maximum structural performance.

The mechanical properties and weight of timber aircraft structures increase with the wood density. When selecting the wood density, a trade-off between weight and mechanical performance is required. In most cases, the optimum trade-off is achieved using wood with an average density of 0.4–0.5 g cm^{-3}.

Only timber from the heartwood of a mature tree should be used in aircraft construction. The timber must be high quality with straight grains and free of defects such as knots.

Wooden aircraft components must be sealed and protected against moisture. Because wood is hygroscopic, the weight rises and mechanical properties drop when moisture is absorbed. For this reason, wood must be sealed with a water-resistant coating.

17.7   Terminology

Cambium: A layer which is one or two cells thick where new wood is formed in trees.

Extractives: Compounds that give colour and sometimes provide decay and insect resistance to wood.

Fibre (wood): A single grain of wood composed of microfibrils aligned in approximately the same direction.

Grain (wood): The direction, arrangement or appearance of the fibres in wood.

Hardwood: General term used to describe timber produced from deciduous broad-leafed trees.

Heartwood: The older, harder, nonliving central wood of a tree that serves the sole function of mechanical support.

Hemicellulose: A carbohydrate (polysaccharide) that is attached to the microfibrils of the cell wall in wood. Lignin and hemicellulose bond crystalline cellulose into microfibrils.

Lignin: A chemically complex substance found in trees that bonds the microfibrils of the cell wall in wood.

Microfibrils: Small fibres found in the cell walls of wood which contain crystalline cellulose strands bonded by hemicellulose and lignin.

Sapwood: The area of a tree trunk which contains some living cells that conduct water.

Softwood: General term used to describe timber produced from needle and/or cone bearing trees.

17.8   Further reading and research

Ashby, M. F., Easterling, K. E., Harrysson, R. and Maiti, S. K., 'The fracture and toughness of woods', *Proceedings of the Royal Society of London*, **A398** (1985), 261–280.

Easterling, K. E., Harrysson, R., Gibson, L. J. and Ashby, M. F., 'On the mechanics of balsa and other woods', *Proceedings of the Royal Society of London*, **A383** (1982), 31–41.

Faherty, K. F. and Williamson, T. G., *Wood engineering and construction handbook (3rd edition)*, McGraw–Hill, 1997.

17.9   Case study: *Spruce Goose* (Hughes H-4 Hercules)

The *Spruce Goose* is the largest aircraft ever built and it is made almost entirely of wood (Fig. 17.11). The idea for the *Spruce Goose* was conceived in the United States during World War II as a means of transporting a large number of troops and defence material over long distances. During the war, the Allied Forces were suffering huge losses from the sinking of troop and cargo ships by enemy submarines. To avoid these losses, the concept of a large flying boat (originally called the Hughes HK-1) was developed to carry troops, tanks and other military equipment. The aircraft was designed to transport up to 750 troops or two Sherman-class tanks at a cruising speed of 320 km h^{-1} for a maximum range of 4800 km. The government of the USA requested that the aircraft was built using wood because of the scarcity of aluminium, which was more urgently needed for building fighters and bombers.

The aircraft was designed and constructed by the eccentric aviator and aircraft manufacturer Howard Hughes and his company. When the design of the aircraft was complete the aircraft was called the Hughes H-4. The project

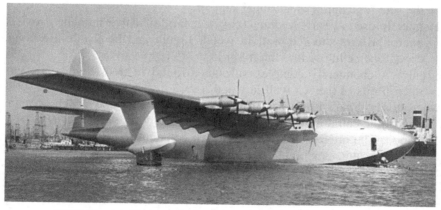

17.11 Spruce Goose – the largest wooden aircraft.

suffered major delays and cost overruns, and the aircraft was not completed until after the war. These problems drew criticism from US government officials, and the press dubbed the Hughes H-4 the *Spruce Goose*. Harsher critics called it the *Flying Lumberyard*. Despite the criticism, the *Spruce Goose* was a remarkable and innovative aircraft. It was a massive aircraft with a wingspan of 97.5 m, length of 66.6 m and height of 24.2 m. For comparison, the wingspan of the *Spruce Goose* is about 50% greater than the Boeing 747-400.

The nickname of *Spruce Goose* is misleading because it was built almost entirely from birch. The airframe and surface structures were made of laminated birch. The laminated sections were bonded together, and the aircraft is almost totally devoid of nails and screws. A major innovation was the construction of the laminated wood structure using a process called 'Durmold'. This involves bonding timber ply sheets together using adhesive and then moulding the plywood into the final shape at high temperature using steam. This allowed curved structures to be made without cutting and shaving the wood, which was the conventional practice of the time.

The *Spruce Goose* is considered by many as a failure in aerospace engineering, despite its many innovations. The aircraft was only flown once. Howard Hughes flew the aircraft off the coast of California on 2 November 1947 at just 20 m above the water at a speed of 130 km h^{-1} for a distance of only one mile. The flight lasted just one minute. Experts realised that the plane was still in ground effect during the flight, and they concluded that the *Spruce Goose* lacked enough power to fly like a conventional aircraft.

Fracture processes of aerospace materials

18.1 Introduction

Sudden fracture of structural materials was one of the most common causes of accidents in early aircraft. Materials for early aircraft were selected for maximum strength and minimum weight, and their fracture resistance was not an overriding consideration. Furthermore, the capability of materials to resist cracking and sudden failure was not well understood in the design and construction of these early aircraft. Many of the materials used in the earliest aircraft, particularly wood, were prone to sudden fracture which gave the pilot no opportunity to avoid crashing (Figure 18.1). The serious injuries and fatalities during the first decades of powered flight forced engineers to consider fracture as a critical factor in the safe design of aircraft structures and, later, jet engine components.

Damage tolerance and fracture resistance became key considerations in the selection of aircraft materials following the de Havilland *Comet* jet airliner accidents in the early 1950s. From this period, fracture toughness became a critical property in the selection of materials for modern aircraft, and it is considered just as important as other material properties such as stiffness and strength. Aerospace engineers do not select a material based solely on its strength qualities, but also consider the ability of a material to withstand

18.1 Fracture of materials caused many accidents in early aircraft.

damage below a critical size (e.g. corrosion and fatigue cracks in metals or delamination cracks in composites) without catastrophic failure.

Fracture is a failure process that involves the initiation and growth of a crack, which can cause the material to break at a stress below its ultimate strength in the crack-free condition. In practice, there is no such thing as a defect-free material or a crack-free aircraft structure. Cracks and crack-like flaws (e.g. voids, corrosion damage) are present in many aircraft components. These cracks are produced during processing of the aerospace material and manufacture of the aircraft. Defects in metal alloys include gas holes, shrinkage, brittle inclusions and stress cracks formed during casting, quenching, heat treatment and shape forming (e.g. rolling, extrusion). Defects in metal components also occur during assembly as a result of poor machining quality or incorrect drilling of fastener holes. Defects in aircraft composite materials include delamination cracks, skin-core interfacial cracks, voids and dry spots. The aircraft industry uses manufacturing practices with stringent procedures to minimise defects in metals and composites, but it is still virtually impossible to produce a defect-free material. However, the defects are often microscopic in size and few in number and, therefore, do not normally pose an immediate problem to the structural integrity of the aircraft component.

In addition to defects created during manufacturing and processing of the material being the cause of cracking, cracks can also initiate at regions of high stress within the material. Regions of high stress occur when there is an abrupt change in the shape of the material, and this is often called a stress raiser. Examples of stress raisers are fastener holes, the corners of windows and doors, and the root of turbine blades. Careful design and materials selection can reduce the stress build-up in such regions, but it cannot be completely avoided. Stress raisers are a common cause for the formation of cracks within materials which are otherwise free of defects.

Cracks may grow by overstressing, fatigue cycling or corrosion damage during the operation of an aircraft, possibly leading to catastrophic damage. For example, Fig. 18.2 shows severe damage to a Boeing 737 fuselage owing to cracking caused by the combined effects of fatigue and corrosion. Cracks can also occur because of poor design, incorrect materials selection, and damage during normal flight operations from bird impacts, lightning strikes, large hail impact or other adverse events. Components are thoroughly inspected for cracks after fabrication and throughout the service life of an aircraft. Inspection and maintenance costs represent a high percentage (20% or more) of the direct operating costs for commercial airliners. The cost is also high for maintaining airframe and engine components of military aircraft against cracks. To minimise the cost, there is a need for more damage tolerant structural materials which resist crack growth under normal flight conditions, and which therefore require less maintenance.

18.2 Fuselage damage to an Aloha Airlines B737 aircraft.

Because cracks and flaws cannot be completely avoided, the important thing is to ensure they remain harmless and do not grow to be large enough so that the material fails at below the design ultimate load. The aerospace industry uses the so-called 'damage tolerance design' philosophy to ensure the safety of aircraft structures containing cracks. Damage tolerance is the ability of structures to withstand the design load and maintain their function in the presence of cracks and other types of damage. The goal of this requirement is to ensure the continued safe operation of an aircraft over a specified period of time. Safe operation must be possible until the crack is detected during routine maintenance or, if undetected, for some multiple of the design life of the structure such as two times or four times (depending on the component and aircraft type).

The fracture behaviour of aerospace materials is examined in this chapter and chapter 19. This chapter deals with the fracture mechanics and fracture mechanisms of materials whereas chapter 19 deals with their fracture toughness properties. This chapter examines the types of fracture processes that occur in metals and fibre–polymer composites used in aerospace structures. The fracture mechanisms of other materials such as polymers and ceramics are briefly mentioned, but are not considered in detail because they are not used in large quantities in aircraft structures. The fracture mechanics of brittle and ductile materials are described in this chapter, including models for predicting the failure strength and critical crack size. The fracture toughness properties of the various types of aircraft materials used in the aircraft structures and engines are discussed in the next chapter, including methods to improve the fracture strength of metals and fibre–polymer composites.

18.2   Fracture processes of aerospace materials

18.2.1 Modes of fracture

There are two types of fracture: brittle fracture and ductile fracture. Brittle fracture involves crack growth with little or no ductile deformation of the material around the crack tip. This is an undesirable mode of fracture because brittle cracking can lead to complete failure of the material very rapidly when a critical load is reached. Ductile fracture, in contrast, involves plastic deformation of the material at the crack tip. This often results in a stable and predictable mode of fracture in which crack growth can only occur under an increasing applied load; when the load is reduced the crack stops growing. As a result, ductile fracture is the preferred failure mode for damage-tolerant materials. Whether brittle or ductile fracture occurs, the mode of fracture depends on many factors, including the stress level, type of loading (static, cyclic, strain rate), presence of pre-existing cracks or defects, material properties, environment and temperature.

Aerospace structural metals including aluminium, magnesium, titanium, high strength steel and nickel-based superalloys usually fail by ductile fracture processes which involve a certain amount of ductility. In contrast, crack growth in fibre–polymer composites and ceramic aerospace materials (such as heat shields) occurs with less ductility and therefore is a more brittle fracture process. This chapter considers the mechanisms of ductile and brittle fracture of aerospace metals and the brittle fracture of fibre–polymer composites.

18.2.2 Ductile fracture processes

Crack tip stress and plastic zone

Ductile fracture is the most common failure mode in aerospace metal alloys and polymers (including structural adhesives). A ductile crack in a metal usually starts at an existing flaw, such as a brittle inclusion within a grain, a precipitate at a grain boundary, or a void. The stress condition ahead of a crack in a ductile material loaded in tension is shown schematically in Fig. 18.3. The stress ahead of the crack is not distributed evenly. Instead, the stress in the region immediately ahead of the crack tip is much higher than the nominal (applied) stress. Figure 18.3 shows that the closer one approaches the crack tip, the higher the local stress becomes until, at some distance r_y from the crack the stress reaches the yield strength σ_y of the material. The material over a distance between the crack tip ($r = 0$) and r_y is plastically deformed, and this region is called the plastic zone.

The size of the plastic zone is dependent on the yield strength of the material, the applied stress level, and the load conditions (e.g. tension, shear).

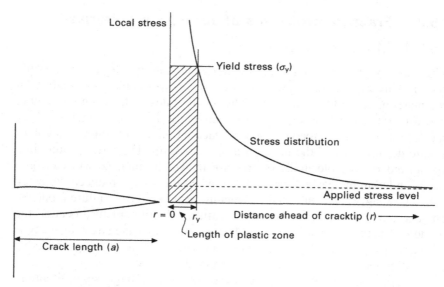

18.3 Stress field ahead of the main crack front in a ductile material, showing the plastic zone (shaded).

The plastic zone can range in size from a few tens of micrometres in high-strength metals to many millimetres in soft plastics. The material outside the plastic zone is not stressed above the yield strength and, therefore, does not plastically deform. Owing to the formation of the plastic zone, which absorbs energy and thereby resists crack growth, the stress needed to initiate a crack in ductile materials is lower than the stress needed to grow the crack. In other words, it is easier to form a crack than to get the crack to grow and, therefore, the applied stress needed to cause crack growth increases with the length. This behaviour provides ductile materials with an intrinsic amount of damage tolerance because, as the crack becomes larger, it also becomes more difficult to grow to the critical size necessary to cause complete failure.

Ductile crack-growth mechanisms

Crack growth in ductile metals can occur in several ways, depending on the type of material and the applied stress conditions (Fig. 18.4). In some cases, complex dislocation interactions occur in the plastic zone that lead to the formation of microscopic cracks. This process involves the formation and movement of large numbers of dislocations within the plastic zone. The dislocations become entangled into a high density from which tiny cracks develop ahead of the main crack front. The cracks link up with the main crack to advance the fracture process. Another important fracture process is the

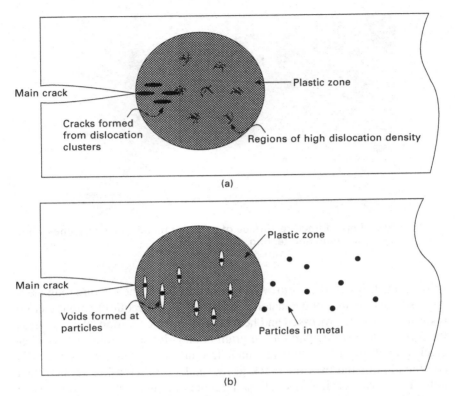

18.4 Schematic representations of the ductile fracture process involving crack development from (a) local regions of high dislocation density, and (b) particles in the metal.

formation, growth and coalescence of micrometre-sized voids in the plastic zone. The voids often develop at fractured inclusion particles or particle-matrix interfaces, and then grow in size under increasing local stress. The growing voids eventually link up with the main crack, with the ligaments between voids behaving like miniature tensile test specimens, promoting crack growth by ductile tearing. The fracture surface of a ductile material is often characterised by a dimpled texture, as shown in Fig. 18.5.

Transgranular and intergranular fracture

After the crack has initiated in a metal it grows through the grains, which is called transgranular fracture, or along the grain boundaries, known as intergranular fracture, or by a combination of transgranular and intergranular fracture (Fig. 18.6). Intergranular fracture often occurs in metals that contain a high concentration of brittle particles at the grain boundaries. These particles

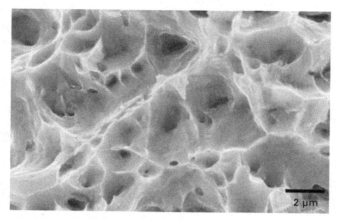

18.5 Dimpled fracture surface of a ductile metal caused by fine-scale plastic yielding and tearing.

provide a pathway for crack growth and thereby lower the fracture toughness and damage tolerance of the material. Structural alloys requiring high fracture resistance must be processed and heat treated under conditions that suppress the formation of brittle particles at grain boundaries. Regardless of whether crack growth occurs via transgranular fracture, intergranular fracture or a combination of the two, the crack grows under an increasing applied load until it reaches a critical size where the remaining uncracked section of the material can no longer support the applied stress at which point complete fracture occurs.

18.2.3 Brittle fracture of materials

Brittle fracture describes the failure of a material that involves little or no plastic deformation at the crack tip. There is always some plastic deformation around the crack tip in many materials, but, in more extreme cases of brittle fracture, the size of the plastic zone is extremely small and has no significant influence on the fracture process. In the absence of the plastic zone, the force needed to grow the crack decreases with its length. In other words, once the critical load for brittle crack growth is reached, the crack grows quickly through the material.

Brittle fracture occurs in two stages: (i) initiation of the crack and (ii) rapid propagation of the crack leading to complete fracture. A brittle crack often starts at a pre-existing defect, such as a void or inclusion. A crack can also initiate in a defect-free material in a region of high stress concentration, such as at the edge of a drilled hole or notch. The stress needed to initiate a brittle crack is higher than the stress needed to grow the crack. It is this

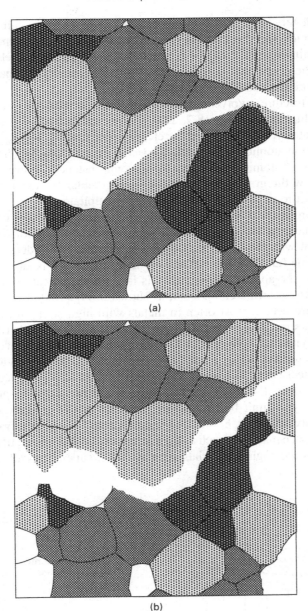

18.6 (a) Transgranular fracture through the grains and (b) intergranular fracture along the grain boundaries.

behaviour that distinguishes brittle fracture from ductile fracture, in which the stress to initiate a crack is lower than the stress to grow a crack.

Brittle fracture involves the sudden failure of the material by rapid crack growth immediately after crack initiation. This behaviour is called fast fracture. The crack speed approaches the speed of sound for the material, which for aluminium and titanium is about 5 km s^{-1} and for steel is 4.5 km s^{-1}. Once the crack has started to grow there is virtually no chance of stopping it before the material completely breaks.

Fully brittle fracture involves the rupture of interatomic bonds ahead of the crack, as illustrated in Fig. 18.7. This produces a fracture surface that can be close to atomically flat, called the cleavage fracture. A cleavage crack grows between the atomic planes along a specific crystallographic direction that has the lowest atomic bond strength. In some instances, however, the crack may follow the grain boundaries (i.e. intergranular fracture) that are weakened by brittle inclusions or intermetallic precipitates.

Brittle fracture is the worse type of failure for aircraft materials because it is fast and catastrophic, with no visible signs of damage or prior warning that the material will break. Brittle fracture can normally be identified by smoothness of the failed surface. When the fracture surface is smooth and perpendicular to the applied stress then this is a strong indication that the fracture has a strong brittle component, as shown in Fig. 18.8.

Brittle fracture occurs most often in metals with high strength and low ductility. A high-strength, low-plasticity metal that is prone to brittle fracture generally has a yield strength of $\sigma_y > E/150$, which for steels is above 1400 MPa. The steels used in highly loaded aircraft structures, such as 4340 steel and maraging steels used in landing gear, have a yield strength exceeding 1400 MPa. However, the steels are heat-treated and processed to provide sufficient ductility to limit brittle fracture. A metal that is unlikely to fail by brittle fracture has $\sigma_y < E/300$, and failure occurs by ductile fracture. Structural aerospace materials such as high-strength aluminium and titanium alloys as well as the engine materials such as nickel superalloys all fail by ductile fracture.

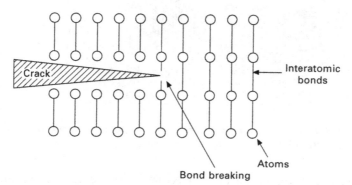

18.7 Schematic of the brittle fracture process involving breaking of interatomic bonds.

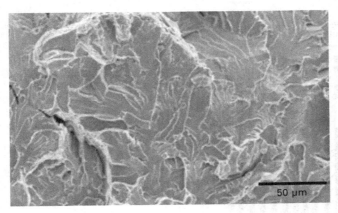

50 µm

18.8 Metal fracture surface with smooth appearance that is a characteristic of brittle (cleavage) failure.

Section 18.9 at the end of the chapter presents a case study of the role of fracture in the space shuttle *Columbia* disaster.

18.2.4 Fracture of fibre–polymer composite materials

Fracture of fibre–polymer composites involves a multiplicity of failure modes that are different from the ductile and brittle processes described for metals. Most composites show no significant sign of plastic deformation before failure and, therefore, their fracture behaviour is often described as being brittle. However, the brittle fracture of composites does not occur in the same way as brittle metals. The operative fracture modes are dependent on the microstructure of the composite, with the volume fraction, strength, toughness and dimensions of the fibres; the volume fraction, strength and ductility of the polymer matrix; and the fibre–matrix interface all influencing the fracture process. The failure modes are also dependent on the loading direction because of the anisotropic microstructure of composite materials.

The two basic modes of fracture for composite laminates are in-plane fracture and interlaminar fracture and they are illustrated in Fig. 18.9. In-plane fracture involves crack growth in the direction normal to the fibres whereas interlaminar fracture involves cracking parallel to the fibre ply layers. The in-plane fracture process is more resistant to crack extension than the process of interlaminar fracture. For example, the amount of energy required for crack growth in the in-plane direction of a carbon–epoxy composite is anywhere from 20 to 1000 times greater than in the interlaminar direction. For this reason, composite aircraft structures must be designed such that their damage tolerance is achieved by in-plane fracture resistance.

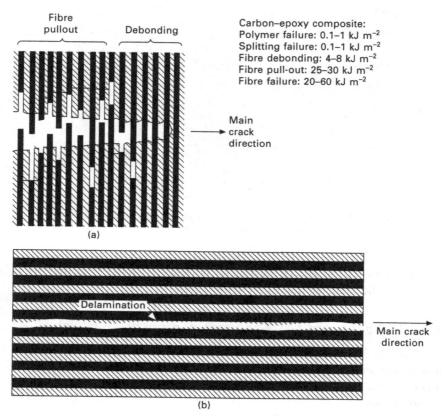

Fibre pullout   Debonding

Carbon–epoxy composite:
Polymer failure: 0.1–1 kJ m^{-2}
Splitting failure: 0.1–1 kJ m^{-2}
Fibre debonding: 4–8 kJ m^{-2}
Fibre pull-out: 25–30 kJ m^{-2}
Fibre failure: 20–60 kJ m^{-2}

Main crack direction

(a)

Delamination

Main crack direction

(b)

18.9 Schematic representations of (a) in-plane and (b) interlaminar fracture of composite laminates.

An interesting feature of carbon–epoxy and other composite materials is that their in-plane fracture toughness is much higher than the toughness of the fibres and polymer resin on their own. The high in-plane fracture toughness is the result of crack growth being resisted by various failure processes that occur near the crack tip and along the crack wake. In-plane fracture involves crack extension by plastic deformation and rupture of the polymer matrix and failure of the fibres. Fibres do not always break at the crack tip, but can fail a distance away at a pre-existing flaw in the fibre where the strength is low. As the two faces of the crack separate the broken fibres detach from the polymer matrix and pull out. The pull out of fibres can extend many millimetres behind the crack front. As the crack grows, its tip is often deflected along the fibre direction (i.e. 90° to the crack direction) as it follows weak interfaces between the fibres and matrix. This produces splitting cracks that can grow many millimetres from the main crack path.

The main crack is repeatedly deflected at the fibre/matrix interfaces into splitting cracks thus resisting crack extension.

The damage processes involved in in-plane fracture absorb different amounts of energy which resist the crack growth process. Figure 18.9 gives the approximate amount of energy absorbed by the fracture processes in carbon–epoxy composite. Fracture and pull-out of the fibres absorb the greatest amount of energy and thereby provide the greater resistance against in-plane fracture whereas the failure of the polymer matrix and fibre–matrix splitting results in less resistance.

The interlaminar fracture process involves cracking in the fibre direction, usually between the ply layers which is called delamination fracture. The crack grows through the polymer-rich regions between the plies without the rupture, debonding and pull-out of the fibres experienced with in-plane fracture. The amount of energy needed for delamination crack growth is relatively low, and therefore interlaminar fracture occurs much more easily than in-plane fracture. Interlaminar fracture resulting in delamination cracking is a major problem with composite structures used in aircraft. Bird strike and other impact events can cause extensive delamination damage in aircraft structural composites because of their low interlaminar fracture resistance.

Section 18.10 at the end of the chapter presents a case study of the fracture of an aircraft composite radome.

18.3   Stress concentration effects in materials

18.3.1 Geometric stress concentration factor

Cracks often initiate and grow from sites of high stress in a material owing to a change in the shape (geometry) of the component. The change in geometry is called a 'stress concentration' or 'stress raiser', and is often the cause for cracking in aircraft components (Fig. 18.10). Common stress raisers in aircraft structures are corners, holes, fillets and notches. As examples, spars and ribs in the wings and fuselage contain cut-outs and notches to reduce the structural weight, and skin panels and many internal parts contain drilled holes for rivets, bolts and screws. Small stress raisers also reside within the material, such as micrometre-sized gas holes in metals or voids in composites.

The study of fracture must consider a parameter known as the geometric stress concentration factor (also called the theoretical stress concentration factor). This factor defines the magnitude of the local stress at the stress raiser compared with the applied stress. When stress is measured at the edge of a stress raiser it is not at the same level as the nominal (or applied) stress, but is much higher. This is because the stress that would normally be supported by the material where the stress raiser occurs is concentrated at its edge. Thus, a stress concentration occurs at the edge of a stress raiser. Cracks in

18.10 Crack growth from a machined notch which acts as a stress raiser under load.

aircraft materials are most likely to develop in regions where there is high stress concentration. Therefore, knowledge of geometric stress concentrations is essential to understand the initiation of cracks which lead to fracture.

Figure 18.11 shows the stress distribution across the load-bearing section (between points $x - x'$) of an infinitely wide plate containing a circular hole. A constant stress is applied to the ends of the plate σ_{app}. The stress in the plate is greatest at the hole edge and drops off rapidly with distance away from the hole. The higher stress at the hole edge is described by the geometric stress concentration factor K_t which is the ratio of the maximum stress at the hole σ_{max} to the applied stress σ_{app}:

$$K_t = \frac{\sigma_{max}}{\sigma_{app}} \qquad [18.1]$$

The geometric stress concentration factor for a noncircular (elliptical) hole can be calculated using:

$$K_t = \frac{\sigma_{max}}{\sigma_{app}} = 1 + \left(\frac{2a}{b}\right) \qquad [18.2]$$

where a and b are the half-width and half-height of the hole.

For an isotropic material containing a circular hole ($a = b$), K_t is 3.0, which means the maximum stress at the hole edge is three times greater than the applied stress. The stress concentration factor is greater than three when the stress raiser has an elliptical shape (i.e. $a > b$) elongated across the plate width.

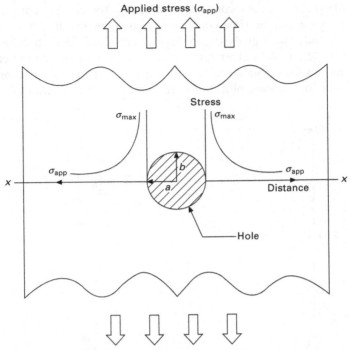

18.11 Stress distribution in a plate containing a stress raiser in the form of a circular hole.

The end radius of a hole (ρ) is determined from:

$$\rho = \frac{b^2}{a} \qquad\qquad [18.3]$$

By combining the two previous equations, the maximum stress at the hole edge is:

$$\sigma_{max} = \sigma_{app} \left[1 + 2\sqrt{a/\rho}\right] \qquad\qquad [18.4]$$

In most cases $a \gg \rho$, and so the approximation is:

$$\sigma_{max} \approx 2\sigma_{app} \sqrt{a/\rho} \qquad\qquad [18.5]$$

where the term $2\sqrt{a/\rho}$ is the geometric stress concentration factor K_t.

The calculation of the stress concentration factor given in equation [18.2] and that for maximum stress given in equation [18.4] are only valid for a plate of infinite width. In reality, of course, components have a finite size and can have shapes other than a flat plate. In these instances, the stress-concentration factor is dependent on the sizes and shapes of both the stress raiser and component.

Calculating the stress concentration factor for components having a complex shape can be time-consuming and often requires finite-element analysis or advanced analytical methods. To avoid detailed calculations, engineering textbooks and aircraft design manuals give the theoretical stress concentrations for components with a wide range of shapes, often presented as a stress concentration plot. For example, Figure 18.12 shows the effects

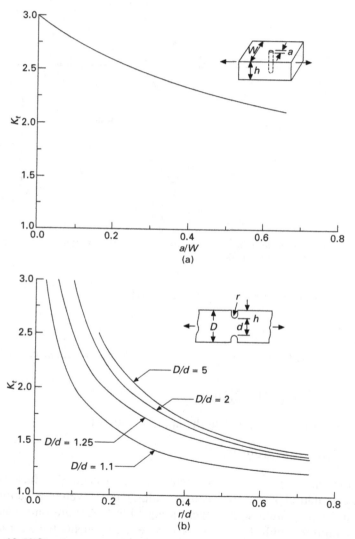

18.12 Stress concentration factor plots for three geometries: (a) axial loading of a flat plate with a circular hole, (b) axial loading of a notched bar, (c) axial loading of a T-section.

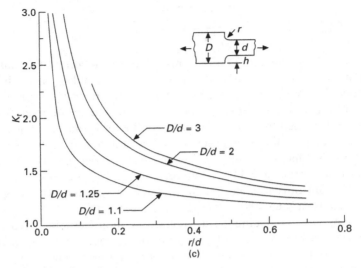

18.12 Continued

of notch shape (*a/b*) and part geometry on the stress concentration factor for three simple cases: a plate with a circular hole, a plate with a double notch, and a T-section.

The stress concentration factor is only valid for elastic stress conditions; that is for brittle materials at any stress level and ductile materials at any stress below the yield strength. In a brittle material, a crack develops at the edge of a stress raiser when the local stress reaches the failure stress of the material. Once the crack has initiated, it grows rapidly (fast fracture) through the remaining material causing complete fracture. Ductile materials plastically deform at the edge of a stress raiser when the local stress exceeds the yield strength. The ability of a ductile material to plastically deform at a stress raiser avoids sudden fracture.

18.3.2 Stress concentration factors for anisotropic (composite) materials

The stress concentration factor for isotropic materials, such as metals and plastics, is determined using the procedure described in 18.3.1. The situation is somewhat different with anisotropic materials, such as fibre–polymer composites, because the stress concentration factor is dependent on the elastic modulus in different directions. The stress concentration factor K_t for an anisotropic material containing a circular hole is given by:

$$K_t = \frac{\sigma_{max}}{\sigma_{app}} = 1 + \sqrt{2\left[\sqrt{\frac{E_x}{E_y}} - v_{xy}\right] + \frac{E_x}{G_{xy}}} \qquad [18.6]$$

where E_x and E_y are the Young's moduli in the loading and transverse directions, v_{xy} is the Poisson's ratio in the x–y plane, and G_{xy} is the in-plane shear modulus. When there is a high degree of anisotropy, such as when all of the fibres in a polymer composite are aligned in the load direction (i.e. $E_x \gg E_y$), then the stress concentration factor is very high at the hole edge. For example, the geometric stress concentration for a unidirectional carbon–epoxy panel containing a circular hole is about 6.6, whereas for an isotropic material it is only 3.0. Therefore, considerable care must be exercised when using anisotropic materials in aircraft components containing stress raisers such as fasteners holes, windows and cut-outs.

One approach to reducing the stress concentration factor is increasing the percentage of ±45° and other off-axis plies in the composite. This increases the shear modulus G_{xy} and brings the ratio of E_x/E_y closer to unity. Table 18.1 shows the effect of reducing the number of axial (0°) plies and increasing the number of ±45° plies on the stress concentration factor for a circular hole in a composite panel. A composite with all ±45° plies has the lowest stress concentration factor of 2, but it also has the lowest strength because of the absence of 0° plies. The best compromise is a laminate that has a sufficient number of 0° plies to carry the load, but also ±45° plies to reduce the stress concentration factor. The two fibre lay-up patterns most often used in carbon–epoxy composites, quasi-isotropic [0/±45/90] and cross-ply [0/90], have stress concentration factors of about 3 and 3.5, respectively.

18.4   Fracture mechanics

18.4.1 Introduction to fracture mechanics

Fracture mechanics is the mechanical analysis of materials containing one or more cracks to predict the conditions when failure is likely to occur. It is an important topic for many reasons, and is used to:

Table 18.1 Stress concentration factors for a 24-ply carbon–epoxy panel containing a circular hole

Lay-up		Stress concentration factor K_t
Number of 0° plies	Number of ±45° plies	
24	0	6.6
16	8	4.1
12	12	3.5
8	16	3.0
0	24	2.0

- design aircraft structures and select materials with high resistance against cracking;
- calculate the residual strength of aircraft structures containing cracks;
- determine whether a crack is benign and does not affect the strength of an aircraft structure or whether the crack may lead to complete failure and therefore the structure must be repaired or replaced before failure occurs;
- determine the cause of structural failures in aircraft accident investigations; and
- minimise the need for expensive structural tests on large aircraft components to assess their damage tolerance and residual strength.

Many analytical methods are available to calculate the fracture strength of materials containing cracks, and almost all are based on the principles of linear elastic fracture mechanics (LEFM). LEFM can be simply described as the analysis of materials containing one or more cracks that fail by brittle fracture. It is assumed in LEFM that failure occurs under elastic conditions, and plastic flow at the crack tip is not significant. LEFM can be applied to brittle materials such as ceramics and high-strength, low-plasticity metals. LEFM is the basis for determination of the critical flaw size or critical stress to cause failure and the prediction of fatigue life for isotropic materials.

LEFM is not accurate for ductile materials where a large plastic zone at the crack tip affects the fracture process. The fracture mechanics analysis of ductile materials is performed using elastoplastic fracture mechanics (EPFM).

18.4.2 Linear elastic fracture mechanics

The simplest LEFM model for calculating the fracture strength of a brittle material containing a crack is the Griffith failure criterion. The model is based on work performed during World War 1 by A. A. Griffith on the fracture strength of glass. Griffith considered an apparent contradiction in the breaking strength of glass plate. The tensile stress needed to fracture bulk glass is around 100 MPa. However, the theoretical stress needed to break the atomic bonds (Si—O) in glass is about 10 000 MPa. Griffith made the important observation that the fracture load is not dependent on the load-bearing area of the material, but that cracks within the glass determine the strength.

Griffith showed experimentally that the fracture stress σ_f of glass is inversely related to the square root of the crack length a:

$$\sigma_f = C/\sqrt{a} \qquad [18.7]$$

where C is a constant. The importance of equation [18.7] is that the fracture stress of a brittle material with a known crack size can be calculated.

Alternatively, equation [18.7] can be used to calculate the maximum crack size that causes a brittle material to break under a known operating stress. Equation [18.7] can be applied to predict the failure strength and damage tolerance for any brittle isotropic material.

In the simple case shown in Fig. 18.13, the brittle material contains a through-thickness crack of length 2a lying normal to an applied tensile stress. The fracture stress σ_f of the material is related to the half-crack length a by:

$$\sigma_f = \sqrt{\frac{2\gamma_e E}{\pi a}}$$ [18.8]

where E is the Young's modulus of the material, and γ_e is the elastic surface energy density needed to form a new crack surface in a brittle material, and has the units $J\,m^{-2}$. Because two new opposing surfaces are produced during crack growth the surface energy term is $2\gamma_e$. Equation [18.8] is the same equation as [18.7], where the constant C equals $\sqrt{2\gamma_e E/\pi}$.

18.4.3 Ductile fracture mechanics analysis

Linear elastic fracture mechanics is accurate for calculating the fracture stress of brittle materials in which the stress field at the crack tip is elastic. However, LEFM does not consider plastic flow at the crack tip that occurs in ductile materials. The metals used in aircraft are ductile, and therefore LEFM cannot be used to calculate the fracture strength. With some modification, however, LEFM can be used to calculate the fracture stress of ductile metals. Two scientists named Orowan and Irwin modified LEFM to account for plastic flow at the crack tip. Orowan adapted the Griffith model for ductile materials by including a term to account for the extra work of fracture that occurs in the plastic zone. This is the basis of elastoplastic fracture mechanics.

In mathematical terms, when the work of fracture for both elastic γ_e and

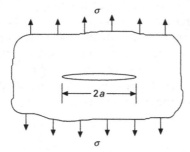

18.13 Through-thickness crack of length 2a in an infinitely wide plate under tension.

plastic γ_p crack growth is considered then the fracture stress is calculated using:

$$\sigma_f = \sqrt{\frac{2E(\gamma_e + \gamma_p)}{\pi a}}$$ [18.9]

The plastic surface energy density γ_p for ductile metals is usually in the range 100–1000 J m^{-2}, which is much higher than the energy density for brittle materials γ_e which is only about 1–20 J m^{-2}. This is because the amount of energy needed to plastically deform the material at the crack tip and then extend the crack by the joining of microvoids and tiny cracks is orders of magnitude higher than the energy needed for elastic deformation. Because the energy needed to grow a crack through the plastic zone is much greater ($\gamma_p \gg \gamma_e$), the elastic fracture energy term can be ignored in plastic fracture analysis. Therefore, the fracture stress for a ductile material containing a small crack can be approximated using:

$$\sigma_f \approx \sqrt{\frac{2E\gamma_p}{\pi a}}$$ [18.10]

It is not easy to measure or calculate γ_p and, therefore, it is difficult to determine the fracture stress of a ductile material using equation [18.10]. To avoid the problem, Irwin showed that when the plastic zone size is very small compared with the size of the metal component, then the fracture stress can be determined by:

$$\sigma_f = \sqrt{\frac{EG_c}{\pi a}}$$ [18.11]

where G_c is the critical strain energy release rate or potential energy release rate of the material, $G = 2(\gamma_e + \gamma_p)$. G_c is one measure of the fracture toughness of a material, and has the units J m^{-2}.

Equation [18.11] shows that increasing the fracture toughness G_c increases the fracture stress σ_f of a material containing a crack. Chapter 19 describes the fracture toughness properties of materials, and how G_c can be increase to improve the fracture strength.

The critical strain energy release rate G_c is a material property that is proportional to the amount of plastic deformation that occurs at the tip of a growing crack. However, the fracture toughness of materials is more often defined by its critical stress intensity factor K_c which has the units of Pa m$^{1/2}$. It is useful at this point to compare the critical stress intensity factor K_c and the geometric stress concentration factor K_t. The meaning of these two factors is easily confused, although they are different parameters used in fracture mechanics. K_c (similarly to G_c) is the fracture toughness of a material, and describes how easily a crack grows under an externally applied stress. This is different from the definition of K_t, which defines the magnification of the applied stress at the edge of a stress raiser. K_c is used more often than the G_c to define the fracture toughness of materials, although both are a direct measure of toughness and are related by the expression:

$$G_c = \frac{\alpha K_c^2}{E}$$ [18.12]

where $\alpha = 1$ for plane stress and $\alpha = (1-v)^2$ for plane strain conditions for crack growth. Plane stress defines the condition where a two-dimensional stress state occurs in the plastic zone at the crack tip whereas plane strain involves a three-dimensional stress state. Plane stress is often dominant in thin materials and plane strain in thick materials. The effect of plane stress and plane strain conditions at the crack tip on the fracture toughness of materials is explained in chapter 19.

18.5   Application of fracture mechanics to aerospace materials

Fracture mechanics is indispensable for the design of damage-tolerant aircraft structures. The application of fracture mechanics in materials selection for new aircraft and the calculation of residual strength for existing aircraft components depend on three parameters:

- fracture toughness K_c of the material;
- critical crack length a_c; and
- operating stress σ.

The equation relating these parameters is:

$$K_c = \beta\sigma\sqrt{(\pi a_c)}$$ [18.13]

The parameter β is a geometry factor that depends on the crack location and the shape of the component. Table 18.2 shows equation [18.13] for three different cases of a single crack within a material. Values for β for a wide variety of crack conditions can be found in engineering handbooks.

Equation 18.13 can be used in several important ways to design damage-tolerant aircraft structures. The engineer decides what is the most important about the design: certain material properties (e.g. E, σ_y), the design stress level (σ), or the critical crack length (a_c) that must be tolerated for safe operation of the component. Equation [18.13] is then used to design the structure, including the selection of material, based on the main design requirement. For example, if a specific type of aluminium alloy is selected with a known toughness K_c for an aircraft component that is required to support a certain

Table 18.2 Stress intensity factors for various geometries

Crack type		Stress intensity equation
Centre crack of length 2a in infinite plate		$K_I = \sigma_{app}(\pi a)^{1/2}$
Centre crack of length 2a in a plate of width W		$K_I = \sigma_{app}\left[W \tan\left(\dfrac{\pi a}{W}\right)\right]^{1/2}$
Edge crack of length a in semi-infinite plate		$K_I = 1.12\sigma_{app}(\pi a)^{1/2}$

stress σ, then equation [18.13] can be used to calculate the critical crack length (a_c) that can be tolerated without causing fracture. Alternatively, if a crack of a known size is detected within an aircraft component, the residual strength of the component can be calculated.

18.6   Summary

Structural materials used in the airframe and engine must be damage tolerant, which means that they retain their strength, fatigue life and other properties in the presence of cracks below a critical length. A key property that ensures high damage tolerance in materials is fracture toughness.

The initiation of cracks often occurs in metals at microstructural imperfections such as large and brittle particles and gas holes or in fibre–polymer composites at voids. These defects must be avoided by careful processing of the material, including the heat treatment of metals. Cracks also initiate at geometric stress raisers, such as fastener holes, sharp corners, and abrupt changes in section thickness or shape. Careful design can eliminate or minimise the concentration of stress at these sites.

The metals used in aircraft fail by ductile fracture processes, which involves plastic yielding of the material ahead of the main crack. Crack growth occurs along the grain boundaries (intergranular fracture) or through the grains (transgranular fracture) or a combination of both. Ductile metals can display brittle-like fracture properties when a high concentration of brittle particles occurs at the grain boundaries, which promotes intergranular fracture. Structural metals must be processed under conditions which eliminate or minimise the presence of brittle particles at grain boundaries.

Brittle fracture is an unstable failure process that occurs in fibre–polymer composite materials, metals with high strength and low ductility, and in some metal types at low temperature (i.e. below the ductile/brittle transition temperature). Aerospace metals used in structural and engine applications should not display brittle fracture properties to ensure sufficient damage tolerance.

Fracture mechanics, based on linear elastic and elastoplastic fracture, is used to calculate the damage tolerance of materials. Fracture mechanics is used to determine the fracture toughness, operating stress and maximum crack size of materials to avoid fracture.

18.7   Terminology

Brittle fracture: A fracture event that involves little or no plastic deformation of the material. Typically, brittle fracture occurs by fast crack growth with less expenditure of energy than for ductile fracture.

Cleavage fracture: Fracture event that involves cracking along specific planes in the crystal resulting in a smooth fracture surface.

Critical strain energy release rate: The fracture toughness of a material expressed in the units $J\ m^{-2}$.

Critical stress intensity factor: The fracture toughness of a material expressed in the units $Pa\ m^{1/2}$.

Delamination: Cracking between the ply layers in composite laminates.

Ductile/brittle fracture transition: Change in the failure mode of certain types of metals from ductile tearing to brittle (cleavage) fracture when cooled below the transition temperature.

Fast fracture: A fracture event in which the crack in the material grows rapidly (usually near the speed of sound of the material) and leads to catastrophic failure. Stress acting on a material when fast fracture occurs is less than the yield strength.

Fracture toughness: A generic term to describe the resistance of a material to crack extension.

In-plane fracture: Crack growth normal to the fibre direction in composite materials.

Intergranular cracking: Cracking along the grain boundaries, and not through the grains. Usually occurs when the phase in the grain boundary is weak and brittle.

Interlaminar fracture: Cracking between the ply layers along the fibre direction in composite materials. Also called delamination.

Linear elastic fracture mechanics: A method of fracture analysis that can determine the stress (or load) required to induce brittle fracture in a material or structure containing a crack-like flaw of known size and shape.

Plane strain: The stress condition in linear elastic fracture mechanics in which there is zero strain in a direction normal to both the axis of applied tensile stress and the direction of crack growth (that is, parallel to the crack front). Usually occurs in loading thick plate along a direction parallel to the plate surface.

Plane stress: The stress condition in linear elastic fracture mechanics in which the stress in the thickness direction is zero. Usually occurs in loading thin sheet along a direction parallel to the sheet surface.

Plastic zone: The region ahead of a crack tip where a ductile material is plastically deformed owing to the local stress field exceeding the yield strength.

Stress concentration factor: A parameter that defines the magnification of the applied stress at the crack tip, that includes the geometrical parameter. Ratio of the greatest stress in the region of a notch or other stress raiser to the nominal (applied) stress. It is a theoretical indication of the effect of stress concentrations on the fracture strength of materials.

Transgranular cracking: Cracking through the grains of the material.

18.8   Further reading and research

Dieter, G. E., *Mechanical metallurgy*, McGraw–Hill, London, 1988.

Dimatteo, N. D., *ASM handbook, volume 19: fatigue and fracture*, ASM International, 1996.

Gordon, J. E., *The new science of strong solids or why you don't fall through the floor*, Penguin Science, 1991.

Hertzberg, R. W., *Deformation and fracture mechanics of engineering materials*, John Wiley & Sons, 1996.

Knott, J. F., *Fundamentals of fracture mechanics*, Wiley, 1973.

18.9   Case study fracture in the space shuttle *Columbia* disaster

One of the most high profile accidents involving fracture was the space shuttle *Columbia* disaster that occurred during re-entry into the Earth's atmosphere on February 1, 2003. The seven crew members of flight STS-107 were killed when *Columbia* broke up while travelling at about Mach 18.5 at an altitude of 64 km. Following an exhaustive investigation it was concluded that the loss of *Columbia* was the result of damage sustained to the thermal protection system. The leading edges of the space shuttle wings are covered with a brittle reinforced carbon–carbon composite to provide thermal protection to the underlying aluminium structure. (Chapter 16 provides details about the ceramic materials used in the thermal protection system).

During take-off a piece of foam insulation broke away from an external fuel tank. The foam, which was about the size of a small briefcase, smashed into the leading edge of the left-side wing of *Columbia*. The reinforced carbon–carbon composite, which is a brittle material with low fracture toughness, broke under the impact force. Tests performed as part of the accident investigation showed that the foam insulation could breach the thermal protection system, leaving a large hole that exposed the underlying aluminium structure (Fig. 18.14). The extremely high temperatures experienced during re-entry caused the exposed aluminium structure to melt which subsequently caused *Columbia* to break up. The reinforced carbon–carbon material has low resistance against fracture because no plastic deformation occurs during crack growth. This accident tragically highlights the risk involved in using brittle materials, even in accidental load cases such as the foam impact on *Columbia*.

18.10   Case study: fracture of aircraft composite radome

An example of the fracture of fibre–polymer composites was the sudden, catastrophic failure of the radome on an F-111C aircraft (Fig. 18.15). In April

18.14 Brittle fracture of reinforced carbon–carbon panel under a simulated test of the foam impact to *Columbia* (photograph supplied courtesy of NASA).

18.15 Damage to an RAAF F-111C radome made of fibreglass composite (photograph supplied courtesy of the Australia Department of Defense).

2008, the aircraft operated by the Royal Australian Air Force was flying at more than 550 km h^{-1} at an altitude of 900 m when the fibreglass radome suddenly fractured as the result of a collision with a large bird (pelican). The low interlaminar fracture toughness of the radome material caused extensive delamination cracking and splintering. However, the high in-plane toughness of the composite stopped the radome from completely breaking away from the forward fuselage section.

19

Fracture toughness properties of aerospace materials

19.1   Introduction

High resistance to fracture is essential to ensure high damage tolerance for the materials used in aircraft structures and engines. In chapter 18 it was explained that an important way to increase the damage tolerance is by raising the fracture toughness. The fracture toughness of materials is defined by two related properties: critical stress intensity factor K_c and critical strain energy release rate G_c. In this chapter we examine these fracture toughness properties for aerospace and other materials. We describe how the fracture toughness of metals is controlled by factors such as strength, ductility, microstructure, alloy content and heat treatment and explain the factors controlling the toughness of composite materials, such as the fibre orientation. Methods used to increase the fracture toughness of metals and composites are described as means for improving the damage tolerance of aircraft components.

19.2   Fracture toughness properties

Fracture toughness properties for aerospace and other engineering materials are given in Fig. 19.1 and Table 19.1. As mentioned in chapter 18, the fracture toughness is expressed as the critical stress intensity factor K_c or critical strain energy release rate G_c, with the former being more commonly used. In simple terms, the higher K_c and G_c then the greater the fracture resistance of the material, which then potentially translates into higher damage tolerance. The fracture toughness varies over an extraordinarily wide range (five orders of magnitude), from the extremely tough to the very brittle. Tough materials have high K_c (100–200 MPa m$^{1/2}$) because they develop a large plastic zone that absorbs energy and thereby resists crack extension. These materials are generally characterised by low yield strength and high ductility, such as pure ductile metals. Most high strength metal alloys, including those used in aircraft structures, have moderately high K_c (20–120 MPa m$^{1/2}$). The toughness of high strength metals is lower than soft ductile metals because their higher yield strength and lower ductility allows less plastic flow at the crack tip. Polymers have low K_c (0.3–5 MPa m$^{1/2}$) and ceramic materials have the lowest toughness because little or no plastic flow occurs at the crack tip.

454

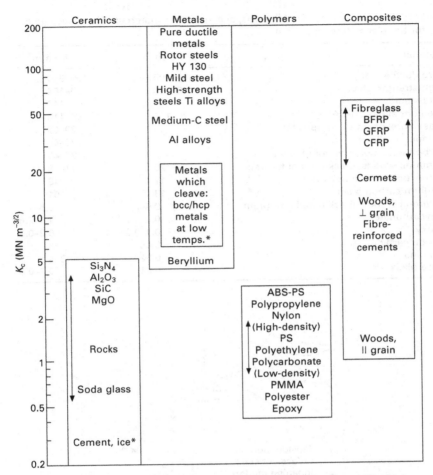

19.1 Range of critical stress intensity factors for materials. Values measured at room temperature unless indicated by* (from M. F. Ashby and D. H. R. Jones, *Engineering materials: an introduction to their properties and applications*, Pergamon Press, Oxford, 1981).

19.2.1 Modes I, II and III of fracture toughness

The fracture toughness of a material may be dependent on the direction of crack growth relative to the load direction. There are three basic modes of crack growth: mode I (tension), mode II (shear) and mode III (tearing), and they are shown in Fig. 19.2. The mode I condition is when a tensile stress is applied normal to the crack, and this is known as crack opening. Mode II is when a shear stress is applied in the crack plane, and this is called crack sliding. Mode III is when a shear stress is normal along the crack plane, and this is described as crack tearing.

Table 19.1 Fracture properties for aerospace and other materials. The values given are for the plane strain condition

Material	G_c (kJ m^{-2})	K_c (MPa m$^{1/2}$)
Pure ductile metals	100–1000	100–350
High strength steel	15–118	50–154
Mild steel	100	140
Titanium alloys (Ti–6Al–4V)	26–114	55–115
Glass fibre–polymer composites	40–100	20–60
Aluminum alloys	8–30	23–45
Carbon fibre–polymer composites	5–30	32–45
Common woods, crack normal to grain	8–20	11–13
Boron–polymer composite	17	46
Medium carbon steel	13	51
Common woods, crack parallel to grain	0.5–2	0.5–1
Polycarbonate	0.4–1	1.0–2.6
Epoxy	0.1–0.3	0.3–0.5
Silicon carbide	0.05	3
Alumina	0.02	3–5
Soda glass	0.01	0.7–0.8

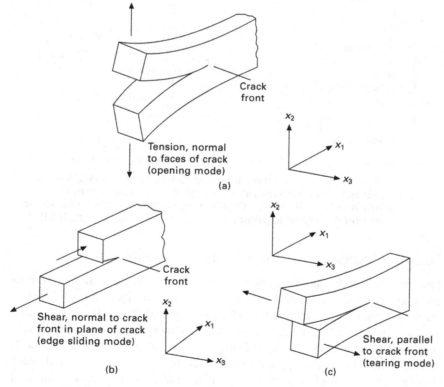

19.2 Crack growth: (a) mode I, (b) mode II and (c) mode III.

The three modes of loading may result in different values for the critical stress intensity factor for some materials. For example, the fracture toughness of a 7000 aluminium alloy in modes I, II and III is 27, 21 and 24 MPa m$^{1/2}$, respectively. The differences in toughness can be much greater in anisotropic materials such as composites. For this reason, the critical stress intensity factor has the subscript number I, II or III (i.e. K_{Ic}, K_{IIc}, K_{IIIc}) to distinguish between the three modes of loading.

19.2.2 Fracture toughness properties of anisotropic materials

The fracture toughness of anisotropic materials, such as composites and wood, is highly directional. The toughness of composites is dependent on the direction of crack growth relative to the fibre orientation. Figure 19.3 shows the effect of loading angle on the fracture toughness of unidirectional, cross-ply and quasi-isotropic composite materials. For unidirectional material, when the direction of crack growth is normal to the fibre direction (i.e. ϕ = 0°) then the toughness is high. This is the in-plane fracture condition described in chapter 18, and the toughness is high because crack growth is resisted by failure processes that absorb a large amount of energy, such as

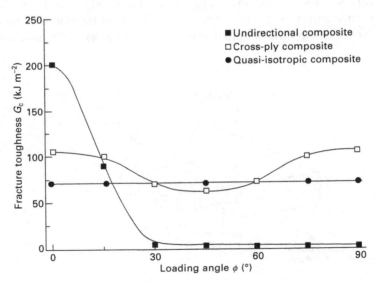

19.3 Effect of loading angle on the fracture toughness of unidirectional, cross-ply and quasi-isotropic glass fibre–epoxy composites (adapted from B. Harris, *Engineering composite materials*, Institute of Metals, London, 1980).

fibre fracture, splitting and fibre pull-out. The toughness of the unidirectional composite drops sharply with increasing load angle away from the fibre direction because crack growth occurs through the low toughness polymer matrix phase between the plies, with little or no fibre fracture. The fracture toughness of a composite loaded at high angles is similar to the fracture toughness of the polymer matrix phase. The anisotropy in fracture toughness of composites is minimised with cross-ply and quasi-isotropic fibre patterns, and this is an important reason for using [0/90] and [0/±45/90] composites in aircraft structures rather than [0] materials.

The fracture toughness properties of woods used in aircraft construction are also anisotropic (Fig. 19.4). The toughness is highest when crack growth occurs in the direction normal to the alignment of the wood cell fibres. The critical stress intensity factor of many woods in this direction is about 12 MPa m$^{1/2}$. The toughness is much lower in the direction parallel to the wood grains because the crack grows easily along the weak and brittle phase that binds the fibres. The fracture toughness in the parallel direction is only about 1–2.5 MPa m$^{1/2}$. In some respects, the anisotropic fracture properties of wood are similar to those of a unidirectional composite material.

The fracture toughness of metals can also be anisotropic when there is an alignment of the grain structure owing to rolling, extrusion or other directional forming process. As described in chapter 7, the plastic working of ductile metals stretches their grain structure in the forming direction (see Fig. 7.9). As a result, mechanical properties such as yield strength and fatigue life become anisotropic (Fig. 7.10). Table 19.2 gives the fracture toughness for several aerospace aluminium alloys in the three principle directions: L,

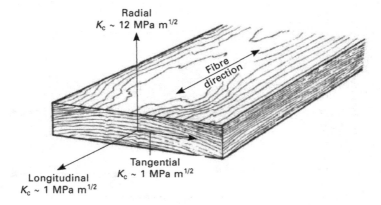

Radial
$K_c \sim 12$ MPa m$^{1/2}$

Fibre direction

Tangential
$K_c \sim 1$ MPa m$^{1/2}$

Longitudinal
$K_c \sim 1$ MPa m$^{1/2}$

19.4 Anisotropic fracture toughness properties of wood. Typical values for the critical stress intensity in the different directions are given. Toughness is highest when the direction of crack growth is in the radial direction.

Table 19.2 Anisotropic fracture toughness values for aluminium alloys in different grain directions.

Aluminium alloy	Longitudinal (L) K_{Ic} (MPa m$^{1/2}$)	Long transverse (LT) K_{Ic} (MPa m$^{1/2}$)	Short Transverse (ST) K_{Ic} (MPa m$^{1/2}$)
2014 Al-T651	23	24	20
2024 Al-T351	32	34	24
7075 Al-T7451	31	36	27
7075-T6	23	32	21
7178-T651	22	26	16

longitudinal; LT, long transverse; and ST, short transverse. The LT toughness (i.e. a crack growing on the plane normal to the L direction, in the T direction) is highest whereas a crack along the plane normal to the S direction has the lowest toughness. As mentioned, the fracture toughness is an important consideration in the design of damage tolerant aircraft structures. The grains in the metal should be aligned across the expected plane of crack growth in the aircraft structure to maximise the toughness.

19.2.3 Fracture toughness for plane stress and plane strain conditions

The critical stress intensity factor is used to calculate the fracture strength of a material containing a crack. Unlike some other material properties such as elastic modulus, the critical stress intensity factor of a ductile material is not a constant property but changes with the thickness of the material. The fracture toughness of a ductile material can be higher when used in thin structures compared with thicker components.

For example, the effect of material thickness B on the critical stress intensity factor for two aircraft-grade aluminium alloys (2024 Al and 7075 Al) is shown in Fig. 19.5. In thin materials, a condition called the plane stress occurs at the crack tip, resulting in the toughness being relatively high. Plane stress is dominant near the surface of the material because the principal stresses have to be zero at the surface. Figure 19.5 shows that the critical stress intensity factor drops with increasing thickness as a progressive transition occurs from plane stress to plane strain. The plane-strain condition dominates towards the centre of the specimen where deformation through the thickness is constrained by the bulk of the material. The stresses in the mid-thickness region are elevated as a result. Over the range of thickness where the material changes from fully plane strain (i.e. thick) to fully plane stress (i.e. thin) the condition at the crack tip can be a complex mix of plane stress and plane strain, with the latter becoming more dominant with increasing thickness. Eventually, a thickness is reached when the plane

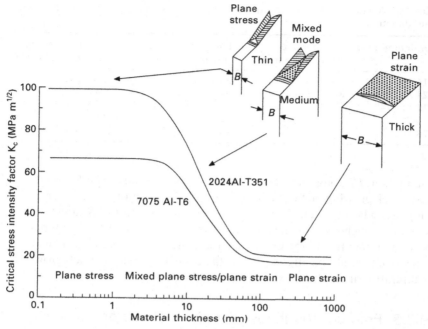

19.5 Effect of thickness B on the critical stress intensity factor K_c of two aircraft-grade aluminium alloys.

stress condition is negligible, and a pure plane strain state exists at the crack tip. When this state is reached the critical stress intensity factor reaches a minimum constant, known as the plane strain fracture toughness K_{Ic}.

As a general rule, the plane strain condition prevails in ductile materials when the thickness is:

$$B \geq 2.5 \left(\frac{K_{Ic}}{\sigma_y} \right)$$ [19.1]

Plane stress conditions occur when the ratio of the radius of the crack tip plastic zone (r_p) to the material thickness (t) is greater than 0.5. When the material becomes sufficiently thin (i.e. $r_p/t > 0.5$) then plastic yielding at the crack tip is constrained in the through-thickness direction, and there is a two-dimensional stress state in the plastic zone. Plane strain occurs when the material is thick compared with the plastic zone size ($r_p/t < 0.02$). In between the conditions when neither plane stress nor plane strain is dominant then mixed plane stress/plane strain occurs at the crack tip. As the material thickness is increased relative to the size of the plastic zone, the fracture toughness falls because the constraint of the material around the plastic zone prevents deformation in the through-thickness direction, and this raises the stress at the crack tip.

The large change in the critical stress intensity factor with material thickness must be an important consideration in the design of damage-tolerant aircraft structures with high fracture toughness. It should not be assumed that a material has a constant critical stress intensity factor, but rather the correct value for the given thickness and yield strength of the material must be used in any damage tolerance and structural design analysis of aircraft structural components. It is usual to assume the worst case condition for fracture toughness by using the plane strain stress intensity factor K_{Ic}.

19.2.4 Fracture toughness of high-strength structural metals

For many materials, their fracture toughness depends on both the yield strength and ductility. Within a single group of materials (such as aluminium alloys or steels) an increase in yield strength almost always corresponds with a loss in fracture toughness. Low strength metals (with high ductility) often have high toughness because a large plastic zone develops at the crack tip which absorbs energy and thereby resists crack growth. High-strength metals have lower toughness because of their inability to form a large plastic zone and, as a result, high stresses occur at the crack tip. Figure 19.6 shows

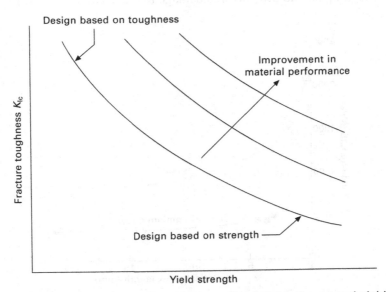

19.6 Generalised relationship between fracture toughness and yield strength of ductile materials.

the relationship between fracture toughness and yield strength for ductile materials. Toughness decreases with increasing yield strength because of the reduction in the size of the plastic zone.

The relationship between fracture toughness and plastic zone size can be described mathematically using fracture mechanics. For an infinitely wide plate containing a through-thickness crack, then the radius of the plastic zone r_y is related to the yield strength σ_y of the ductile material by the plane stress condition:

$$r_y = \frac{1}{2\pi}\left(\frac{K_c}{\sigma_y}\right)^2 \tag{19.2}$$

and the plane strain condition

$$r_y = \frac{1}{6\pi}\left(\frac{K_c}{\sigma_y}\right)^2 \tag{19.3}$$

These equations show that the size of the plastic zone shrinks rapidly when the yield strength of the material is increased. The inverse relationship between fracture toughness and yield strength implies that the maximum critical crack size in metals decreases with increasing strength. This relationship is shown in Fig. 19.7 for aluminium alloy and steel. In practical terms, this relationship means that soft, ductile materials capable of a large amount of plastic flow at the crack tip should provide high damage tolerance to structures. However,

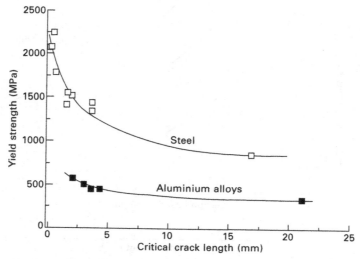

19.7 Relationship between yield strength and allowable crack size for steel and aluminium alloys used in aircraft.

this is offset by the reduced strength of the material. Alternatively, when a high-strength material is used to maximise the load capacity and minimise the weight of an aircraft component, then the maximum crack size that can be tolerated becomes smaller.

19.3   Ductile/brittle fracture transition for metals

Another consideration in the selection of materials for damage-tolerant aircraft structures is the ductile/brittle transition effect. This effect describes the change in the fracture behaviour of metals from ductile cracking at or above room temperature to brittle cracking at low temperatures. The change in the fracture mode often occurs over a temperature range known as the transition temperature. Figure 19.8 shows the change to the critical stress intensity factor of steel with temperature resulting from the ductile/brittle transition effect. In this material, the fracture toughness decreases with temperature over the range of +5 to –20 °C owing to the change in the failure mode from ductile to brittle fracture.

The ductile/brittle transition effect occurs because the development of the plastic zone in some types of metals is a temperature-dependent process. At high temperatures, there is sufficient thermal energy in the crystal structure to aid the movement of dislocations under an externally applied stress. This allows the plastic zone to develop at the crack tip which then allows cracking to proceed by ductile fracture. The thermal energy to assist dislocation slip drops with temperature, and this makes it harder to develop the plastic

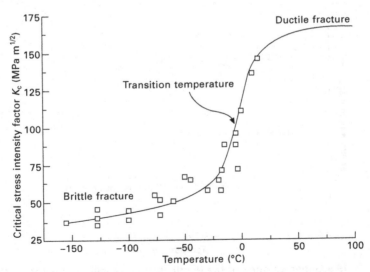

19.8 Ductile/brittle transition curve for medium-strength steel.

zone. Dislocation mobility and, hence, the size of the plastic zone decrease rapidly at the transition temperature which results in a large loss in fracture toughness. Dislocation slip virtually stops below the transition temperature, which causes the metal to fracture by brittle crack growth.

The ductile/brittle transition behaviour of a wide range of metals falls into three categories determined by their yield strength and crystal structure, as shown in Fig. 19.9. Metals with a face centred cubic (fcc) crystal structure do not undergo the transition and retain their ductility at low temperature. This is because fcc metals have a large number of slip systems in their crystal structure which allows dislocation slip to occur, even at very low temperature. Aluminium is an fcc metal and, therefore, does not become brittle at low temperature. Most hexagonal close packed metals (hcp), including magnesium and α-titanium alloys, also do not undergo the transition effect. Metals with a body centred cubic (bcc) crystal structure often display ductile/brittle transition properties. Fracture in many bcc metals occurs by brittle cleavage at low temperatures and by ductile tearing at high temperature. The transition temperature is dependent on the alloy content, thickness and yield strength of the metal, and for steel can be as high as 0 °C.

The operating temperature range of most aircraft structures is between +40 and –50 °C. Although the ductile/brittle transition effect is not a problem for aluminium alloys, other structural metals such as steel may be susceptible. It is essential that any aircraft material does not become brittle at temperatures above –50 °C, otherwise the fracture toughness and damage tolerance of the structure are seriously compromised.

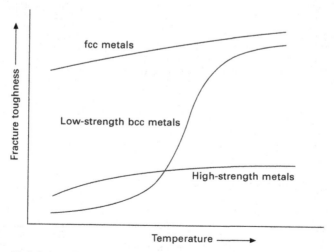

19.9 General trends of the ductile/brittle transition effect for different groups of metals.

19.4   Improving the fracture toughness of aerospace materials

19.4.1 Toughening of metals

The aerospace industry is always seeking new ways to increase the fracture toughness of structural metal alloys without significant loss in strength, fatigue resistance and other important mechanical properties. This is usually achieved by alloying, processing and heat treatment (including ageing). As an example, Fig. 19.10 shows the increases in fracture toughness and yield strength of aircraft-grade aluminium by changes to the alloy composition and heat treatment. The development of aluminium alloys with both high fracture toughness and high strength has focused on composition and processing controls which minimise intergranular cracking by reducing the presence of grain boundary precipitates and increasing the concentration of precipitates within grains.

Control of impurities (such as iron and silicon) within aluminium has resulted in improvements to both fracture toughness and yield strength. Figure 19.11 shows the effect of Fe+Si impurity content on the toughness of 7075 aluminium sheet, and similar changes are achieved with 2000 and 8000 series aluminium alloys. By minimising the impurity content it is possible to reduce the concentration of brittle precipitates that lower the fracture

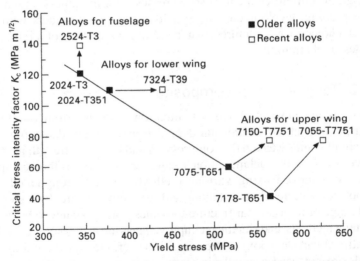

19.10 Improvements to fracture toughness and strength of aluminium alloys by alloy control and heat-treatment (adapted from C. Peel, 'Advances in aerospace materials and structures', in *Aerospace materials*, ed. B. Cantor, H. Assender and P. Grant, Institute of Physics Publishing, Bristol, 2001).

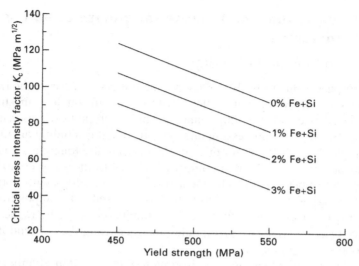

19.11 Effect of Fe+Si content on the fracture toughness and yield strength of 7050 aluminium alloy (adapted from E. A. Starke and J. T. Staley, 'Application of modern aluminium alloys to aircraft', *Progress in aerospace science,* **32** (1996), 131–172).

toughness. Reducing the grain size by thermomechanical processing and the use of grain-refining elements also increases both toughness and strength. These on-going developments are providing aircraft designers with better material choices for structures that need a combination of high fracture toughness and strength.

19.4.2 Toughening of composites

Composites are susceptible to delamination cracking from overloading, bird strike, impact, environmental degradation and other damaging events, which significantly weaken the compression and bending strength properties. Improvements to the delamination toughness of carbon-fibre composites have been achieved using various methods, such as toughened resins, thermoplastic interleaving, stitching and z-pinning. Table 19.3 shows the typical range of interlaminar fracture toughness values achieved by different toughening methods, and significant improvements are achieved compared with conventional carbon–epoxy composite. However, most of these toughening methods are not used currently in aircraft composites structures, and only toughened epoxies are widely accepted.

Table 19.3 Delamination fracture toughness values for various toughening mechanisms for carbon–epoxy composite

Toughening method	Mode I interlaminar fracture toughness G_{Ic} (J m^{-2})
Conventional (brittle) epoxy resin	200–500
Chemical toughened epoxy	800–1400
Rubber toughened epoxy	1000–2500
Nanoparticle reinforced epoxy	1500–4000
Thermoplastic interleaving	1500–3000
Through-thickness stitching	1500–8000
Through-thickness weaving	1500–8000
Z-pinning	2000–10000

19.5   Summary

The fracture toughness as measured by the critical stress intensity factor K_c and critical strain energy release rate G_c define the resistance of materials against crack growth. The fracture toughness properties of materials vary over a wide range, about five orders of magnitude. High fracture toughness in metals is generally achieved by increasing the ductility, but this often comes at the expense of lower yield strength.

The fracture toughness properties of fibre–polymer composites are anisotropic, with the highest fracture resistance occurring with in-plane fracture that involves breakage and pull-out of the fibres and the lowest fracture resistance occurring by interlaminar (delamination) cracking. The fracture toughness properties of wood and metals with a directional grain structure are also anisotropic.

The fracture toughness of metals and other ductile materials is dependent on their thickness. The toughness is highest when plane stress conditions exist at the crack tip, and this occurs when the material is thin (typically less than several millimetres). The fracture toughness falls over a thickness range (between several millimetres and several tens of millimetres) as plane strain conditions become more influential on the plastic yielding process at crack tip. The fracture toughness is lowest in thick materials when fully plane-strain conditions occur at the crack tip. Large differences in fracture toughness exist between thin and thick materials, and this must be considered in the selection of structural materials and the design of damage tolerant aerospace structures.

The fracture toughness of metals can be improved without significant loss in strength in several ways, including minimising the impurity content, reducing the grain size, and reducing the amount and size of intermetallic particles at the grain boundaries.

The fracture toughness of fibre–polymer composites can be improved by using toughening resins, thermoplastic interleaving, through-thickness reinforcement by stitching, pinning or orthogonal weaving, as well as other processes.

19.6   Terminology

Ductile/brittle fracture transition: Change in the failure mode of certain types of metals from ductile tearing to brittle (cleavage) fracture when cooled below the transition temperature.

Transition temperature: The temperature when the fracture behaviour of a material transforms from ductile to brittle.

19.7   Further reading and research

Dieter, G. E., *Mechanical metallurgy*, McGraw–Hill, London, 1988.

Dimatteo, N. D., *ASM handbook, volume 19: fatigue and fracture*, ASM International, 1996.

Gordon, J. E., *The new science of strong solids or why you don't fall through the floor*, Penguin Science, 1991.

Hertzberg, R. W., *Deformation and fracture mechanics of engineering materials*, John Wiley & Sons, 1996.

Knott, J. F., *Fundamentals of fracture mechanics*, Wiley, 1973.

Fatigue of aerospace materials

20.1   Introduction

Fatigue is the most common cause of damage to aircraft structures and engine components. It is estimated that fatigue causes over one-half of all metal component failures, and is responsible for more damage than the combined effects of corrosion, creep, wear, overloading and all the other failure sources on aircraft. For example, a study performed on repairs made to the aluminium fuselage of 71 Boeing 747 aircraft with an average life of nearly 30 000 flight hours revealed that the most common type of damage was fatigue cracking (58% of all repairs), followed by corrosion (29%) and then impact damage from bird strike (13%). Fatigue is also damaging to composite structures, although the incidence of fatigue failures is less than with metals. The damaging effect of fatigue increases with the age of the aircraft. Older aircraft contain many small cracks caused by fatigue, and the inspection and repair of these aircraft to rectify fatigue problems is a major maintenance cost. As aircraft become older the problem of fatigue rises to the forefront of safety and structural reliability.

Fatigue is defined as the deterioration to the structural properties of a material owing to damage caused by cyclic or fluctuating stresses. A characteristic of fatigue is the damage and loss in strength caused by cyclic stresses that are below (often well under) the yield strength of the material. In other words, the repeated application of elastic stresses can damage and weaken a material. Another characteristic of fatigue is that the material often shows no visible sign of damage before it fails. The damage caused by fatigue is often a single crack in a metal component and ultrafine cracks in a composite structure, and these are not easily seen during visual inspection of aircraft. Fatigue damage can grow undetected until the material fails suddenly and without warning.

Aerospace materials can experience several types of fatigue.

- Cyclic stress fatigue is the most common form of fatigue, and occurs by the repeated application of loads to the material. Of the many ways that fatigue can occur, cyclic stress fatigue is responsible for more failures of aircraft components, more aircraft crashes, and requires the highest maintenance inspection of ageing aircraft.
- Corrosion fatigue is another common form of fatigue experience by metallic materials (particularly high-strength aluminium alloys) used

469

in aircraft, and occurs by the combined effects of corrosion and cyclic stress loading. Corrosion fatigue is described in the next chapter on the corrosion of aircraft metals.

- Fretting fatigue is the progressive deterioration of materials by small-scale rubbing movements that cause abrasion of mating components. This type of fatigue occurs most often in joints containing bolts, rivets or screws which have become loose, thereby allowing movement between parts.
- Acoustic fatigue is caused by high-frequency fluctuations in stress caused by noise. The pressure waves of the noise impinge on the material thus inducing fatigue effects. Common sources of acoustic fatigue are jet or propeller noise.
- Thermal fatigue is caused by fluctuating stresses induced by the thermal expansion and contraction of materials owing to thermal cycling (i.e. repeated heating and cooling). Thermal fatigue usually occurs in materials required to operate over a large temperature range, such as jet engine components, heat shields and rocket motor nozzles.

It is essential that aerospace engineers understand the fatigue resistance of materials. Fatigue resistance is the ability of structural materials to maintain an acceptable level of strength under fluctuating stress conditions. Fatigue is a major factor affecting the design of aircraft structures. Aerospace engineers must have the know-how to design components and select materials that have high fatigue damage tolerance (i.e. resistance to failure) over the aircraft life.

In this chapter, we examine the fatigue properties of the metals and fibre–polymer composite materials used in aircraft structures and engine components. Most of the chapter is devoted to cyclic stress fatigue because, as mentioned, it is the most common type of fatigue. The nature of fatigue stress loading, the development of fatigue damage in materials, and the final failure of materials caused by fatigue is explained. In addition, techniques used by aerospace engineers to control the stress fatigue problem in aircraft materials are described. Fatigue caused by fretting, acoustic or thermal effects is also briefly explained in this chapter, whereas corrosion fatigue is described in chapter 21.

20.2   Fatigue stress

20.2.1 Fatigue stress loading on aircraft

Fatigue damage to aircraft structural materials is caused by repeated fluctuating loads. The loads cause cracking in metals and composites that can lead to complete failure if left unrepaired. Figure 20.1 shows typical stress fluctuations on an aircraft wing during a single flight. When the aircraft is stationary the wing is stressed to about minus one g as it deflects downwards under

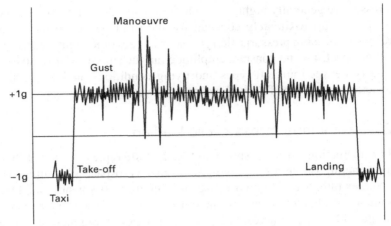

+1g

−1g

Manoeuvre

Gust

Take-off

Taxi

Landing

20.1 Typical fatigue stress profile of an aircraft wing during one complete flight cycle.

its own weight and the weight of any wing-mounted engines. Low-level fatigue stresses are produced during the ground roll and taxi of the aircraft before take-off. The magnitude of the fatigue stress increases rapidly during take-off and ascent of the aircraft until it reaches the cruise altitude. The stress becomes positive when the aircraft becomes airborne because the wing deflects upwards under the pressure needed to support the weight of the fuselage. During the cruise phase, the fatigue stresses fluctuate randomly owing to wind gusts and manoeuvres. High fatigue loads are generated by high speed, tight manoeuvres (otherwise known as 'high g' turns, up to 6g to 9g) in military fighter and aerobatic sports aircraft.

All the stresses generated during a single flight contribute to the progressive fatigue deterioration of the aircraft over many flights. The magnitude and frequency of the fatigue stresses are different for different structures. For example, consider the fatigue stresses applied on the pressurised fuselage of an aircraft. The fuselage skin is loaded in tension when the cabin is pressurised during take-off. The fuselage expands like a balloon owing to cabin pressure being higher than the external atmospheric pressure. The fuselage contracts when depressurised during descent. This expansion and contraction of the fuselage represents one fatigue load cycle per flight. The fuselage is also subjected to fatigue loading during flight from gusts and manoeuvres that repeatedly stress the skin. The fatigue loads applied to the wing are different from those on the fuselage, and involve bending and torsion stresses many thousands of times in a single flight. Therefore, the aircraft fuselage needs to be constructed using material capable of withstanding low cycle fatigue loads (from cabin pressurisation), whereas wings require materials that can sustain high cycle fatigue loads (from gusts and manoeuvres). The cyclic stresses are generally highly variable for an aircraft wing whereas the stress cycles are approximately constant for the fuselage where the main stress results from cabin pressurisation. These two types of loading are, respectively, called spectrum and constant amplitude fatigue stresses. The loads also vary widely between aircraft types and flying conditions, making it difficult to generalise about the fatigue loading on aircraft.

20.2.2 Fatigue stress cyclic loading

The fluctuations in stress shown in Fig. 20.1 are representative of the fatigue loads acting on an aircraft wing and other aerospace structures. This is a complex fatigue condition because the magnitude, frequency and duration of some of the loading changes randomly. The first step in understanding the fatigue properties of aerospace materials is determining their response under simple cyclic loading conditions that do not involve random stress cycles. Figure 20.2 shows two constant-stress amplitude conditions that are used to evaluate the fatigue properties of materials. These conditions are called fully-reversed cycle fatigue and repeated stress cycle fatigue. Fully-reversed fatigue loading involves the material being loaded in tension (which is considered positive stress) and compression (negative stress) within a single cycle. The magnitude of the alternating stress σ_a is the same in tension and compression. The repeated stress cycle involves the load remaining positive or negative for the entire cycle, and the magnitude of the stress varies between a constant maximum σ_{max} and a constant minimum σ_{min}.

The fatigue cycle has several important stress parameters that can affect the fatigue properties of materials, and these are:

- Maximum fatigue stress σ_{max}.
- Mean (or steady-state) fatigue stress σ_m, which is the average of the maximum and minimum stress in the cycle: $\sigma_m = (\sigma_{max} + \sigma_{min})/2$.
- Fatigue stress ratio R, which is the minimum divided by the maximum fatigue stress: $R = \sigma_{min}/\sigma_{max}$.
- Stress frequency f, which is the number of load cycles per second.

The susceptibility of materials to fatigue damage and failure increases with all of the stress parameters. That is, the rate of damage growth increases with the maximum fatigue stress, mean fatigue stress, fatigue stress ratio and, in some materials, the stress frequency.

Fatigue properties are also dependent on the type of fatigue stress, with the repeated application of tension loads being more damaging than cyclic compression loads for metals, whereas fully reversed (tension–compression) loads are more detrimental to fibre–polymer composites than repeated tension–tension or compression–compression loads. The fatigue properties of materials are usually different under cyclic tension, compression, shear,

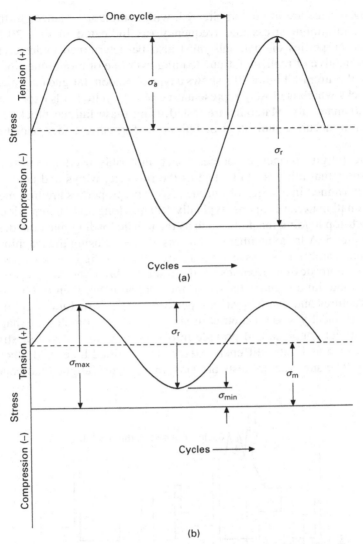

20.2 Fatigue stress profiles for (a) fully reversed and (b) repeated stress cycling.

torsion and other loading states, and, for this reason, it is important that the fatigue properties of aerospace materials are determined for the cyclic load conditions that best represent the actual loads on the aircraft structure.

Fatigue tests on materials are initially performed for simple cyclic stress loading conditions involving a constant load cycle repeated many times until failure. However, the fatigue stresses experienced by aircraft materials during service are more complicated, with continuous changes to the peak stress, minimum stress and frequency (as indicated in Fig. 20.1). The fatigue properties of materials must also be determined under conditions representative of realistic fatigue loading over one or more times the design life of the aircraft. Figure 20.3 shows a typical random fatigue loading applied in blocks which is closely representative of the cyclic loads experienced by aircraft materials. Materials are tested using these fatigue block cycles to obtain an accurate indication of fatigue performance under realistic flight conditions.

The fatigue properties of aerospace materials used in safety-critical structures that must not fail, such as the fuselage, wings and landing gear, are determined in a series of fatigue tests. The properties are first measured with small material coupons (typically 100 mm long and 10 mm wide) using a bench-top loading machine, such as the tensile loading machine described in chapter 5. A large number of coupons are tested using the machine under constant amplitude stress conditions as well as fatigue block loading for two or more design lifetimes of the aircraft. After the fatigue properties of the material are characterised at the coupon level, then fatigue tests on substructures and components are performed. The structural specimens are larger (typically several metres in size) and more complex in shape than the material coupons, and provide information on the fatigue of structural details such as joints, stiffeners, attachments, drilled holes and other stress raisers. The substructure tests are performed under fatigue load conditions

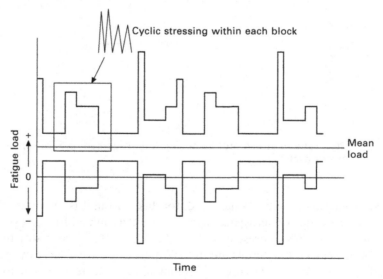

20.3 Fatigue block cycle.

that replicate, as closely as possible, the actual in-service fatigue loads. These tests provide data which help in the design of the aircraft. Finally, a full-scale test on a single aircraft is usually performed to validate the design and to assess the overall fatigue properties of the materials and structures. Because the tests are technically difficult and very expensive, they are not performed on all aircraft types but only on selected types, such as large passenger aircraft or fighter aircraft that have significant new design features or experience extreme fatigue conditions.

20.3   Fatigue life (S–N) curves

The basic method for determining the fatigue resistance of materials is the fatigue life (S–N) graph, which is a plot of the maximum fatigue stress S (or σ_{max}) against the number of stress cycles-to-failure of the material N. The graph is usually plotted with the fatigue stress as a linear scale and load cycles-to-failure as a log scale. S–N graphs are used in aircraft design to determine the number of cycles of stress that a material can endure before failure.

There are two basic shapes for S–N graphs, as shown in Fig. 20.4. One type slopes steadily and continuously downwards with increasing number of load cycles (curve A). Materials that display this type of S–N curve fail in fatigue with a sufficient number of load cycles. Many nonferrous alloys, including aluminium and magnesium, have this type of S–N curve. This is undesirable because failure may eventually occur at low fatigue stress levels. The other type (curve B) becomes horizontal at a limiting fatigue stress, which is called the endurance limit or fatigue limit. When the maximum fatigue stress is

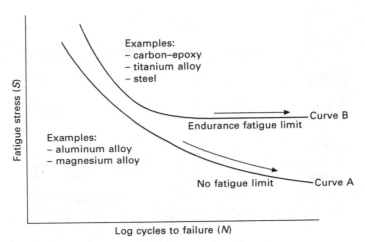

20.4 Two basic shapes of the S–N curve.

below the endurance limit then the material can endure an infinite number of load cycles. This is a desirable fatigue property for aerospace materials because an infinite life is assured when the fatigue stress is kept below the endurance limit. Examples of aerospace materials that have an endurance limit include titanium alloys, steels and carbon fibre–epoxy composites.

S–N curves for several aerospace structural materials are given in Fig. 20.5. The S–N curves show that large differences exist in the fatigue performance between aerospace structural materials. In general, carbon fibre–epoxy composites, titanium alloys and high-strength steels have better fatigue resistance than aluminium and magnesium alloys under cyclic tension loading. The S–N curve is useful for determining the fatigue life, which is the number of repetitions in stress that a material can withstand before failure. The number of cycles-to-failure can be easily determined for any fatigue stress level using the S–N curve.

An S–N curve is only valid for a specific set of fatigue conditions (e.g. R ratio, load frequency, temperature), and the graph may be different when the conditions are changed. The S–N curves for metals are dependent on many factors, with the most important being the alloy composition, microstructural features (e.g. grain size, type and size of precipitates), and the thermomechanical treatment. The graphs for composites are also dependent on the composition; with the type, volume content and orientation of the fibres and the type of polymer matrix being important. Another important feature is that fatigue of both metals and composites is a stochastic process, with considerable scatter even in a controlled environment. Figure 20.6 shows the scatter in the fatigue life data for an aluminium alloy, and the variability in the number of load cycles-to-failure is as high as one order of magnitude. Other aerospace

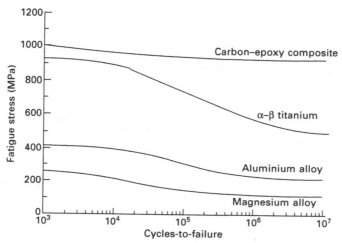

20.5 S–N curves for aerospace structural alloys and composite.

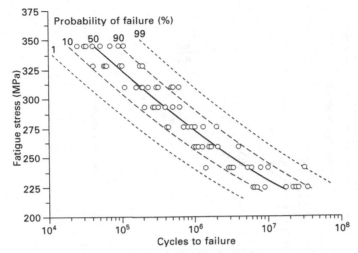

20.6 Scatter in the fatigue life of 2024 aluminium alloy.

materials also experience scatter in their life, and this must be considered in assessing the fatigue properties. A- and B-basis fatigue allowables (as described in chapter 5) are used to define the fatigue limits on aerospace materials.

20.4   Fatigue-crack growth curves

S–N curves are useful for determining the number of load cycles-to-failure for a material, but they do not provide information on the amount of fatigue damage the material sustains before failure. Fatigue-crack growth curves are used, in combination with *S–N* curves, to determine the fatigue resistance of metals. The curves are used to determine the fatigue stress conditions under which cracks initiate, grow and cause complete failure.

Figure 20.7 shows the general form of the fatigue-crack growth curve for metals. The stress intensity factor range, ΔK is used to represent the variation in the fatigue stress acting on the crack within a single load cycle. Expressed mathematically, $\Delta K = (\sigma_{max} - \sigma_{mm})Y\sqrt{\pi a}$, where a is the crack length and Y is a correction factor that is dependent on the shape and geometry of the crack. The fatigue-crack growth rate, da/dN, is the distance a that the crack propagates in one load cycle N.

A fatigue-crack growth curve is divided into three regions. Region I is the condition when no fatigue-crack growth can occur because the stress intensity range is too low. That is, the peak fatigue stress applied to the metal is too small to cause crack growth. It is only when the stress intensity factor range rises above a threshold, ΔK_{th}, that a crack begins to grow. It is important

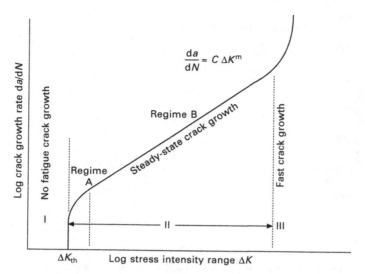

20.7 Fatigue-crack growth curve.

that metal aircraft structures are designed with adequate section thickness to ensure the stress intensity factor range is kept below ΔK_{th}, thereby avoiding fatigue cracking problems.

Region II covers the stress intensity range over which fatigue cracks grow stably cycle-by-cycle. This region is subdivided into two regimes. Regime A occurs over a narrow stress intensity range immediately above ΔK_{th}. The fatigue-crack growth rate is very slow in this regime ($<10^{-8}$ m/cycle), with the rate changing continuously with ΔK. Steady-state crack growth rate occurs in regime B, with the rate covering the range about 10^{-6} to 10^{-8} m/cycle for most metals. The graph in this regime is approximately rectilinear when plotted on log–log scales, and has a slope m. The crack growth rate is related to the stress intensity factor range in the linear regime by:

$$\frac{da}{dN} = C\Delta K^{m} \qquad [20.1]$$

where C is a material constant. This is known as the Paris law for fatigue-crack growth. When the range of stress intensity factors, material constants (C, m), and initial (a_o) and final (a_f) crack lengths are known, then the number of load cycles that the material can safely withstand can be estimated by integrating the expression into:

$$N_f = \int_0^{N_f} dN = \int_{a_0}^{a_f} \frac{da}{A(\Delta K)^m} \qquad [20.2]$$

This equation is often used to estimate the fatigue life of materials containing cracks.

When the peak fatigue stress is high enough to cause general plastic deformation (i.e. low cycle fatigue), then the expressions described above are not accurate. Instead, the number of load cycles to failure is related to the plastic strain by the Coffin–Manson relation:

$$\Delta\varepsilon_p = 2\varepsilon_f(2N)^c \qquad [20.3]$$

where $\Delta\varepsilon_p$ is the change in the plastic strain during one fatigue load cycle, ε_f is the static failure strain, and c is an empirical constant known as the fatigue ductility coefficient, which is typically –0.5 to –0.7 for most metals.

Region III of Fig. 20.7 is the most severe fatigue condition because it involves rapid crack growth ($>10^{-6}$ m/cycle) that quickly leads to fracture. When the stress intensity range approaches the fracture toughness (K_c) of the material, the speed of the fatigue crack accelerates rapidly and failure occurs quickly. The material fails within a few hundred load cycles owing to the rapid advance of the fatigue crack.

Fatigue-crack growth curves for metals are dependent on their alloy composition and microstructure as well as the cyclic loading condition. For example, Fig. 20.8 shows the curves for three aluminium alloys used in aircraft structures, and there is a large difference between their fatigue-crack growth rates. Crack growth in the 8090 Al alloy is about ten times slower than in the 2024 Al alloy and slower still compared with the 7075

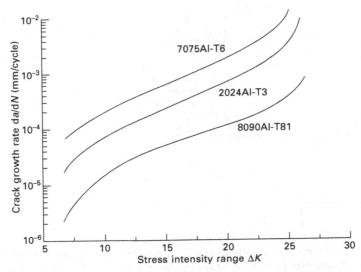

20.8 Fatigue crack growth rate curves for three aircraft-grade aluminium alloys.

Al alloy. This information allows the designer to select the best material for aircraft structures with the highest resistance against fatigue cracking. For example, pressurised cabins and lower wing skins are two areas prone to fatigue through the application and relaxation of tensile stresses. For this reason, 2024 Al is often preferred over 7075 Al because of its slower fatigue-crack growth rate; 8090 Al is not used because of its higher cost. The upper wing skin, which has to withstand compressive stresses as the wing flexes upwards during flight, is not prone to metal fatigue because cracks do not grow in compression. For this reason, the higher strength 7075 Al is often used because its lower fatigue resistance is not a problem.

20.5   Fatigue of metals

20.5.1 Fatigue-crack growth in metals

The total fatigue life, from the first load cycle to the last load cycle that causes final failure, of metals is divided into three stages:

(i)   fatigue crack initiation,
(ii)   crack growth under cyclic loading and
(iii)   final failure.

The initiation and growth of a fatigue crack under repeated tensile loading is illustrated in Fig. 20.9. A fatigue crack (also called Stage 1) in metals often initiates as a shear crack and then grows as a tensile crack (Stage 2

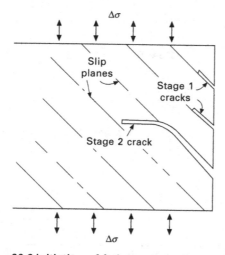

20.9 Initiation of fatigue cracks in metals (reproduced from M. F. Ashby and D. R. H. Jones, *Engineering materials 1: an introduction to properties, applications and design*, Elsevier Butterworth–Heinemann, 2005).

crack). Crack initiation can occur in a defect-free region or in a pre-existing defect in the material. In a defect-free region, any microstructural feature that concentrates the stress is a potential site for the formation of a fatigue crack. Crack initiation usually occurs at coarse slip bands with a high dislocation concentration or at localised soft regions such as precipitate-free regions near grain boundaries. Figure 20.10 shows the initiation and growth of a fatigue crack in an aircraft-grade aluminium alloy. A small crack develops in the

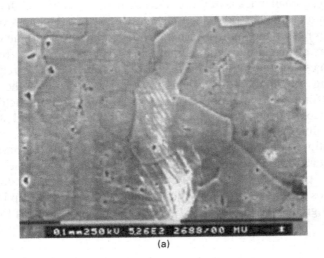

(a)

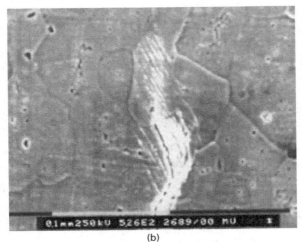

(b)

20.10 (a) Initiation and (b) growth of a fatigue crack in a 2024 Al-T351 alloy (from X. P. Zhang, C. H. Wang, L. Ye and Y.-W. Mai, *Fatigue and Fracture of Engineering Materials and Structures*, **25** (2002), 141–150).

defect-free region after repeated loading, even when the maximum fatigue stress is less than the yield strength of the metal.

More often, fatigue cracks initiate at pre-existing defects, such as voids, large inclusions or surface flaws that act as stress raisers. Common examples of pre-existing surface defects that initiate fatigue cracks are machining burrs, scratches, corrosion pits, and sharp corners; and these must be avoided to delay the initiation of fatigue cracks. Under cyclic loading, the metal near the pre-existing defect is plastically deformed owing to the stress concentration and, eventually, a small crack is initiated. Defects that concentrate high levels of stress, such as scratches or sudden changes in section thickness of the component can reduce the number of load cycles to initiate a fatigue crack by many orders of magnitude. For example, Fig. 20.11 shows the effect of surface roughness on the fatigue life of steel. The fatigue life is reduced with increasing roughness which act as stress concentrators at the steel surface. Therefore, care is required in the design, manufacture, machining and maintenance of metal components to avoid surface damage and abrupt changes in section that act as stress raisers.

Once the fatigue crack has formed, it propagates away from the initiation site under repeated loading, and this is the second stage of fatigue. This stage involves two modes of fatigue-crack growth (Fig. 20.9). Stage 1 growth involves the extension of the initial crack along specific slip bands within a metal grain. The rate of crack growth can be very slow; as low as

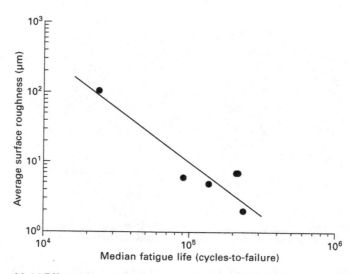

20.11 Effect of surface roughness on the fatigue life of a steel part (data from P. G. Fluck, *Proceedings American Society for Testing and Materials*, **51** (1951), 584–592).

several angstroms (10^{-10} m) per load cycle. Fatigue cracks grow by a series of opening and closing motions at the tip of the crack. The crack grows slowly along the slip bands for a few grain diameters, which may be less than a few micrometres, before it changes to Stage 2. This second stage involves the crack growing in the direction normal to the applied fatigue load. Stage 2 fatigue cracks grow in discrete steps, typically advancing at several micrometres per load cycle, with the step length increasing with the fatigue stress level. Fatigue cracks advance in steps with each load cycle until the load-carrying capacity of the remaining (uncracked) portion of the metal component is reached. When the residual load capacity of the fatigued metal reaches the maximum fatigue load level then sudden failure occurs. The crack grows through the remaining section causing complete failure.

Fatigue cracking in metals, which can involve a single dominant crack, is difficult to detect without the aid of nondestructive inspection or structural health monitoring techniques (which are described in chapter 23). It is easy then for fatigue cracks to grow undetected in metal aircraft structures and engine components, sometimes over many years, without displaying any obvious sign of damage until sudden and catastrophic failure. The insidious nature of fatigue cracks is a major factor compromising the structural integrity and safety of aircraft, particularly ageing aircraft approaching the end of their design life.

Cracks in metals are often classified as occurring by low cycle fatigue or high cycle fatigue. Low cycle fatigue occurs when the maximum fatigue stress exceeds the yield strength of the metal. Fatigue cracks initiate easily and grow quickly when plastic deformation occurs with each load cycle, and many metals fail in less than 1000–10 000 cycles. High cycle fatigue occurs when the peak fatigue stress is below the yield strength, and the fatigue process occurs under apparently elastic conditions. There is no general plasticity of the metal; only local plasticity where a pre-existing stress raiser such as notch or hole concentrates the stress. The fatigue cracks that develop under this condition have a plastic zone at the crack tip, but the surrounding material is elastically deformed with each load cycle. The number of load cycles to failure under high cycle fatigue is generally above 10 000.

Section 20.11 at the end of the chapter presents a case study of aircraft fatigue in Japan Airlines flight 123.

20.5.2 Surface analysis of fatigued metals

The fractured surfaces of metals that fail owing to fatigue are used to determine the nature of the crack growth process. Figure 20.12 shows the fracture surface of a fatigued metal component, and on close examination shows a pattern of ripples (often called 'beach marks'). The rippled region is the area over which the fatigue crack grows in discrete steps. Each ripple

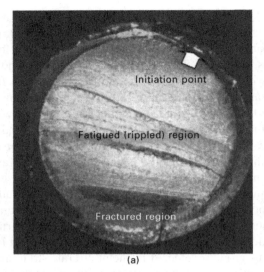

(a)

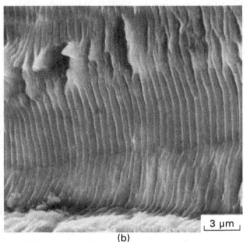

3 μm

(b)

20.12 Fracture surface of fatigued metal: (a) low magnification view showing fracture surface; (b) ripples caused by fatigue crack growth, each ripple indicating an incremental advance in the fatigue crack length.

is a fatigue fracture striation showing the distance the crack has advanced in one load cycle. The distance between the ripples is usually very small, typically less than a few micrometres, with the spacing increasing with the fatigue stress intensity range. The ripples usually radiate outwards from a single point which is the site of crack initiation. As mentioned, the crack usually initiates at a point of stress concentration such as a notch or sharp corner. The crack grows outwards from the initiation point with each load cycle above the threshold stress intensity range ΔK_{th} leaving behind a ripple with every advancing step. When the load capacity of the fatigued metal is reduced to the maximum fatigue load, then sudden fracture occurs through the remaining uncracked region. This is shown by the rougher region on the fracture surface, where ductile tearing has occurred during the rapid growth of the crack.

The appearance of the fracture surface is used to analyse the fatigue failure of metals, including aircraft accident investigations into structural failures. The initiation point shows the cause of the start of fatigue cracking, such as inferior quality metal processing producing large inclusions; poor machining producing surface notches; incorrect design resulting in sharp corners; or in-service damage causing surface dents or scratches. The initiation point is used in accident investigations to determine the original cause of aircraft structural failures. The ripples or more prominent markings which appear when the fatigue loading changes significantly (called progression marks) can be used to determine the rate of fatigue crack growth. The number of ripples (or progression marks) in the smooth region provides information on the time period that the fatigue crack spent growing through the metal component before final failure. This allows accident investigators to assess whether the crack was present during a routine maintenance inspection of the aircraft or whether the crack initiated and grew to final failure between inspections. The spacing between the ripples is used to determine whether the component was overstressed. The ripple spacing (da/dN) is related to the fatigue stress intensity range (ΔK), as shown in Fig. 20.7, and therefore the distance between ripples is used by aircraft accident investigators to determine whether the metal was loaded above its design limit during service. The information gained by careful examination of the fracture surface of fatigued metal components has solved the cause of many aircraft accidents.

Section 20.12 at the end of the chapter presents a case study of the role of metal fatigue in the *Comet* aircraft accidents.

20.5.3 Improving the fatigue properties of metals

Many techniques are used to resist the initiation and slow the growth of fatigue cracks in aircraft structures. As mentioned, fatigue normally starts at a stress raiser, such as a fastener hole, although internal stress concentrations can also initiate cracks. Therefore, any method that removes stress concentration, such as smoother surfaces or blended radii, delays or prevents the initation of fatigue cracks. Methods include ensuring the structure is free from stress concentrations such as sharp corners and sudden changes in section thickness. When stress concentrations cannot be avoided, such as cut-outs for windows, doors and access panels, the structure should be reinforced with additional material to increase the section thickness and thereby reduce the fatigue stress. For the same reason, the material should be made thicker around fasteners holes and other small cut-outs. The material must also be free from surface scratches, machine marks and other stress raisers inadvertently caused by poor quality manufacturing and finishing.

It is common practice to shot peen metal components prone to surface fatigue. Shot peening involves blasting the metal with a high velocity stream of hard particles which introduce a residual compressive stress into the surface region. The applied tensile fatigue stress must first overcome the residual compressive stress on the surface before this region actually experiences any net tensile strain. This then resists the initiation and initial growth of tensile fatigue cracks from surface stress raisers. Figure 20.13 shows the large improvement in the fatigue life achieved using shot peening. Similarly, fastener holes in metal components are often cold worked to introduce residual compressive stress that resists the development of fatigue cracks at the hole edge.

Control of the microstructural properties of metals is an effective way of improving their fatigue properties. The metal must be cast, processed and heat treated using processes that avoid the formation of microstructural defects such as voids and large inclusions, which can initiate fatigue cracks. The grain size also affects the fatigue properties; fine-grained metals generally possess a longer fatigue life than coarse-grained materials. Furthermore, increasing the yield strength of a metal, by appropriate alloying and heat-treatment, often provides greater resistance to fatigue cracking. The fatigue resistance is also dependent on surface protective coatings to resist corrosion, erosion

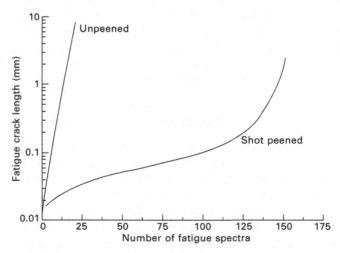

20.13 Fatigue crack growth curves for aluminium alloy which has or has not been shot-peened.

and other environment effects that may damage the material and thereby create a site for fatigue cracking.

20.6   Fatigue of fibre–polymer composites

20.6.1 Fatigue damage in composites

Fatigue of fibre–polymer composites is different to that of metals. As mentioned, fatigue of metals often involves the growth of a single dominant crack. Fatigue cracks in metals grow under repeated tension (but not compression) loads in the direction which is approximately perpendicular (90°) to the load direction. Fatigue damage in composites is different to that encountered in metals. Rather than the single dominant crack that occurs with metals, various types of fatigue damage occur at many locations throughout composite materials. Fatigue of composites is characterised by a multiplicity of damage types, which includes cracks in the polymer matrix, debond cracks between the fibres and matrix, splitting cracks, delamination cracks, and broken fibres. The damage types initiate at different times and grow at different rates over the fatigue life of the composite material. As a result, composites often fail progressively through a series of damage events rather than because of a single, large crack. Despite the many types of damage, continuous fibre–polymer composites, such as carbon–epoxy, often exhibit a fatigue life which is much longer and a fatigue endurance limit which is higher than aerospace-grade aluminium alloys.

Aerospace engineers are reliant on S–N curves to determine the fatigue life and fatigue strength of composites. Figure 20.14 shows the S–N curve and the fatigue damage states for a quasi-isotropic carbon–epoxy composite material under cyclic tension loading. At high fatigue stress and short life, the fatigue damage process is dominated by fibre breakage. The fatigue strains in the high-stress regime approach the failure strain of the fibres and, therefore, they rapidly break leading to early failure of the material. At intermediate stress levels, the fatigue process involves many types of damage, including matrix cracks, fibre–matrix debonding and delamination cracks. Fibre breakage also occurs in the intermediate stress range, but it occurs slowly and therefore the fatigue life of composites is prolonged. At the lowest fatigue stress regime, the fatigue life is infinite and failure does not occur. This is the fatigue endurance limit of the composite, and is determined by the fatigue limit of the polymer matrix.

20.6.2 Fatigue life of composites

The development of fatigue damage and, as a result, the fatigue life is dependent on the composition and microstructure of the composite material.

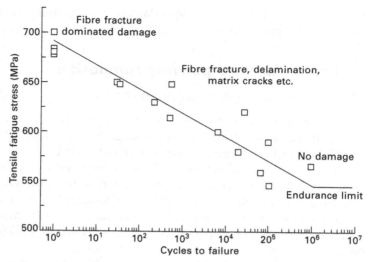

20.14 S–N curve and damage modes to a quasi-isotropic carbon–epoxy composite for tensile fatigue.

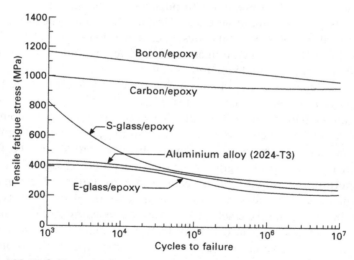

20.15 S–N curves for unidirectional fibre–polymer composites and aluminium alloy under cyclic tensile loading.

The fibre type, fibre volume percent, fibre lay-up pattern, and matrix properties all influence the fatigue life. For example, the effect of the type of fibre reinforcement on the fatigue performance of epoxy matrix composites is shown in Figure 20.15. The S–N curve for an aerospace-grade aluminium alloy (2024-T3) is shown for comparison. The fatigue resistance of composites generally improves with their elastic modulus and strength; with materials containing high stiffness, high-strength carbon fibres having excellent fatigue resistance. The fatigue performance of carbon–epoxy is much superior to that of aluminium alloy, which is a key reason for the increased use of these composites in aircraft structures. Glass fibre composites have inferior fatigue performance compared with carbon fibre materials because of their lower fibre stiffness. The lower stiffness of fibreglass causes the composite to undergo greater strain (change in shape) under cyclic loading, which causes more fatigue damage and thereby reduces the fatigue life. The low fatigue performance of glass fibre composites is an important reason for their exclusion from primary aircraft structures.

The fatigue life of composites is dependent on their fibre lay-up pattern. Figure 20.16 shows S–N curves for carbon–epoxy with unidirectional [0], cross-ply [0/90], quasi-isotopic [0/±45/90] and angle-ply [±45] fibre patterns. The fatigue life decreases with a reduction in the percentage of load-bearing fibres (which in this case are 0° fibres) in the composite. Lowering the amount of load-bearing fibres reduces the stiffness of the composite, which increases the amount the composite is strained under cyclic loading. This creates a greater amount of damage with each load cycle that lowers the fatigue life.

The cyclic loading conditions have a major influence on the fatigue life of composites. As mentioned, fatigue-crack growth in metals occurs under repeated tension loading but not under cyclic compression. Fatigue damage in

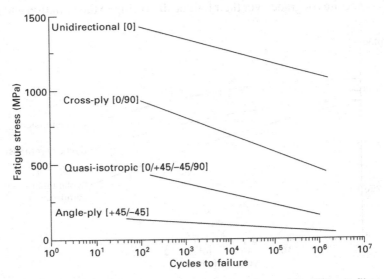

20.16 S–N curves for carbon–epoxy composites with different fibre patterns.

composites, on the other hand, can occur under both tension and compression loads. Figure 20.17 shows S–N curves for carbon–epoxy subjected to repeated tension–tension, compression–compression and reversed tension–compression loads. Fatigue occurs with all three load cases, but tension–compression loading is the most severe followed by compression–compression and then tension–tension. This is an important issue in the use of composites in aircraft structures subjected to fatigue loading. Composites used in structures that predominantly experience fluctuating tension loads, such as the underside of wings and pressurised fuselages, have the highest fatigue performance. The fatigue performance is worse under compression loading because the fibres are prone to buckling and kinking. Aramid fibre composites are particularly susceptible to fibre buckling, and should not be used in compression-loaded structures.

A final point on the effect of loading is the influence of the frequency of the fatigue cycles. The fatigue life of composites is often reduced when the load frequency is above about 20 Hz. This is because heat builds up in the material at high frequencies, thus softening the polymer matrix and thereby hastening the onset and development of fatigue-induced damage. The fatigue performance of metals, on the other hand, is not affected by the frequency of loading until extremely high levels, which are rarely encountered in aircraft.

20.6.3 Mechanical properties of fatigued composites

When the fatigue stress exceeds the endurance limit, the mechanical properties of composites degrade over their fatigue life owing to the initiation and spread

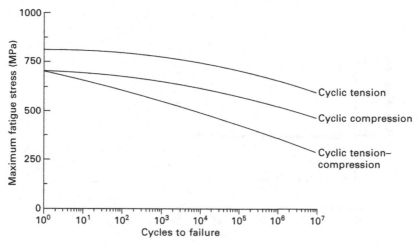

20.17 S–N curves for carbon–epoxy composite under different cyclic loading conditions.

of damage. Figure 20.18 shows the reductions to the Young's modulus and tensile strength that can occur over the fatigue life. The stiffness decreases early in life, but usually by only a few percent, and then remains reasonably steady for most of the fatigue life. Towards the end of life, the stiffness drops sharply owing to the failure of fibres and, within a short time, the material fails completely. The change in strength follows a different trend to stiffness. The strength remains unchanged for a large period of the fatigue life, and it is only when fibres begin to break well into the life-span of the composite that the strength begins to fall. The strength drops as more and more fibres are broken under cyclic loading until eventually the composite is completely broken.

20.6.4 Improving the fatigue properties of composites

The fatigue performance of composites can be improved in various ways. The fatigue life generally increases with the stiffness and strength of the composite, which can be improved in several ways including maximising the volume percentage of load-bearing fibres and using high stiffness, high strength fibres. The fibre volume content of the composite used in aircraft materials structures is usually already high (55–65%), and there is little opportunity to increase it further. The maximum possible fibre content that can be achieved is about 70%, which is the upper limit of fibre packing within composites. Therefore, there is little scope in practice to greatly increase

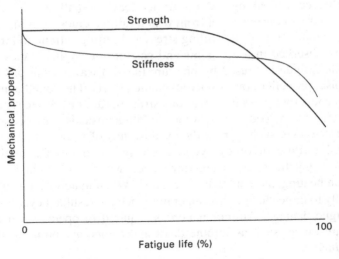

20.18 Reduction in stiffness and strength of carbon–epoxy with increasing number of tensile load cycles.

the fatigue properties of composite by raising the fibre content above the current level. Improving the fatigue life of composites by increasing the volume fraction of load-bearing [0°] fibres comes at the cost of reducing the amount of fibres in the other directions (e.g. ±45°, 90°). Therefore, for aircraft structures subjected to multidirectional loading it may not be possible to increase the volume fraction of load-bearing fibres.

Without doubt, the most practical method of improving the fatigue properties is by the use of high-stiffness, high-strength fibres. The fatigue life improves with the fibre stiffness, and this demonstrates that high modulus carbon fibres are best suited for aircraft composite structures that require high fatigue resistance. There are many types of carbon fibres that can be used in composite aircraft structures, but those fibres with the highest stiffness generally provide the greatest fatigue resistance.

20.7   Fretting, acoustic and thermal fatigue

The most common source of fatigue damage in aircraft is cyclic stress loading. However, fatigue can also be caused by fretting, noise and temperature changes. The crack growth process caused by these types of fatigue is similar to cyclic stress fatigue; the main difference is the source of the cyclic load.

Fretting fatigue involves the progressive deterioration of a material by small-scale rubbing that causes abrasion and cracking. Fretting occurs when two materials, pressed together by an external load, are subjected to transverse cyclic loading that causes one surface to slide back and forth against the other surface. The small sliding displacements cause wear, with small pieces breaking off the surface leaving shallow pits. The pits are sites of stress concentration from which fatigue cracks can initiate and grow under the action of the sliding stresses. Fretting fatigue in aircraft is most often observed in loose joints and gas turbine engine blades.

Acoustic fatigue is caused by pressure (sound) waves from gas turbine engine noise, propeller noise or aerodynamic effects. The sound consists of high-frequency, low-pressure waves that strike the aircraft surface and cause cracking after many load cycles. Acoustic fatigue usually only occurs close to the noise source, such as near the exhaust duct of turbine engines.

Thermal fatigue involves cyclic stressing owing to fluctuations in temperature. Cyclic stresses are introduced when a restrained material expands on heating and contracts on cooling. When materials are subjected periodically to large changes in temperature then the resultant cyclic stresses cause fatigue damage. Aircraft materials required to operate over a wide temperature range, such as turbine discs and rotors, are prone to thermal fatigue damage.

20.8 Summary

Fatigue is the most common cause of damage to aircraft metal structures, and causes more damage than the combined effects of corrosion, wear, impact and other damaging events. The incidence of fatigue damage increases with the age of the aircraft, and old aircraft can require expensive maintenance and repair to combat fatigue.

The most common type of fatigue damage to aircraft is cyclic stress fatigue. Other types of fatigue can also occur: corrosion fatigue (also called fatigue corrosion cracking), fretting corrosion, acoustic fatigue and thermal fatigue.

The fatigue properties of aircraft materials are determined by fatigue tests performed on coupons, structural details, full-scale components and, if necessary, the entire aircraft. The fatigue behaviour is dependent on the cyclic loading condition, such as the type of stress (tension, compression, torsion, etc.), stress level, loading frequency, R ratio, and environmental conditions. Therefore, the fatigue properties of aerospace materials should be determined under test conditions that closely replicate the operating condition.

The existence of a fatigue endurance limit makes it easier to design carbon–epoxy composites, titanium alloys and high-strength steels for a long fatigue life than is the case for aluminium alloys, which lack a clearly defined limit. The inferior fatigue performance of aluminium alloys is a major reason for the increasing use of composites and titanium alloys in military and civil aircraft.

Fatigue cracks often initiate in metals at stress raisers, such as notches or corrosion pits at the surface or voids or large intermetallic particles within the material. Therefore, the fatigue performance of metals is improved by eliminating stress raisers by careful design, manufacturing and maintenance. Other ways of improving the fatigue resistance of metals include shot peening to introduce surface compressive stresses and heat treatment to produce a fine-grained microstructure that is free of large, brittle particles.

The fatigue performance of fibre–polymer composites generally increases with the elastic modulus, strength and volume content of the load-bearing fibres. Because carbon fibres have higher stiffness than glass fibres, their composites have superior fatigue properties. The high fatigue resistance of carbon-fibre composites is a key reason for their use in primary aircraft structures.

The mechanical properties of composites decrease with increasing number of load cycles owing to the initiation and spread of fatigue-induced damage that occurs when the fatigue stress exceeds the endurance stress limit. The Young's modulus decreases by a small amount (usually less than a few percent) early in the fatigue life and then remains stable until towards the end of life when large losses in stiffness occur owing to fibre fracture. The tensile strength is unaffected by fatigue damage until well into the fatigue life (about 50% of remaining life), when it drops as a result of fibre damage.

The fatigue properties of composites are dependent on the cyclic loading condition. The fatigue performance of composites decreases in the order: cyclic tension, cyclic compression, reversed tension–compression. The fatigue performance also decreases with increasing loading frequency above about 20 Hz owing to internal heating effects softening the polymer matrix.

20.9   Terminology

Acoustic fatigue: Fatigue caused by cyclic stress loading from acoustic (noise) waves.

Constant amplitude fatigue load: Cyclic stress loading of constant amplitude with the maximum and minimum loads remaining constant with each cycle.

Corrosion fatigue: Fatigue caused by the combined effects of cyclic stress loading and corrosion.

Cyclic stress fatigue: Fatigue caused by cyclic stress loading.

Endurance limit: The maximum fatigue stress (or strain) that a material can withstand without failing.

Fatigue stress ratio R: The ratio of the minimum to maximum fatigue stress within a load cycle.

Fretting fatigue: Fatigue caused by cyclic stress loading (rubbing) between two contacting surfaces.

Fully reversed cycle fatigue: Repeated loading involving both positive (i.e. tensile) and negative (i.e. compressive) stresses of equal magnitude within one load cycle.

High cycle fatigue: Fatigue loading where the maximum stress is below the yield strength of the material and therefore global plasticity effects are absent which results in long fatigue life (typically in the range 10^4–10^8 load cycles).

Low cycle fatigue: Fatigue loading where the maximum stress exceeds the yield strength of the material and involves plasticity, resulting in short fatigue life (typically under 10^3–10^4 load cycles).

Maximum fatigue stress σ_{max}: The peak stress applied to a material within a load cycle.

Mean (or steady-state) fatigue stress σ_m: The average stress applied to a material within a load cycle; the mid-stress value between the maximum and minimum stress.

Repeated stress cycle fatigue: Repeated loading involving the same type of stress (i.e. always tension or always compression) and the magnitude of the peak stress is the same with each load cycle.

Spectrum fatigue load: Cyclic stress loading of varying amplitude.

Stress frequency *f*: The number of load cycles applied each second.

Thermal fatigue: Fatigue caused by repeated expansion (tension) and contraction (compression) of the material owing to cyclic variations in temperature.

20.10   Further reading and research

Dimatteo, N. D., *ASM handbook, volume 19: fatigue and fracture*, ASM International, 1996.

Harris, B. (editor), *Fatigue in composites*, Woodhead Publishing Limited, 2003.

Pook, L., *Metal fatigue*, Springer, 2007.

Suresh, S., *Fatigue of materials (2nd edition)*, Cambridge University Press, 1998.

Talreja, R., *Fatigue of composite materials*, Technomic Press inc., 1987.

20.11   Case study: aircraft fatigue in Japan Airlines flight 123

Japan Airlines flight 123 was a domestic flight of a Boeing 747 from Tokyo International Airport to Osaka International Airport on August 12, 1985. The aircraft was cruising towards Osaka when suddenly the rear pressure bulkhead, which was made of high-strength aluminium, failed. The resulting explosive depressurisation of the main cabin ripped the vertical stabiliser and other sections from the empennage. A photograph taken from the ground shows the damage (see Fig. 20.19). The explosive event severed all four of the hydraulic systems causing a complete loss of flight control. The pilots struggled to control the aircraft by adjusting engine power, but loss of the stabiliser and control systems caused the aircraft to plunge and ascend uncontrollably in wild oscillations in what is known as a phugoid cycle. The aircraft collided at a speed of 630 km h^{-1} into a mountain, killing all 15 crew members and 505 of the 509 passengers. This is the deadliest single-aircraft accident in aviation history, and comes second in the highest number of fatalities after the twin aircraft disaster in Tenerife when 583 people died.

A complex series of events led to the crash of flight 123, although it can be traced to fatigue damage in the rear pressure bulkhead. Seven years earlier, the aircraft was involved in a tail-strike incident that damaged the rear pressure bulkhead, aft fuselage frames and skin. Aircraft maintenance engineers failed to repair the damaged bulkhead according to the Boeing approved repair method. Testing performed by the accident investigators showed that the incorrect repair performed on the aircraft reduced the fatigue life by about 70%. The aircraft performed over 12 000 flights in the seven-year period between the repair and final accident. Over this period, fatigue cracks initiated and propagated in the repair under the cyclic stress loads applied to the rear bulkhead from the pressurisation/depressurisation of the

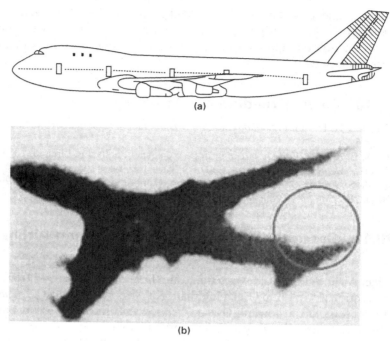

20.19 Japan Airlines flight 123: (a) schematic and (b) photograph showing the damage.

cabin with every take-off and landing. The cracks remained undetected and grew to the critical length to cause catastrophic failure of the bulkhead. Although the accident was caused by incorrect repair rather than the material itself, it tragically demonstrates the importance of fatigue in aircraft structural reliability and safety.

20.12   Case study: metal fatigue in *Comet* aircraft accidents

There have been many aircraft crashes caused by metal fatigue, but the *Comet* aircraft accidents brought fatigue to the forefront of aviation safety. The *Comet* aircraft was designed by the de Havilland company (UK) and first put into service by the British Airways Corporation in 1952. The *Comet* was a revolutionary airliner in an era of major advances in aerospace technology (Fig. 20.20). The *Comet* had a pressurised cabin built with aluminium alloy that allowed cruising at 35 000 ft. Compared with older-style aircraft, the cabin had extra large windows to allow the passengers a wider view. Sectors of the aerospace industry lauded the *Comet* as the aircraft of the future until

20.20 Comet aircraft.

two crashes raised serious concerns about its safety. The first *Comet* crash occurred in early 1954, less than two years after the first flight, killing 29 passengers and 6 crew. The cause of this accident was not immediately obvious because the aircraft seemed to breakup during flight and crashed into the sea. Only four months later another *Comet* broke apart in-flight, this time killing 14 passengers and 7 crew.

Wreckage from the second aircraft was recovered and the damaged fuselage was reconstructed by accident investigators. Failure of the fuselage was traced to a crack that had started from an antenna aperture which, because of its design, was a stress raiser. Closer examination revealed ripples on the fracture surface of the fuselage material, which indicated that fatigue was the problem. It was concluded that the stress concentration at a fastener hole at the antenna aperture caused a fatigue crack to form. The crack grew under the cyclic stressing of the fuselage owing to pressurisation on take-off and depressurisation on landing. The crack grew undetected until eventually a large section of the skin suddenly broke away from the fuselage during flight. The *Comet* accidents highlighted, in the most tragic way, that careful design and fatigue-resistant structural materials are critical to avoid fatigue failure.

Corrosion of aerospace metals

21.1 Introduction

Corrosion of metals used in aircraft structures and engines is a large and expensive problem for the aviation industry. The yearly cost to the industry is over $2.2 billion, which includes the expense of designing and manufacturing aircraft components to resist corrosion ($0.2 billion); downtime when aircraft are inspected for corrosion ($0.3 billion); and corrosion maintenance of aircraft ($1.7 billion). Despite the large sums of money spent on corrosion prevention, it remains a common cause of damage to metal components. Corrosion accounts for about 25% of all metal component failures on aircraft; only fatigue is responsible for more failures than corrosion. The risk and cost of corrosion damage increases with the age of the aircraft, with the hours spent on corrosion maintenance often higher than the actual flight hours for many old aircraft.

Corrosion is simply defined as the chemical attack of metals that results in deterioration and loss of material. A corrosive fluid is usually involved, with the most common being water containing reactive chemicals (such as chloride ions). Moisture condenses on metal surfaces and can seep into and drip down the inside surface of the fuselage and around the lavatories and galley, causing corrosion in often hard-to-access areas. The low humidity (under 5–8%) inside pressurised aluminium fuselages helps minimise condensation, although the dry air affects the comfort of passengers.

Corrosion of metal aircraft components can range in severity from superficial discoloration to severe pitting and cracking that can cause sudden, catastrophic failure. Figure 21.1 shows the many causes and sources of corrosion during the design/manufacturing stage and in-service operation of aircraft. Common examples of corrosion damage to aircraft include corrosion thinning at fastened joints owing to water intrusion; pitting of exterior skins; stress-corrosion cracking at drilled holes, cut-outs and other geometric stress raisers; and corrosion of fuel tanks. Figure 21.2 shows examples of corrosion damage to metals used in aircraft.

The type of corrosion and the rate that corrosion takes place is determined by many factors. Important factors include the:

- composition, metallurgical properties and heat treatment of the metal alloy;
- type of surface films and protective systems on the metal;

498

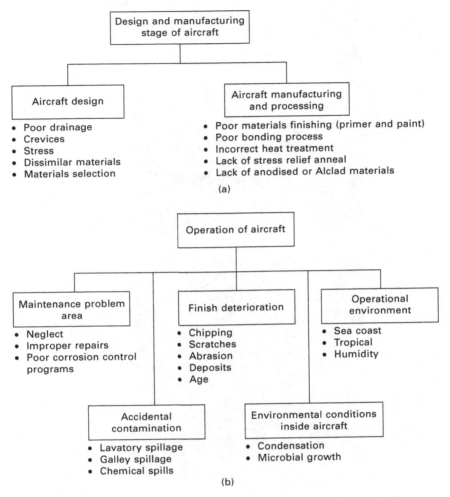

21.1 Common sources of corrosion during (a) the design and manufacturing stage of aircraft production and (b) in-service operation of aircraft (adapted from information provided by The Boeing Company).

- presence of stresses, voids and other defects in the metal;
- composition and concentration of the corrosive liquid or gas; and
- temperature and humidity of the environment.

Routine inspections for corrosion damage must be performed over the entire life of the aircraft using the nondestructive inspection methods described in chapter 23. The time between inspections becomes shorter as the aircraft becomes older which increases the maintenance cost. For example, Fig. 21.3 shows the increase in the maintenance cost of Boeing 727 aircraft with

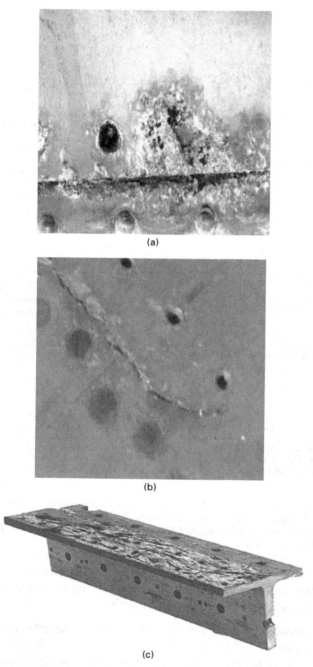

(a)

(b)

(c)

21.2 Examples of corrosion damage to metal aircraft components:
(a) surface corrosion; (b) corrosion cracking; and (c) exfoliation
corrosion.

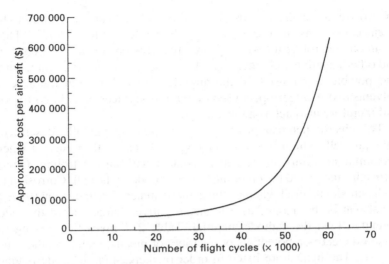

21.3 Total cost of maintenance per aircraft for Boeing 727s with increasing number of flight cycles.

increasing number of flights. The cost rises at an increasing rate with the age of aircraft owing to many factors, including corrosion inspection and repair. Therefore, managing corrosion is an important factor in the cost-effective operation of aircraft.

The aerospace engineer must understand the process of corrosion in order to design and manufacture metal components that are resistant to corrosion, specify the protection method which best resists corrosion, and identify and assess the severity of corrosion damage found in aircraft. In this chapter, we study the corrosion of aerospace metals. The next section provides a general description of the electrochemical process of metal corrosion. This is followed by a description of the different types of corrosion damage to metal aircraft components. Techniques used by the aircraft industry to prevent or slow corrosion are then described. It is worth noting that corrosion of fibre–polymer composites does not occur and, therefore, is not described. The excellent corrosion resistance of composites is an important reason for their use in aircraft structures.

21.2   Corrosion process

Corrosion is an electrochemical process involving two dissimilar materials that are placed in electrical contact with each other in the presence of an electrolyte. The two dissimilar materials can be two metals (e.g. aluminium and titanium), metal and composite (e.g. aluminium and carbon fibre–epoxy laminate) or, on a smaller scale, the boundary and core of a metal grain. The electrolyte is usually a corrosive liquid such as water containing charged atoms called ions, such as negatively charged chloride (Cl^-) ions. The most common electrolyte involved in aircraft corrosion is water that contains Cl^- and other negatively charged ions. The water can come from rain, humidity, and potable water used on the aircraft. Certain aviation fluids, cleaning solvents and paint strippers used on aircraft surfaces contain corrosive ions and therefore also act as an electrolyte.

The electrochemical process of corrosion is generally described as a galvanic cell, which is shown in Fig. 21.4. The cell is created when two dissimilar materials are placed in contact with an electrolyte. One metal forms the anode and the other metal the cathode to the cell. Corrosion occurs to the anode material whereas the cathode material remains unaffected. The metal that is the anode has a higher negative charge, called the electrode potential, than the cathode material. Table 21.1 gives the ranking for the standard electrode potentials of a variety of metals, including those used in aircraft. The metals are listed in order of increasing electrode potential to form the electromotive force (emf) series. When two dissimilar metals form a galvanic cell then the metal higher in the emf series is the cathode and the lower metal in the series is the anode. For example, aluminium alloy is higher (or more cathodic) than magnesium and, therefore, when these two metals are in contact in the presence of an electrolyte then the galvanic cell which is created causes the magnesium to corrode whereas the aluminium

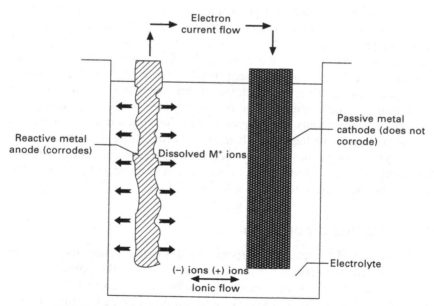

21.4 Schematic of the galvanic cell.

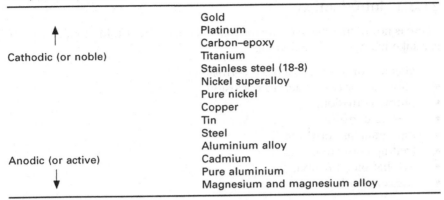

Table 21.1 Standard electrode potential rankings for selected metals and composites in seawater

Cathodic (or noble) ↑	Gold Platinum Carbon–epoxy Titanium Stainless steel (18-8) Nickel superalloy Pure nickel Copper Tin Steel Aluminium alloy
Anodic (or active) ↓	Cadmium Pure aluminium Magnesium and magnesium alloy

is unaffected. As another example, carbon fibre–epoxy composite causes aluminium to corrode because of its higher position in the emf series. In general, the greater the difference between the electrode potential of the two materials the faster the anodic metal corrodes.

Corrosion occurs in a galvanic cell by the movement of electrons from the anode to cathode. The anode and cathode must be electrically connected, usually by being in contact, to allow the flow of electrons. The liquid electrolyte, which is also conductive owing to the presence of ions, must be in contact with both the anode and cathode to complete the electric circuit. In the corrosion process, the loss of electrons from the anode creates positively charged metal ions, which leave the anodic metal surface and dissolve in the electrolyte fluid. For example, when aluminium corrodes it undergoes the anode reaction: $Al \rightarrow Al^{3+} + 3e^-$. The aluminium ions ($Al^{3+}$) created by this reaction pass into the electrolyte, causing the loss of material from the metal surface. The electrons produced by the anode reaction flow to the cathode where they combine with ions already dissolved in the electrolyte. The ions present in water are positively charged hydrogen (H^+), which react with the electrons to form hydrogen gas: $2H^+ + 2e^- \rightarrow H_2$. The metal of the cathode does not corrode, but simply conducts the electrons received from the anode to the positively charged ions in the electrolyte.

Three conditions must exist simultaneously for corrosion to occur:

1. Two dissimilar materials or two regions of different electromotive potential within a metal in order to form the anode and cathode.
2. A conductor (usually a metal) between the anode and cathode.
3. An electrolyte such as water.

Stopping corrosion of metal aircraft components involves removing one or more of these conditions.

21.3   Types of corrosion

21.3.1 Introduction

There is no single type of corrosion that occurs in aircraft. Instead, corrosion can take many forms, including:

- general (or uniform) surface corrosion,
- galvanic (or two-material) corrosion,
- pitting corrosion,
- crevice corrosion,
- intergranular corrosion,
- fretting corrosion,
- exfoliation corrosion, and
- stress corrosion.

21.3.2 General surface corrosion

General (also called uniform) surface corrosion is the most common type of corrosion. It involves an electrochemical reaction that proceeds uniformly over the entire exposed surface of the metal, as illustrated in Fig. 21.5. General corrosion is responsible for the greatest destruction of metal on a tonnage basis, although it is not usually a serious corrosion problem for aircraft owing to the surface protection measures described later in this chapter. General corrosion of metals used in the airframe only occurs when the surface protection is damaged or incorrectly applied. When it does occur, general corrosion occurs at the surface and is detected during maintenance inspection by the presence of grey or white powdery deposits. These deposits are the residual solid by-product of the corrosion process, such as the electrochemical degradation of aluminium:

$$Al \rightarrow Al^{3+} + 3e^-$$

to form metal cations (Al^{3+}) that react with oxygen in the atmosphere to form aluminium oxide powder:

$$4Al^{3+} + 3O_2 \rightarrow 2Al_2O_3$$

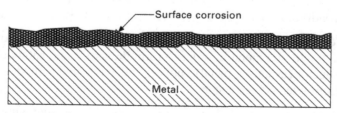

21.5 General corrosion.

21.3.3 Pitting corrosion

Pitting is one of the most destructive and insidious types of corrosion. Without appropriate protection, metals such as aluminium, steel and magnesium used in aircraft are susceptible to pitting corrosion, which is a form of extremely localised attack that results in small holes, as shown in Fig. 21.6. Pitting can start at precipitates at the surface of certain alloys when the particle has a different electrochemical potential from the surrounding metal matrix. Pitting can also occur in surface regions where the corrosion protective layer is absent. When surface protection is used small gaps in the layer can occur because of incorrect application or in-service damage by abrasion, erosion or some other event. The corrosion forms as a hole at the gap in the protective layer, which then develops into a wider cavity below the surface (Fig. 21.7).

Pitting requires an incubation period before it becomes visible, which can be months or years depending on the type of metal and electrolytic

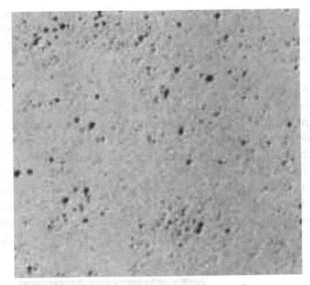

21.6 Metal surface showing pitting corrosion.

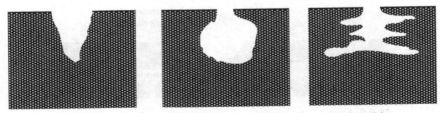

21.7 Examples of different shapes of pitting corrosion cavities.

fluid. Once started, however, the pit tends to penetrate the metal at an ever increasing rate forming a large cavity beneath the surface. Most pits grow downwards from horizontal surfaces, such as upper wing and horizontal stabiliser surfaces. Pits are less likely to develop on vertical surfaces, and only rarely grow upwards from the bottom of horizontal surfaces. Therefore, the inspection of aircraft components for damage caused by pitting corrosion should focus on upper horizontal surfaces. However, pits are difficult to detect by visual inspection because their surface opening is very narrow and is often covered with corrosion products, even though the underlying metal is severely corroded. It can take many months or years for pitting corrosion to appear as visible holes, by which time the component can be damaged beyond repair. Another problem caused by pitting corrosion is that the cavities are potential starting points for the growth of fatigue cracks.

21.3.4 Crevice corrosion

Crevice corrosion (also called concentration cell corrosion) is the most common type of corrosion damage found on many older aircraft which have not been adequately maintained. It is an aggressive form of corrosion that occurs locally inside crevices and other shielded areas of metals exposed to corrosive fluid. The process of crevice corrosion is shown schematically in Fig. 21.8. Crevice corrosion occurs in shielded regions where a small volume of stagnant corrosive fluid is trapped between two surfaces, such as under loose paint, within a delaminated bond-line or in an unsealed joint. Oxygen molecules in the stagnant fluid have low solubility. As a result, the fluid inside the crevice becomes depleted of oxygen when it remains stagnant and shielded from the atmosphere. The lower oxygen content in the crevice helps to form an anodic region at the metal surface. The metal surface in contact with the trapped moisture exposed to air forms a cathode. An excess of positively charged ions occurs in the crevice causing the stagnant solution to become acidic. To compensate for the excess of positive ions, chloride

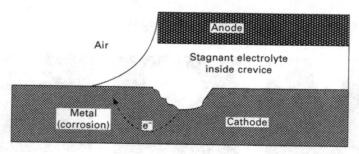

21.8 Schematic of crevice corrosion within a fastened joint.

ions in the fluid migrate into the crevice causing a local and small galvanic cell to be created. Because crevice corrosion can only occur when the oxygen level in the corrosive fluid inside the crevice is different from that outside the crevice, it is also known as 'differential aeration corrosion'. The most effective way to eliminate crevice corrosion is to keep water out of joints and tight spaces between surfaces.

Crevice corrosion can quickly develop into pitting or exfoliation corrosion when left untreated, depending on the types of metal and corrosive fluid. It is for this reason that crevices within aircraft, such as fastener holes and joints, must be sealed with a durable protective coating that stops the ingress of corrosive fluid. Crevice corrosion is one of the most common types of corrosion found in aircraft, and it usually occurs in crevices under fastener heads, under loose paint, within delaminated bonded joints, or in unsealed joints.

21.3.5 Intergranular corrosion

Intergranular corrosion involves the localised attack of the grain boundaries in a metal. The intergranular corrosion process is shown in Fig. 21.9. Corrosion damage occurs along the grain boundaries whereas the core of the grains is unaffected. Intergranular corrosion occurs because the electrochemical potential of the grain boundary is different from the grain core and, therefore, the grain boundary and grain core form, respectively, the anode and cathode of a tiny galvanic cell.

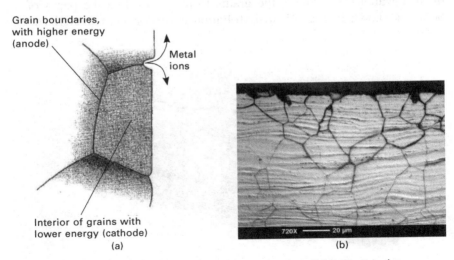

21.9 (a) Schematic of integranular corrosion. (b) Intergranular corrosion below a metal surface.

The grain boundary is anodic because its chemical composition is different to the grain core. The difference can be caused by a higher concentration of impurity elements at the grain boundaries, depletion of alloying elements in the grain boundary region, or some other chemical difference between the boundary and core of a grain. For example, in 2000 series aluminium alloys (Al–Cu) the precipitation of $CuAl_2$ particles within the grains depletes the amount of copper at the grain boundaries. The lower copper content makes the grain boundaries anodic with respect to the grain core. The formation of other types of precipitates, either along the grain boundaries or inside the grains, can also cause a difference in the electrochemical potential between the outer and inner regions of grains.

In the presence of a corrosive fluid, a small-scale galvanic cell is created between the core and boundary of a grain and this leads to intergranular corrosion. The corrosion process causes the gradual disintegration of the metal by grains breaking away from the surface after the boundaries have been corroded. Intergranular corrosion is a potential problem for many types of heat-treatable aluminium and magnesium alloys used in aircraft, although it may be avoided by the use of surface protection.

A special type of intergranular corrosion is called exfoliation corrosion, which involves the lifting of surface grains by the force of corrosion products at the grain boundaries. Exfoliation corrosion starts at the surface but it mainly involves subsurface attack that proceeds along narrow paths parallel with the surface. The attack is usually along the grain boundaries (intergranular corrosion). When the grain boundaries corrode they form corrosion products that exert pressure on the surface grains which forces them upwards. This causes the grains to peel back like the pages of a book, as shown in Fig. 21.10. Exfoliation corrosion is characterised by

21.10 Exfoliation corrosion. Photograph supplied courtesy of L. Hahara, Hawaii Corrosion Laboratory.

leafing of thin layers of uncorroded metal between layers of corrosion product. The grains are removed from the surface by abrasion or other mechanical action, which allows the underlying grains to then lift up and the exfoliation process to continue. High-strength aluminium alloys are susceptible to exfoliation corrosion, and their resistance to this type of corrosion is improved by over-ageing during heat treatment.

21.3.6 Fretting corrosion

Fretting corrosion, which is also called friction oxidation or wear oxidation among several other terms, involves the deterioration of contacting metals subjected to vibration or slip. The fretting (or rubbing) action results in fine particle fragments being abraded from one or both materials. These fragments oxidise into hard, abrasive particles which wear and destroy the metal surface. For example, the fretting of aluminium produces aluminium oxide (Al_2O_3) particles that are many times harder than the metal surface. The process is considered corrosive because the metal particles must oxidise, which is a form of dry corrosion.

Scratching and abrasion of the metal by the hard oxide particles causes a loss in dimensional tolerance between contacting surfaces such as structural joints. In extreme cases, it can cause seizing and galling of moving parts. Fretting corrosion is not usually a common problem with aircraft, although it has contributed to several aircraft accidents. The most notable case involved fretting corrosion between electrical contacts in the fuel control system of the F-16 *Flying Falcon*. Fretting damage between the contacts caused the control system to automatically shut off the valves to the main fuel supply without warning, resulting in at least six F-16 crashes before the problem was identified and fixed.

21.3.7 Stress-corrosion cracking

The inspection of airframes during routine maintenance often includes looking for signs of stress-corrosion cracking (Fig. 21.11). Although stress-corrosion cracking is not the most common form of corrosion, it does account for between 5 and 10% of all aircraft component failures. Stress-corrosion cracking, which is also called environmentally assisted stress corrosion, is caused by the combination of stress and corrosion. The stress acting on the metal may arise from an external applied stress such as structural or aerodynamic loads or an internal stress that comes from a variety of sources during metal processing, with the most common being metal working (such as rolling or bending), nonuniform cooling during heat treatment, and machining without proper stress relief. Internal residual stresses often provide the driving force for stress corrosion in many metal components. Another potential source of

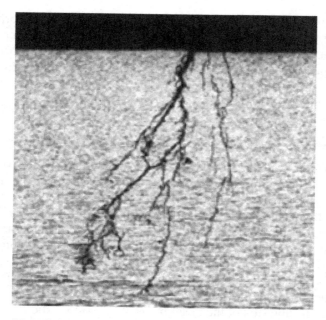

21.11 Subsurface view of stress-corrosion cracking. Photograph supplied courtesy of Corrosion Testing Laboratories Inc., Newark DE, USA.

stress in aircraft structures is fastener heads which have been overtightened, thus causing stress-corrosion cracking in the material underneath the fastener head. Failure by stress-corrosion cracking can occur at stress levels well below the yield strength of the metal.

Stress-corrosion cracks often initiate at pits, notches or other stress raiser sites on the metal surface in the presence of a corrosive fluid. The specific nature of stress corrosion is complex and difficult to describe via a single mechanism. It is generally believed that when the stress is high enough then the passive metal oxide film, such as the protective oxide (Al_2O_3) layer on aluminium alloys, ruptures. A corrosive fluid attacks the underlying stressed metal by anodic dissolution, thus causing a crack to grow into a branched structure, as shown schematically in Fig. 21.12. At the same time, the applied or residual stress causes local plastic tearing at the crack tip, thus increasing the crack size. Stress-corrosion cracking in most metals occurs by this mechanism of anodic dissolution and plastic tearing at the crack tip. An alternative mechanism involves the absorption of corrosive chemicals at the crack tip; these break the strained metal bonds, thereby forcing crack growth.

Stress-corrosion cracks often grow along the grain boundaries or (less

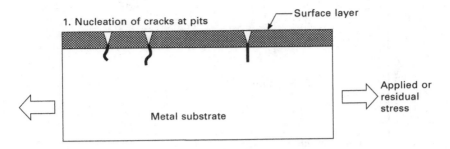

1. Nucleation of cracks at pits

Surface layer

Applied or residual stress

Metal substrate

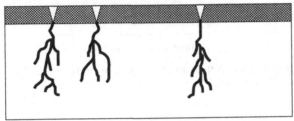

2. Stress corrosion cracks form branched structure

21.12 Stress-corrosion cracking.

often) through the grains via a brittle-type mode of fracture. This is one of the problems associated with stress-corrosion cracking; metals that normally fail by ductile processes in a noncorrosive environment can fracture in a brittle-type mode. A major problem with the stress-corrosion cracking process is that the cracks are difficult to detect by visual inspection of the metal surface. Large cracks can be present inside aircraft components, but be virtually impossible to observe by the eye. Careful examination of the aircraft using nondestructive inspection methods such as radiography is essential, and these methods are described in chapter 23.

Stress-corrosion cracking only occurs when the applied or residual stress is above a certain threshold, as shown in Fig. 21.13. Below this threshold, the driving force for crack growth is too low. Ideally, all aircraft metal components should operate in this low stress regime. The threshold may be increased by annealing the metal component to relieve the residual stresses thickening the section. When the stress is above the threshold, the time-to-failure drops rapidly with increasing stress owing to faster crack growth. Cracking often occurs move quickly when the metal is subjected to alternating stresses rather than constant stress. This is a special case of stress-corrosion cracking called corrosion fatigue, which occurs under the combined actions of cyclic stressing and corrosion. The crack growth rate in corrosion fatigue is faster and, in some cases, many times faster than the sum of the rates of

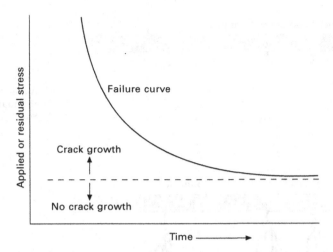

21.13 Effect of stress on the time-to-failure of metal caused by stress-corrosion cracking.

corrosion and fatigue when each act alone. The damage process is faster because cyclic stressing tends to remove or dislodge corrosion products at the crack tip. Corrosion products often slow the cracking process by acting as a barrier between the corrosion fluid and crack tip. When the products are removed by fatigue the crack growth rate is increased.

Stress-corrosion cracking causes a loss in failure strength, and is a potential cause of component failures in aircraft. The susceptibility of metals to stress-corrosion cracking or corrosion fatigue is determined by several factors, including alloy composition of the metal; types and distribution of precipitate particles; amount of strain hardening; and orientation and size of grains.

High-strength aluminium alloys are susceptible to stress-corrosion cracking when exposed to water and many other corrosive fluids. Aluminium alloys that contain solute elements such as copper, magnesium, zinc and lithium are prone to stress-corrosion cracking. These alloying elements are present in the 2000, 7000 and 8000 alloys used in aircraft components. The susceptibility of aluminium to stress-corrosion cracking generally increases with the solute content. The resistance of heat-treatable aluminium alloys to stress-corrosion cracking is also affected by the age-hardening treatment used to promote precipitation hardening. Figure 21.14 shows the change to the stress-corrosion resistance and strength owing to the age-hardening treatment of an aluminium alloy. The resistance to stress-corrosion cracking is lowest for the heat-treatment condition needed to achieve maximum strength. This occurs because the formation of $CuAl_2$ and other precipitate particles during

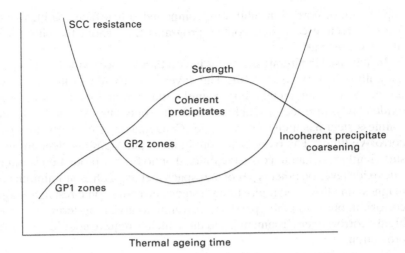

21.14 Effect of heat-treatment time on the stress-corrosion cracking (SCC) resistance and strength of age-hardening aluminium alloys.

age-hardening reduces the corrosion resistance of aluminium. Therefore, any improvement in strength gained by the age-hardening of aluminium alloys comes at the expense of lower resistance against stress-corrosion cracking. The T7 heat treatment process, which involves thermally ageing aluminium beyond the point of maximum strength, is often used to improve resistance against stress corrosion cracking. In addition, it is essential that aluminium structures are treated to resist stress corrosion by corrosion protective methods, such as Alclad, which are described in the next section. High-strength steel components are also susceptible to stress-corrosion cracking, and must be protected using a surface coating such as cadmium or chromium plating.

Section 21.8 at the end of the chapter presents a case study of the role of corrosion in the incident that occurred on Aloha Airlines flight 243 in April 1988.

21.4   Corrosion protection of metals

Commercial airliners are designed and built to have adequate corrosion protection to last the design service life, which is usually more than 20 years. Corrosion protection and corrosion control is achieved by proper materials selection, drainage of water and other moisture away from the aircraft, sealants to stop the ingress of corrosive fluids into crevices and other potential corrosion 'hot spots', selection of durable surface finishes, application of corrosion-inhibiting compounds in surface coatings, and the use of effective corrosion control programs throughout the entire service life of the aircraft.

Metals used in aircraft structures have a thin surface oxide layer that forms naturally when exposed to air. This layer can provide sufficient corrosion protection in some metals. For example, titanium and its alloys have a thin oxide (TiO_2) film that is stable in most environments. As another example, stainless steel has a chromium oxide (Cr_2O_3) layer that is resistant to most corrosive fluids. The oxide skins on titanium and stainless steel are usually sufficiently resistant to corrosive fluids that no further protection is required. The oxide layer on other types of aerospace metals, such as aluminium alloys, magnesium alloys and high-strength steels, provides some resistance against corrosion, but is too thin, pervious or fragile to give long-term protection in highly corrosive environments, and these metals require additional corrosion protection.

Various methods are available to protect aerospace metals from the damaging effects of corrosion. Most methods involve applying a durable, impervious coating that is resistant to corrosive fluids. Common coating methods are painting, cladding and anodising, and often several methods are used together to provide adequate corrosion protection. In addition, careful design of metal components is essential to ensure corrosion resistance. This section examines the most common methods for protecting aerospace metals against corrosion, with emphasis given to the corrosion protection of high-strength aluminium alloys used in fuselage and wing skins.

The simplest method of corrosion protection is painting the metal surface with a thin organic coating that provides a barrier against the corrosive fluid. The coating most often used is moisture-resistant paint, which shields the metal surface from corrosive fluids and humidity. Paint protects more metal components on aircraft on a tonnage basis than any other method for combating corrosion. A primer is applied directly to the metal surface, and is then coated with paint. The primer also provides protection by releasing corrosion inhibiting chemicals in the presence of moisture. For example, the primer zinc chromate ($ZnCrO_4$) was for many years used as a corrosion inhibitor for aluminium used in aircraft. The primer contains an active CrO_4^{2-} anion which impedes corrosion by neutralising chloride and other reactive ions present in corrosive fluids. However, zinc chromate is toxic and is now rarely used in modern aircraft. New types of primers contain strontium chromate or other corrosion-inhibiting compounds.

Figure 21.15 shows where corrosion-inhibiting compounds are commonly applied to aluminium structures. Paints and primers are only effective when the metal surface is properly prepared to ensure good bonding. Surface preparation involves removing dirt, fluids and other contaminants from the metal and then roughening the surface using grit blasting or chemical

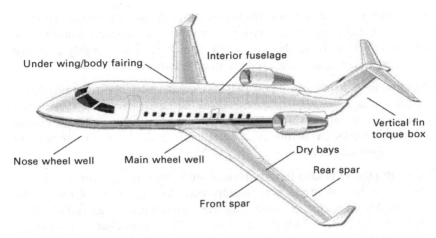

Under wing/body fairing

Interior fuselage

Vertical fin
torque box

Nose wheel well   Main wheel well

Dry bays

Rear spar

Front spar

21.15 Applications of corrosion-inhibiting compounds on aluminium commercial aircraft (adapted from information provided by The Boeing Company).

treatments before the primer is applied. Good corrosion protection also requires that the primer adheres strongly to the top coat of paint, again to avoid peeling. Some paints also protect the aircraft structure from erosion damage by stone chips thrown up from the runway and ice particle impacts during flight, although their primary function is corrosion protection.

High-strength aluminium alloys are vulnerable to corrosion, particularly stress-corrosion cracking and corrosion fatigue. In addition to painting and priming, the aluminium alloys used in aircraft are often protected with surface cladding called Alclad, which is a thin coating of pure aluminium or aluminium alloy (such as Al–1%Zn) applied over the surface. The coating is applied by hot rolling onto the aluminium alloy, which gives a durable, strong bond. The thickness of the cladding is typically in the range of 1.5 to 10% of the component thickness. The cladding works by being anodic relative to the underlying aluminium alloy, and thereby corrodes preferentially. Clad aluminium sheet and plate is used where weight and function permit, such as fuselage skins.

Another important method of corrosion protection for aluminium alloys is anodising. This process involves forming a thick layer of aluminium oxide (Al_2O_3) on the aluminium metal surface. The oxide layer is formed by immersing the aluminium component in a bath of chromic acid or sulfuric acid. During immersion an electrolytic cell is created, with the acid bath serving as the cathode and the aluminium being the anode. The electrolytic reaction causes the growth of a thick oxide surface layer that provides excellent corrosion protection for the underlying aluminium metal.

High-strength steel components used in aircraft are often protected from corrosion by priming and applying cadmium or chromium plating. The plating is very thin, typically only 5 to 15 μm thick, but is highly effective at protecting steel. Cadmium protects steel in two ways:

(1) cadmium is a passive metal that has good resistance to atmospheric attack, and
(2) cadmium is anodic (higher in the galvanic series) to steel alloys and thereby works by acting as a sacrificial anode in a similar way to aluminium cladding.

Chromium plating is also used to protect steel parts against corrosion.

Proper design of metal components can slow or stop corrosion. An important design factor is preventing the formation of a galvanic cell at joints by using metals with the same or similar electrochemical potential values. The Boeing Company group materials into four categories of different galvanic properties, as shown in Table 21.2. The objective is to avoid coupling of materials from different groups unless required by economic or weight considerations. If this is not possible, then electrical insulation of the contact between dissimilar materials to impede the flow of electrons is effective. For example, an insulating layer of fibreglass laminate at the joint between aluminium alloy (anode) and carbon–epoxy composite (cathode) panels is often used in aircraft assembly. Another design method is to ensure the area of the anode material is much larger than the cathode area, which reduces the corrosion rate of the anodic metal. Stress-corrosion cracking is slowed or stopped by reducing the load applied to the metal by increasing the net-section and by relieving residual stresses through heat treatment. Crevice corrosion is avoided by ensuring correct tightening of fasteners and that there are no gaps in the joint by which moisture can enter. Joints, connections and potential sites for crevices must be sealed with a durable,

Table 21.2 Grouping of materials based on galvanic properties by The Boeing Company

Reactive end (anodic)	1	Magnesium and magnesium alloys
⇕	2	Aluminium and aluminium alloys, cadmium–titanium plate and cadmium, zinc
	3	Steels (except corrosion-resistant steels)
Passive end (cathodic)	4	Nickel and nickel-based superalloys, cobalt alloys, corrosion-resistant steels, titanium and titanium alloys, carbon-fibre composites, glass-fibre composites

flexible and impervious coating (such as polysulfide sealant) to avoid the ingress of corrosive fluids.

21.5   Summary

Corrosion is an expensive problem for the aviation industry and accounts for about one-quarter of all metal component failures on old aircraft.

Corrosion occurs in many different ways on aircraft, with the most common types being stress corrosion, fatigue corrosion, pitting corrosion and crevice corrosion.

Electrochemical corrosion occurs when two dissimilar materials with different electrode potentials are in close contact in the presence of a corrosive fluid. Corrosion of aircraft components occurs most often in the presence of moisture containing reactive ions (e.g. Cl⁻) which condenses on the metal surface. Other types of fluids may also cause corrosion of aircraft metals, including paint strippers, cleaning solvents and certain types of aviation fuel.

Corrosion between two dissimilar materials is minimised or avoided in several ways, including inserting an insulating medium (such as fibreglass composite) between the materials and increasing the size of the anodic metal.

Stress corrosion and corrosion fatigue are the most common forms of corrosion in high-strength aluminium components. These types of corrosion are insidious to aircraft because the corrosion cracking is difficult to detect visually. This kind of corrosion cracking can be minimised in several ways, including careful control of the heat treatment (age-hardening) process, reducing the applied stress by increasing the net-section thickness of components, and removing residual tensile stresses by stress-relief annealing.

Metallic aircraft components, particularly when made using aluminium alloy, magnesium alloy and high-strength steel, require protection against the damaging effects of corrosion. External metal surfaces should be coated with a durable, moisture-resistant paint which provides a barrier against corrosion fluids. A primer containing corrosion-inhibiting chemicals should be applied between the metal and paint for added protection. High-strength aluminium components should be protected with surface cladding (e.g. Alclad) and/or anodised. Steel components should be plated with cadmium, chromium or other metal protective coating.

21.6   Terminology

Alclad: Common name given to surface cladding of a corrosion-resistant material onto an aluminium alloy substrate.

Anode: Electrode in a galvanic cell at which electrons are released during corrosion.

Anodising: Process involving the deposition of an adherent synthetic oxide film on a metal surface (usually aluminium alloys) to improve corrosion resistance.

Cathode: Electrode in a galvanic cell that accepts electrons from the anode during electrochemical corrosion. The cathode is not damaged by corrosion.

Corrosion fatigue: Process of cracking and failure of materials caused by the combined action of local corrosion at the crack tip and repeated cyclic loading.

Crevice corrosion: Localised corrosion of a metal surface at, or immediately adjacent to, an area that is shielded from the environment. The corrosion process is characterised by the absence or low level of oxygen in the corroding region of the crevice.

Electrode potential: The voltage of the half-reactions occurring in an electrochemical cell. The electrode potential is the potential (voltage) difference of the half-reaction of a material and the half-reaction of a standard hydrogen electrode.

Electrolyte: A substance (usually a liquid), containing free ions, that conducts an electric current.

Exfoliation corrosion: Corrosion process involving the exfoliation (peeling) of a surface material owing to pressure applied from corrosion products formed below the surface.

Fretting corrosion: Accelerated deterioration at the interface between contacting surfaces as the result of corrosion and slight oscillatory movement between the surfaces.

General (or uniform) surface corrosion: Corrosion that occurs at a constant rate over a metal surface.

Galvanic (or two-material) corrosion: Electrochemical process in which one metal corrodes preferentially when it is contact with another material in the presence of an electrolyte.

Intergranular corrosion: Selective preferential attack and penetration of corroding agents along the grain boundaries of a metal.

Pitting corrosion: Localised form of corrosion that leads to the creation of small holes in the surface.

Stress corrosion: Process of cracking and failure of materials caused by the combined action of local corrosion and stress at the crack tip.

21.7   Further reading and research

Fontana, M. G. and Greene, N. D., *Corrosion engineering*, McGraw–Hill Inc., Tokyo, Japan, 1986.

Hagemaier, D. J., Wendelbo, A. H. and Bar-Choen, Y., 'Aircraft corrosion and detection methods', *Materials evaluation*, **43** (1985), 426–437.

Revie, R. W. and Uhlig, H. H., *Corrosion and corrosion control (4th edition)*, John Wiley & Sons, 2008.

Roberge, P. R., *Handbook of corrosion engineering*, McGraw–Hill, 2000.

Wallace, W., Hoeppner, D. W and Kandachar, P. V., *AGARD corrosion handbook. volume 1. Aircraft corrosion: causes and case histories*, 1985.

21.8 Case study: corrosion in the Aloha Airlines flight 243

On April 28 1988, a 19-year old Boeing 737 aircraft operated by Aloha Airlines lost a large piece of the upper fuselage as a result of stress-corrosion cracking (Fig. 21.16). A 4 to 6 m section from the aluminium upper fuselage suddenly broke away when the aircraft was cruising at an attitude of 24 000 feet. The flight crew had no warning before a large piece of the fuselage was torn off the aircraft. A flight attendant was killed and many passengers were injured by flying debris, but miraculously the pilot managed to land the aircraft without further incident on the island of Maui, Hawaii.

Inspection of the aircraft after landing revealed the presence of multiple cracks in the fuselage, with many growing from rivet holes in skin lap joints. The cracks were caused by stress corrosion and corrosion fatigue. Before the accident, the aircraft was operated for many years by flying between

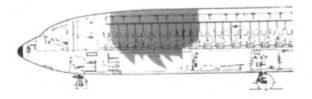

21.16 Stress-corrosion failure of the Aloha Airlines aircraft.

the Hawaiian Islands. Because of the short flight times between the islands, the aircraft performed many take-offs and landings in a single day, with the pressurisation of the cabin after take-off and depressurisation during landing causing fatigue stressing of the fuselage skin. Operating in Hawaii meant the aircraft was exposed to sea mist and salty air. The combination of fatigue stressing and seawater caused corrosion-fatigue cracks to develop at a rapid rate in the fuselage. The fatigue cracks that formed at the fastener holes joined into a single large crack, which subsequently caused catastrophic failure of the fuselage. The Aloha incident dramatically shows the danger of stress-corrosion cracking.

Creep of aerospace materials

22.1   Introduction

Creep is a process that involves the gradual plastic deformation of a material over time. The remarkable thing about creep is that plastic deformation occurs at stress levels below the yield strength of the material. In other words, creep causes a material to plastically deform and permanently change shape over time when subjected to an elastic load. This runs counter to the concept that plastic deformation can only occur when the applied stress exceeds the yield strength of the material.

When most engineering materials, including the metal alloys and composites used in aircraft structures, are elastically loaded then the amount of deformation that occurs does not change with time. However, this is only true when the temperature is moderately low and the elastic load is applied to the material for a short time. When the temperature is raised then 'elastic loads', which give no permanent deformation at room temperature, can cause the material to plastically deform via creep. (Creep of materials does occur at room temperature, but the rate of plastic deformation is usually extremely slow and any permanent deformation is not noticeable.) Most metals undergo creep at temperatures higher than 30–40% of their absolute melting temperature (in Kelvin). Creep of polymers and polymer composites occurs at lower temperatures than in metals, and in some materials is noticeable at only 50–75 °C. Creep deformation of metals, polymers and composites can continue unabated under elastic loading until eventually fracture occurs via a process called stress rupture.

Creep of aerospace materials can be a serious problem when they are required to withstand high elastic loads and elevated temperatures for long periods of time. Aerospace metals must have high resistance to creep and stress rupture, otherwise the aircraft component may be damaged (Fig. 22.1). For example, without high creep resistance the materials used in aircraft jet engines, such as the turbine blades, discs and compressor parts, distort owing to the high operating stress and temperature. Close tolerances are critical in jet engines, and even a small amount of plastic deformation caused by creep can cause an engine to seize. Excellent creep resistance is also essential for structural materials used in the body skins of supersonic aircraft, rocket nose cones and re-entry spacecraft such as the space shuttle. High temperatures are generated by frictional heating from molecules in the atmosphere, and

521

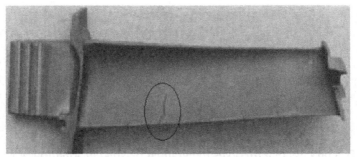

22.1 Cracking (circled) of a turbine blade caused by creep.

this can cause the skin materials to permanently deform and warp when they lack sufficient resistance against creep. There are many other examples when high resistance to creep and stress rupture is needed for the materials used in aircraft, such as engine components, and spacecraft, such as rocket nozzles. It is essential to aircraft safety that aerospace engineers understand the creep behaviour of structural materials.

In this chapter we study the creep and stress rupture properties of metal alloys, polymers and polymer composites. We also discuss ways to improve the resistance of aerospace materials against creep and stress rupture.

22.2   Creep behaviour of materials

When a material is held under a constant stress for a period of time, the process of creep can be divided into three stages of development: (I) primary creep when the process begins at a fast rate, (II) secondary creep when the process proceeds at a steady rate, and lastly (III) tertiary creep that occurs quickly and eventually leads to failure (or rupture). These three stages are observed in the creep curve of a material, which is a plot of increasing strain against time under load (Fig. 22.2).

22.2.1 Primary creep

The initial strain represented by ε_0 occurs when load is first applied to the material and is the result of elastic deformation, whereas the higher strains are caused by time-dependent plastic deformation owing to creep. The creep rate is initially very rapid in the earliest period of the primary stage, but slows over time as the material resists the deformation by strain hardening.

22.2.2 Secondary creep

The second stage of creep, called 'steady-state creep', is a period of nearly constant creep rate defined by the slope $d\varepsilon/dt$. The creep rate is constant because

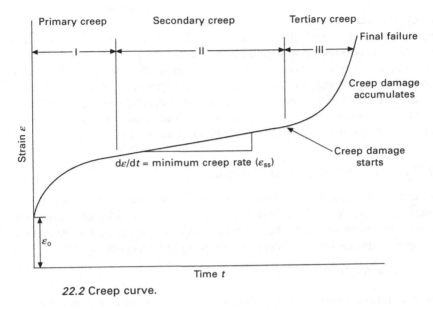

Primary creep   Secondary creep   Tertiary creep

Final failure

I   II   III

Creep damage accumulates

Strain ε

$d\varepsilon/dt$ = minimum creep rate $(\dot{\varepsilon}_{ss})$

Creep damage starts

ε_0

Time t

22.2 Creep curve.

a balance exists between the competing processes of plastic deformation and strain hardening. Secondary creep occurs for most of the creep life.

22.2.3 Tertiary creep

The tertiary creep stage occurs when the creep life is nearly exhausted, and the material specimen begins to neck or develop internal voids which reduce the load capacity. The creep rate accelerates during the tertiary stage as the load capacity drops owing to increased necking or void growth until eventually the specimen fails. By holding the specimen at a constant stress level and temperature until failure, the stress rupture life can be measured.

22.2.4 Steady-state creep

An important property determined from the creep curve is the steady-state creep rate, $\dot{\varepsilon}_{ss}$, which occurs during the second stage. Because this stage lasts for most of the creep life, the steady-state creep rate is used to calculate the change in shape of a material over most of the operating life. The creep rate is calculated using the Arrhenius relationship:

$$\dot{\varepsilon}_{ss} = A\sigma^n e^{-(Q_c/RT)} \qquad [22.1]$$

where T is the absolute temperature, R is the universal gas constant, A and n are creep constants specific to the material, and Q_c is the creep activation energy for the material. The creep constants and activation energy are determined from creep testing (which is explained in chapter 5) and, using the results, it is possible to calculate the creep rate of the material for any operating stress and temperature. This equation allows aerospace engineers to calculate the change in shape of a metal component during service and hence specify its design creep life. Alternatively, the equation can be used to determine the maximum operating stress and temperature for an aerospace material without it suffering excessive creep deformation.

22.2.5 Creep life and creep failure

The creep life, otherwise known as the stress rupture time, of aerospace materials can be determined from their creep curve, provided of course that the test conditions replicate the in-service operating stress and temperature. The stress rupture time (t_r) is calculated using the expression:

$$t_r = K\sigma^m e^{(Q_r/RT)} \qquad\qquad [22.2]$$

where K and m are material constants and Q_r is the activation energy for stress rupture, and these are measured by experimental testing. The stress rupture time decreases rapidly with increasing stress and temperature, as shown in Fig. 22.3 for a nickel-based superalloy used in jet engines. Therefore, it is important to limit the operating stress and temperature to avoid stress rupture as well as excessive creep deformation.

Creep failure of aircraft components can be defined in several ways. The failure modes are known as 'displacement-limited creep', 'stress-limited creep',

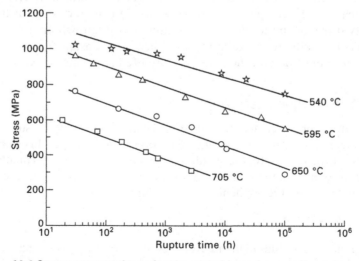

22.3 Stress rupture times for the nickel-based superalloy Inconel 718.

'buckling-limiting creep', and 'stress rupture'. These are now explained in this order:

- Displacement-limited creep failure occurs when the material changes shape beyond a specified limit, such as 0.1% elongation. This type of failure is important for aircraft components that have precise dimensions or when small clearances must be maintained, such as discs and blades for gas turbine engines.
- Stress-limited failure is when the permanent change in shape owing to creep relaxes the initial stress on a material. For example, stress relaxation creep can loosen pretensioned fasteners in aircraft joints.
- Buckling-limited creep occurs in beams, panels and other structures that carry compressive loads. This failure mode involves the buckling or collapse of thin sections owing to creep. For example, an upper wing skin could experience creep-induced softening and buckling as a result of frictional heating when flying at supersonic speeds for a long time. For this to occur in practice, however, the skin material would need to have exceptionally low creep resistance.
- Stress rupture occurs at the end of the tertiary creep stage when the load capacity of a material has dropped to the applied stress level, causing final fracture.

22.3   Creep of metals

Creep of metals occurs from the action of two plastic deformation processes: dislocation slip and grain boundary sliding. It is often assumed that dislocations do not move when the stress acting on a metal is below its yield stress. Strictly, this assumption is only true when the temperature is absolute zero (−273 °C). Above this temperature, the metal atoms have sufficient mobility to cause the dislocations to move. At room temperature, the atomic mobility is low and therefore the movement of dislocations is extremely slow and an extraordinarily long period must pass before plastic deformation is noticeable. For this reason, at room temperature it is assumed that the deformation of a metal is completely elastic when the applied stress is below the elastic limit. Atomic mobility increases with temperature and can, with sufficient time, aid dislocations to move through the crystal structure and thereby cause plastic deformation.

Dislocation movement during creep occurs by two processes: dislocation slip (which is described in chapter 4) and dislocation climb. The former process involves the movement of dislocations along the slip planes of the crystal lattice whereas the latter process is the movement (or climb) of dislocations perpendicular to the slip planes. Dislocation slip occurs at all temperatures above absolute zero when the applied stress is high enough

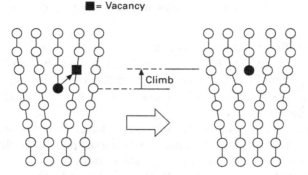

22.4 Dislocation climb mechanism that causes creep in metals.

whereas dislocation climb usually occurs only at high temperature. The dislocation climb process is illustrated in Figure 22.4, and requires atoms to move either to or from the dislocation line by diffusion involving a lattice vacancy. This action allows dislocations to 'climb' around obstacles impeding plastic flow, such as precipitate particles or clusters of solute atoms (e.g. GP zones), and thereby cause creep deformation even at low stress levels. Dislocation movement by slip or climb increases rapidly with temperature, as observed by an increase to the creep rate. At high temperatures, new slip systems become operative in the crystal lattice of some metals, thus further assisting dislocation motion and thereby increasing the creep rate.

The other important deformation process controlling the creep rate of metals is grain boundaries sliding. At high temperature the grains in polycrystalline metals are able to move relative to each other by plastic flow at the grain boundaries. The sliding that occurs between grains increases with temperature, which thereby raises the creep rate. Grain boundary sliding is the dominant creep process in most metals when the applied stress and temperature are low. Dislocation movement by slip and climb are the more dominant creep mechanisms at high stress and temperature.

Dislocation slip and, in particular, grain boundary sliding promote the formation of small voids at the grain boundaries (Fig. 22.5). Voids initiate at grain boundaries which are oriented transverse to the direction of the applied creep load. The voids develop at the start of the tertiary stage of the creep life, and then increase in number and size during this stage until eventually the metal fails by intergranular fracture. The time taken for the metal to fail under a constant stress and temperature is used to define the stress rupture life.

22.4   Creep of polymers and polymer composites

A potential problem with using polymers in aircraft components is viscoelastic creep, which can cause permanent distortion and damage. Polymers exhibit

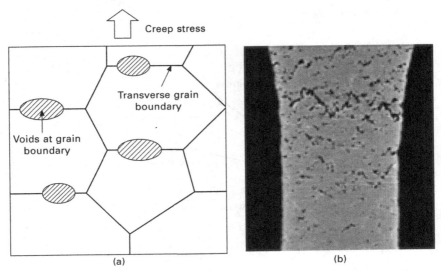

22.5 (a) Schematic and (b) photograph of void formation in metals caused by creep that eventually leads to stress rupture.

both elastic (instantaneous) and viscous creep (time-dependent) deformation when under an applied stress that is below the yield strength. This combination of viscous and elastic deformations is called viscoelasticity. When a polymer is under load there is an immediate elastic response. As explained in chapter 13, this is caused by the elastic stretching of bonds along the polymer chains and the partial straightening of twisted and coiled segments of the chains. When the load is removed the chains relax back into their original position, and this is the elastic component of viscoelasticity. However, when a polymer is held under load for a period of time then a second deformation process known as viscous creep occurs which is time-dependent. The chains have time to unfold and slide relative to one another when load is applied for a sufficient time. This viscous or creep flow is a time-dependent process, which decreases with increasing time until a steady-state condition is reached when the initially folded chains reach a new equilibrium configuration. When the polymer is then unloaded, there is an immediate (elastic strain) recovery followed by a time-dependent recovery; however, a permanent deformation remains.

The viscoelastic effect in polymers is dependent on the loading and environmental conditions, as shown in Fig. 22.6. Permanent deformation caused by viscous flow increases when the loading (or strain) rate is reduced. When load is applied rapidly, the polymer chains do not have sufficient time to uncoil and slide and, therefore, the creep effect can be quite low and the polymer behaves in a brittle manner. When load is applied slowly or

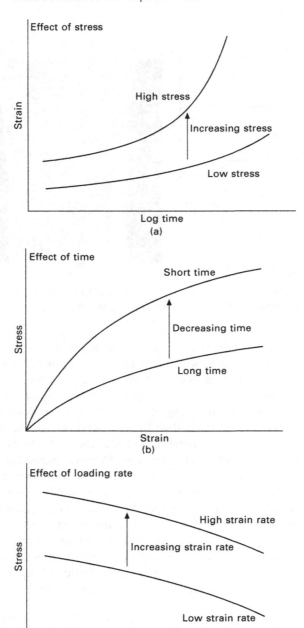

22.6 Creep curves showing the effects of (a) applied stress, (b) loading time and (c) loading rate.

stress is applied for a long period, there is sufficient time for the chains to slip and straighten and, therefore, the viscoelastic creep behaviour is more pronounced. Creep deformation also increases rapidly with temperature because there is more internal energy available for the chains to slide and uncoil. A polymer eventually fails by stress rupture when the load is applied for a sufficiently long time. The stress rupture time decreases with increasing stress and temperature as shown in Fig. 22.7.

The importance of creep is that permanent deformation to loaded plastic aircraft parts (including bonded connections) increases with the loading rate and operating temperature. Creep can occur in some polymers at room temperature, and these materials must be avoided in aircraft. Polymers should only be used when creep cannot occur, such as in lightly-loaded parts. Although creep occurs in all polymers, the creep rate can be controlled by the molecular structures. Creep is reduced by any process that resists the unfolding and sliding of chains, such as increasing the degree of crystallinity in thermoplastics or the amount of cross-linking in thermosets. Polymers that have large side-groups along the chain also have higher resistance to creep.

In chapter 15 we discussed the importance of aligning the fibres in the load direction for high stiffness, strength and fatigue life of polymer composites, and it is also important for high creep strength and resistance to stress rupture. The creep properties of fibre–polymer composites are anisotropic, and depend on the fibre direction. Carbon and glass fibres used in aerospace composites are resistant to creep and, therefore, when the applied load is

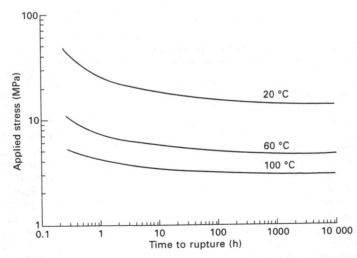

22.7 Creep rupture curves for a polymer showing the effects of stress and temperature.

parallel with the fibres then composites do not experience creep or stress rupture. However, when the load is not acting on the fibres and is carried by the polymer matrix (such as through-thickness loading) then significant creep can occur.

22.5   Creep-resistant materials

There are several methods used to improve the creep resistance and prolong the stress rupture life of aerospace materials. The most important method is selecting a material with a high melting (or softening) temperature. As a general rule, creep occurs when metals are required to operate at temperatures above 30 to 40% of their absolute melting point. Rapid creep of polymers occurs at 30–40% of their glass transition temperature, whereas creep of heat-resistant ceramic materials starts above 40–50% of their melting temperature (in Kelvin). Table 22.1 gives the melting or softening temperatures of various aerospace materials. Most polymers and polymer composites have low softening temperatures (typically under 150–180 °C) and so are not suited for high-temperature service. Aluminium and magnesium alloys have relatively low melting temperatures and not suitable for aircraft components required to operate for long periods at temperatures above about 150 °C. Nickel, iron–nickel and cobalt superalloys have high melting temperatures, which makes them suitable for gas turbine engines and other high-temperature components. Ceramic materials have very high softening temperatures that make them useful for extreme temperature applications, such as rocket nose cones and the heat insulation tiles on the space shuttle.

Table 22.1 Melting or softening (S) temperatures of aerospace materials

Material	Temperature (K)
Carbon–carbon composite	4000
Alumina (Al$_2$O$_3$)	2320
Cobalt alloys	1650–1770
Nickel alloys	1550–1730
Titanium alloys	1770–1940
Carbon steels	1570–1800
Stainless steels	1660–1690
Aluminium alloys	750–930
Magnesium alloys	730–920
Polycarbonates	400 (S)
Epoxy, high-temperature	400–425 (S)
Epoxy, general purpose	340–380 (S)
Carbon fibre–epoxy composite	340–425 (S)
Glass fibre–epoxy composite	340–425 (S)

22.5.1 Creep-resistant metals

The creep resistance of metals is controlled by their alloy composition and microstructure as well as by their melting temperature. The creep resistance of metal alloy systems increases with the concentration of alloying elements dissolved into solid solution. The presence of alloying elements in interstitial crystal sites increases the lattice strain and thereby resists the processes of dislocation slip and climb that drive creep. This is one reason for the high alloy content of nickel-based and iron–nickel superalloys used in jet engines. Creep resistance is also improved by the presence of finely dispersed intermetallic precipitates, which are stable at high temperature. The precipitates resist the movement of dislocations that causes high temperature creep. It is for this reason that many nickel-based superalloys contain small amounts of aluminium and/or titanium which combine with the matrix to form γ and γ' intermetallic precipitates Ni_3Al, Ni_3Ti or $Ni_3(Al,Ti)$. Chapter 12 gives more information on the control of creep using the metallurgical properties of superalloys.

Control of the grain size and structure is also an effective method of reducing creep. Increasing the grain size by thermomechanical processes reduces the creep rate and extends the stress rupture life of metals by lowering the amount of grain boundary sliding. Therefore, metals with a coarse grain texture are often used in creep-resistant components. The elimination of transverse grain boundaries along which sliding occurs provides an even greater improvement to the creep resistance and rupture life. High-pressure turbine blades for jet engines are fabricated using the directional solidification process, which involves chilling the metal casting from one end when removed from the furnace (chapter 6). The sharp temperature gradient used in directional solidification forces the grains to grow continuously from one end of the casting to the other. The final casting has a columnar grain structure with few or no transverse grain boundaries, thus providing high creep resistance. The improved creep properties of turbine blades fabricated from directionaly solidified metal allows them to operate at a temperature about 50 °C higher than the material with a coarse polycrystalline structure, thereby providing greater propulsion efficiency. Even better creep properties are achieved by casting metals using the single crystal process. Because there are no grain boundaries in single crystal metals, they have outstanding creep resistance and prolonged stress rupture life.

22.5.2 Creep-resistant polymer and polymer composites

There are several ways of increasing the creep resistance of polymers, structural adhesives and fibre–polymer composites. The creep resistance of thermoset polymers increases with the amount of cross-linking between the chains. The cross-links increase the glass transition temperature by resisting the sliding and straightening of the chains under applied loading, and this raises the creep-softening temperature. Therefore, thermoset polymers such as epoxy should be fully cured to maximise the amount of cross-linking. The creep resistance of polymers also increases with their molecular weight. Thermoplastics are generally less resistant to creep than thermoset polymers, and their creep properties are controlled by the arrangement of the network structure of the chains; with crystalline and semicrystalline polymers being more creep resistant than amorphous (or glassy) polymers. The most effective method of maximising the creep resistance of composite materials is aligning the fibres in the load direction. Carbon-fibre and glass-fibre composites have good creep resistance when the fibres carry the applied load.

22.6   Summary

Creep is an important deformation process in aerospace materials required to operate at high temperatures and stresses for long periods of time, and it is especially important for jet engine materials.

Creep involves the plastic deformation of material when an elastic load is applied for a long time. The amount of creep increases with time until eventually the material breaks; this is the stress rupture time.

The creep process in metals is controlled by dislocation slip and climb, and grain boundary sliding.

Creep becomes significant in metals when required to operate above 30–40% of their melting temperature (in Kelvin).

Creep resistance of aerospace metal components is improved by:

- using a high melting point material (such as nickel-based superalloys) having a high concentration of interstitial alloying elements dissolved in the crystal structure;
- having thermally stable intermetallic compounds; and
- eliminating transverse grain boundaries.

Polymers are susceptible to plastic deformation when load is applied for a sustained period as a result of viscoelastic creep. The rate of creep deformation increases with the applied stress, temperature or loading rate. The operating load and temperature of polymers used in aircraft components must be sufficiently low to avoid creep deformation which can eventually lead to stress rupture.

Creep resistance of thermoset polymers (including structural adhesives) and fibre–polymer composites is improved by increasing the amount of cross-linking and the molecular weight. Creep resistance of thermoplastics is improved by increasing the molecular weight and the amount of crystalline polymer. Creep resistance of composites is improved by ensuring the fibres carry the applied load, and not the polymer matrix. Carbon-fibre composites have much higher creep resistance in the fibre direction compared with the anti-fibre direction.

22.7   Terminology

Creep: Slow plastic deformation of a material under the influence of stress that is usually below the yield strength.

Dislocation climb: Movement of a dislocation perpendicular to its slip plane, usually assisted by the movement of vacancies through the crystal structure.

Grain boundary sliding: A plastic deformation process that usually occurs at elevated temperature in which grains slide past each other along, or in a zone immediately adjacent to, their common boundary.

Stress rupture: The fracture of a material after carrying a sustained load (usually below the yield stress) for an extended period of time.

Stress rupture life: The time taken for a material to fail under constant stress and temperature conditions by the plastic deformation process of creep.

Viscoelasticity: Combination of viscous and elastic properties in a material with the relative contribution of each being dependent on time, temperature, stress and strain rate.

22.8   Further reading and research

Nabarro, F. R. N. and De Villiers, H. L., *The physics of creep: creep and creep-resistant alloys*, Taylor & Francis, 1995.

Penny, R. K. and Marriott, D. L., *Design for creep (2nd edition)*, Chapman & Hall, 1995.

23

Nondestructive inspection and structural health monitoring of aerospace materials

23.1   Introduction

The high quality of materials used in aircraft structural components is essential for reliability and safety. Structural materials must be free from defects that would reduce the structural properties, failure strength and design life. However, despite careful control of the operations used in processing of aerospace materials, it is difficult to ensure that every structure is completely free from defects. For example, brittle intermetallic inclusions and gas holes form in the casting of metal components and voids and dry spots occur in the manufacture of fibre–polymer composite structures. Damage also occurs during in-service operation of aircraft from poor design, mechanical damage (e.g. impact, fatigue, battle damage) and environmental degradation (e.g. corrosion, moisture absorption, lightning strikes). Tables 23.1 and 23.2 list common manufacturing defects and in-service damage found in metal and fibre–polymer composite components.

Aviation safety authorities such as the Federal Aviation Administration apply strict regulations to the types and amount of damage allowed in structural materials without the replacement or repair of the damaged component. The regulations require that aircraft structures and engine components have

Table 23.1 Common defects and damage in metals

Defect	Description of damage
Manufacturing	
Porosity	Pocket of air or gas trapped in solid material during solidification
Intermetallic inclusion	Hard, brittle particle formed during casting or heat treatment
Shrinkage crack	Cracks formed by shrinkage of metal casting
Cracks, scratches	Defects caused by machining and drilling operations
In-service	
Fatigue	Cracks formed owing to stress, thermal or acoustic fatigue
Corrosion	Material degradation (eg. pitting, cracking) owing to corrosion
Impact	Cracks

534

Table 23.2 Common defects and damage in fibre–polymer composites

Defect	Description of damage
Manufacturing	
Delamination	Separation of ply layers in laminate by cure stresses
Matrix crack	Crack within polymer matrix owing to cure stresses
Unbond	Area between prepreg layers that fail to bond together
Foreign object	Inclusion of a foreign object (e.g. peel ply)
Porosity	Pocket of air or gas trapped during lay-up or cure
Core crush	Crushing of core owing to excessive cure pressure
In-service	
Impact damage	Delamination, matrix cracks and fibre breaks owing to impact loading
Moisture absorption	Softening and cracking owing to moisture absorption
Fatigue	Delamination, matrix cracks and fibre breaks owing to stress, thermal or acoustic fatigue
Core degradation	Corrosion of aluminium core and softening of Nomex core owing to moisture absorption

a minimum level of damage tolerance, which means that the part must retain functionality and design structural properties when damage is present in the material. In commercial aviation, the maximum allowable defect (delamination) size for composites is typically 12.5 mm and the maximum porosity content is about 1.0% by volume. For military aircraft, which usually operate at higher stress levels than civil aircraft, the maximum allowable defect size is generally smaller, depending on the design and function of the structure. Maximum allowable defect sizes are also specified for metal aircraft components. Metal and composite structures containing damage greater than the allowable limits must be repaired or taken out of service.

Most damage is impossible to detect by eye because it is too small and buried below the component surface. The aerospace industry cannot rely solely on visual examination to determine material quality. Instead, the industry is reliant on nondestructive inspection (NDI), also called nondestructive testing (NDT) and nondestructive evaluation (NDE), to assess material integrity. As the name implies, NDI does not inflict damage on the component during the inspection process (as opposed to other methods which require destruction of the component to detect the damage).

Aviation safety regulations require the aerospace manufacturing companies to nondestructively inspect all primary structures before being built into the aircraft. Structures such as wing sections, fuselage panels, control surfaces and landing gear components are thoroughly inspected using NDI methods after fabrication to ensure they are free of defects. Airline companies and other aircraft operators (including the military) must also undertake regular NDI throughout the operating life of their aircraft. Aircraft and helicopters are taken out of service for routine NDI examinations to ensure the structures are damage-free.

Whether for the inspection of as-manufactured or in-service components, NDT is used to determine the type, size and location of damage which is essential information to assess the flight-worthiness and residual strength of the structure. There are various NDI methods, and the types most often used by the aerospace industry are listed in Table 23.3. The methods regularly used are: ultrasonics, dye penetrant, magnetic particle, radiography, thermography and eddy current. Quite often two or more methods must be used in combination to obtain a complete description of the type, size and location of internal damage. The types of damage that can be detected in metals and composites with the various methods are given in Table 23.3.

The NDI of aircraft is often an expensive, labour-intensive and slow process. The inspection of as-manufactured aircraft components adds considerably to the production time and cost (by as much as 50%). Inspection of in-service aircraft requires grounding and downtime, which affects the profitability of airlines and the operational capacity of airforces. For these reasons, the aerospace industry is assessing the use of structural health monitoring (SHM) to detect defects and damage. SHM systems use *in situ* sensor networks and intelligent data processing for the continuous inspection of aircraft structures. The sensor networks are embedded or surface-mounted on the aircraft, and can provide information on the presence of damage during manufacture or in-service operation with little or no human intervention. The concept of SHM is about determining the condition of a structure in real-time with

Table 23.3 Nondestructive inspection technologies

NDE technique	Inspection method	Metal	Composite
Ultrasonics	Acoustic waves	Intermetallic inclusions, cracks, corrosion and porosity	Delamination cracks, porosity and foreign objects
Radiography	X-rays, neutrons	Intermetallic inclusions, cracks, corrosion and porosity	Matrix cracks and porosity
Thermography	Heat	Intermetallic inclusions, cracks, corrosion and porosity	Delamination cracks, porosity, foreign objects and core degradation
Eddy current	Electromagnetic radiation	Cracks, corrosion and porosity	Not applicable
Magnetic particle	Electromagnetism	Cracks, corrosion and porosity	Cracks and voids (carbon-fibre composites only)
Dye Penetrant	Visual	Surface cracks	Surface cracks
Acoustic emission	Noise	Intermetallic inclusions and cracks	Delaminations and cracks

structurally integrated sensing equipment. This way, corrective maintenance can be performed when required, rather than at intervals based on flight times. Many SHM technologies are currently being developed and evaluated by the aerospace industry, with the most promising including Bragg grating optical-fibre sensors, piezoelectric transducers, and comparative vacuum monitoring.

In this chapter, we examine the NDI techniques used to detect manufacturing defects and in-service damage in metals and composites. We study the operating principles of the techniques and learn about their capabilities and limitations. In addition, there is an introduction to SHM for aircraft and several SHM techniques that are emerging as damage-detection methods for aircraft are discussed.

23.2   Nondestructive inspection methods

23.2.1 Visual inspection and tap testing

Two of the simplest inspection methods are visual inspection and tap testing. Visual inspection involves the careful examination of the material surface with the eyes, often assisted with a magnifying glass. Although simple, visual inspection is the first step in any NDI method, and it can identify obvious signs of damage. However, visual inspection is only suitable for detecting surface damage such as large cracks or general or exfoliation corrosion in metals, and it is not suitable when the damage is buried below the surface of metal or composite components.

Tap testing involves the repeated tapping over the component surface using a coin, soft hammer or some other light object to produce a ringing noise (Fig. 23.1). Damage immediately below the surface may be detected using tap testing by a change in the pitch of the noise. For example, tapping the surface of damaged composite material containing large delamination cracks can produce a dull sound compared with the higher-pitch ringing noise of the damage-free material. However, tap testing is not always reliable and can easily fail to detect damage. Instrumented tap testing devices which measure and analyse the noise generated by the tapping are available to eliminate the need for human hearing, which is not always sensitive to small changes in pitch. Both visual inspection and tap testing can be used for the initial inspection of aircraft components, but more sophisticated NDI methods are needed for reliable inspections.

23.2.2 Ultrasonics

Ultrasonics is an NDI method used widely in the aerospace industry to inspect aircraft structures and engine components. Ultrasonics is used for

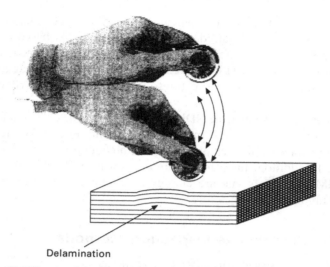

Delamination

23.1 Tap testing of a composite material for delamination damage
(from M. C. Y. Niu, *Composite airframe structure*, Hong Kong
Conmilit Press, 1992).

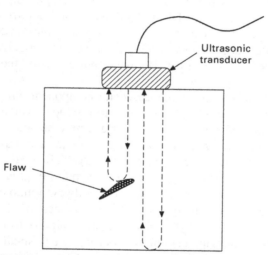

Ultrasonic
transducer

Flaw

23.2 Operating principles of ultrasonics.

the detection of both manufacturing defects and in-service damage. Although
ultrasonics cannot detect every type of damage, it can determine the presence
of common types of damage found in metals (e.g. voids, corrosion damage,
fatigue cracks) and composites (e.g. delamination, porosity).

The operating principle of ultrasonics is shown schematically in Fig. 23.2.
The method involves the transmission of ultrasonic pulses generated by a

piezoelectric transducer through the material. The pulses are high frequency (typically 1 to 15 MHz) compressive or shear elastic waves. When the waves encounter a region with an acoustic impedance value different from the host material, such as cracks or voids, then they are reflected and scattered. The characteristic acoustic impedance (Z) of a medium, such as air, metal or composite, is a material property:

$$Z = \rho c$$

where ρ is the density of the medium and c is the longitudinal wave speed in the medium. The acoustic impedance values for the main types of aerospace materials and air (which is the typical value for a crack) are given in Table 23.4. A large difference in the acoustic impedance value results in the loss in acoustic intensity owing to reflection and scattering, which is called attenuation and is measured using a receiving transducer. The received signal is analysed to determine the location and size of defects and damage.

Ultrasonics is operated in two basic modes: pulse–echo (or back-reflection) and through-transmission (Fig. 23.3). The pulse–echo mode involves the use of a single transducer located at one side of the material to radiate and receive the acoustic waves. When a defect is blocking the wave path then part of the acoustic energy is reflected back to the transducer. The reflected acoustic wave is transformed into an electrical signal by the transducer and is displayed on an oscilloscope. Pulse–echo ultrasonics can accurately measure the size and depth of damage in metallic and composite materials. Through-transmission ultrasonics involves using one transducer to generate the waves and another transducer located on the other side of the material to receive the signal. The waves generated by the transmitting transducer propagate through the material. When the waves encounter a defect with an acoustic impedance value different from the host material they are scattered and back-reflected, which attenuates the transmitted wave signal. The receiving transducer records a weakened signal owing to blocking of the acoustic waves by the damage, which is used to indicate its presence.

Pulse–echo ultrasonics is the preferred method for the inspection of in-service aircraft because the equipment is portable for field use and only

Table 23.4 Acoustic impedance values for aerospace materials and air (crack)

Medium	Acoustic impedance (Pa s m^{-1})
Air	420
Aluminium	17×10^6
Titanium	10×10^6
Nickel	54×10^6
Steel	45×10^6
Carbon–epoxy	9×10^6

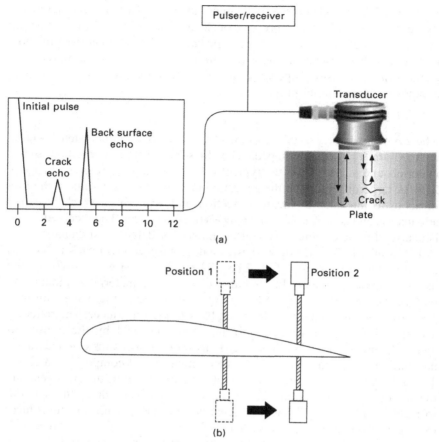

23.3 (a) Pulse–echo and (b) through-transmission ultrasonics.

one-side access is required (no need to remove aircraft components to gain access to both sides). Through-transmission ultrasonics is used more for the inspection of as-manufactured components before they are assembled into the aircraft. This mode of inspection is faster and more easily automated than pulse–echo ultrasonics. Furthermore, ultrasonics can generate a two- or three-dimensional image (called a C-scan) of damage inside the material by measuring the time-of-flight of reflected acoustic waves. Figure 23.4 shows a C-scan image of an aerospace composite material, with the bright zone revealing internal delamination damage, caused by a low-energy impact event, that cannot be observed visually. Other ultrasonic methods are used occasionally by the aerospace industry, such as Lamb waves and laser ultrasonics, although their application is less common than pulse–echo and through-transmission ultrasonics. Ultrasonics is best suited for the detection

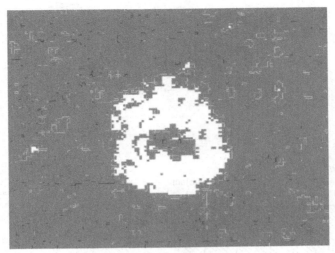

23.4 C-scan image of impact damage (white zone) to carbon–epoxy composite.

of planar damage such as cracks aligned parallel with the surface. The technique is not well suited to detecting damage aligned parallel with the propagation direction of the acoustic waves, although angled probes can be used.

23.2.3 Radiography

Radiography involves the use of radiation, such as x-rays, γ-rays or high-speed neutrons, to detect damage in solids (Fig. 23.5). Radiation is emitted from an energetic source, such as an x-ray tube, and directed to the test component. The radiation energy is absorbed during its passage through the material. However, the absorption rate changes when the radiation passes through a defective region having different absorption properties to the host material. The absorption value for an air gap is much lower than the aerospace materials and, therefore, less energy is absorbed during the passage of radiation through cracks and voids.

The radiation intensity is measured using x-ray film after passing out of the material. Regions of high-intensity radiation (owing to the presence of damage) and low-intensity radiation (pristine material) appear different on the x-ray image (as shown in Fig. 23.6). The size and shape of the defect is measured from the image. Radiography can detect defects such as voids, intermetallic inclusions, corrosion damage, and cracks larger than ~0.5–1.25 mm, which is below the critical damage size in aircraft structures. However, the shape and orientation of the defect affect how easily it is detected. Long

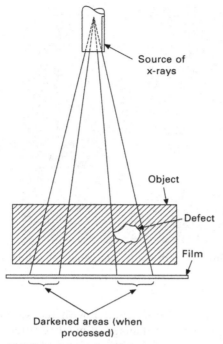

23.5 Principles of radiography.

23.6 Radiographic image of impact damage in carbon–epoxy composite.

cracks aligned parallel with the direction of radiation are more easily detected than cracks running perpendicular to the incident radiation. Therefore, it is necessary to inspect a component with different radiation angles to ensure cracks at different orientations are detected.

23.2.4 Thermography

Thermography is used by the aerospace industry for the rapid, wide-area inspection of components. There are two main types of infrared thermography known as passive and active. Active thermography is the more widely applied of the two, and is shown in Fig. 23.7. The active method involves short duration heating (usually less than 1 s) of the component surface using flash tubes, hot-air guns or some other controllable heating device. The component surface is heated 10–20 °C above ambient temperature, with the heat being absorbed into the material. The method measures the heat dissipated from the heated surface as the component cools. The amount of heat dissipated depends on the thermal properties of the material together with the type, size and location of damage. Any defect which creates an air gap such as a delamination, void or corrosion cavity absorbs less heat than the parent material. Consequently, more heat is dissipated from the surface above the defective region. The damage is then observed using an infra-red (IR) camera as a 'hot spot' on the surface. For instance, Fig. 23.8 shows a hot spot in a thermographic image of a carbon–epoxy composite caused by delamination damage. Differences in surface temperature are recorded using an IR camera to reveal the damaged area, and temperature-data-processing

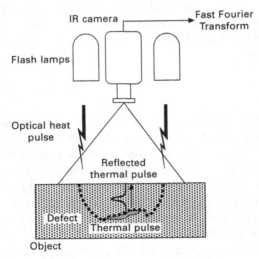

23.7 Principles of active thermography.

23.8 Thermographic image showing damage in a carbon–epoxy composite.

methods are used to determine the damage depth. Thermography can detect delamination cracks, porous regions and foreign objects in composites and intermetallic inclusions and large voids in metals.

The passive thermography method uses internally generated heat, often from damage growth, rather than externally applied heat. Heat generated by damage growth is measured as a hot spot on the component surface. This technique is not as popular as active thermography because the material must be damaged to generate the internal heat and, therefore, it is rarely used to inspect aircraft components.

23.2.5 Eddy current

The eddy current method is widely used to inspect metallic aircraft components for surface cracks and corrosion damage. Figure 23.9 shows the eddy current process of inspection. Eddy current testing equipment contains a conductive metal coil which is electrified with an alternating current. The current generates a magnetic field around the coil. This magnetic field expands as the alternating current rises and collapses as the current drops. When the coil is placed in close proximity to another conducting material, such as a metal component, then an alternating current, called an eddy current, is induced in this material by the magnetic field. The eddy currents are induced by electrical currents that flow in circular paths. Surface and near-surface

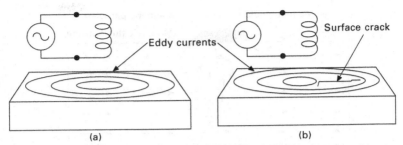

Eddy currents

Surface crack

(a)   (b)

23.9 Eddy current inspection of materials (a) without and (b) with surface cracks.

cracks interrupt the eddy currents, and this is detected by changes in the coil's impedance. By passing the eddy current equipment, which is a hand-held device, above the component surface it is possible to detect cracks, including shallow and tight surface cracks that may be less than 50 μm deep and 5 μm wide. Cracks that cut across the eddy current path are easily detected, although cracks parallel to the current flow can be missed because they do not disturb the eddy currents.

Eddy current equipment is lightweight and portable, thus allowing for inspection of grounded aircraft. However, access to both sides of the component is necessary for complete inspection. Eddy current testing only works on conductive materials such as aerospace metal alloys. It cannot be used to inspect insulating materials with low electrical conductivity, such as fibreglass composites, and difficulties are experienced with ferromagnetic materials.

23.2.6 Magnetic particle

Magnetic particle inspection is a simple NDI method used to detect cracks at the surface of ferromagnetic materials such as steels and nickel-based alloys. The inspection process begins with the magnetisation of the component. The surface is then coated with small magnetic particles, which is usually a dry or wet suspension of iron filings. Surface cracks or corrosion pits create a flux leakage field in the magnetised component, as shown in Fig. 23.10. The magnetic particles are attracted to the flux leakage and thereby cluster at the crack. This cluster of particles is easier to see than the actual crack, and this is the basis for magnetic particle inspection.

23.2.7 Liquid dye penetrant

Liquid dye penetrant is used to locate surface cracks in materials, but cannot detect subsurface damage. The method is shown schematically in Fig. 23.11.

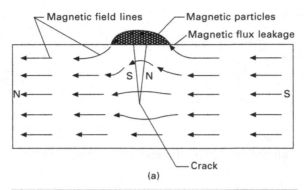

(a)

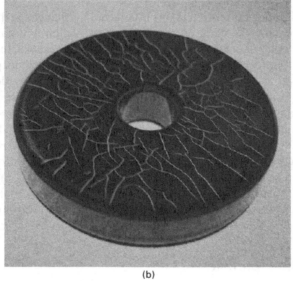

(b)

23.10 Magnetic particle inspection: (a) principles (b) in detection of surface cracks. Photograph supplied courtesy of MR Chemie GmbH.

The component surface is cleaned before a visible or fluorescent liquid dye is applied using a spray, brush or bath. The dye seeps into surface cracks by capillary action. Excess dye retained on the surface is wiped off leaving only the dye that has seeped into the cracks. Chemical developer is then applied and it reacts with the dye, drawing it from the crack on to the surface. The dye can then be observed, either because it changes the colour of the developer or because it fluoresces under ultraviolet light.

Liquid dye penetrant is a popular inspection method because it is simple, inexpensive and can detect cracks to a depth of about 2 mm. The main drawback is that the method can only detect surface breaking cracks. Despite

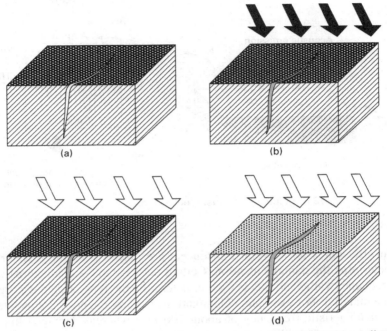

23.11 Dye-penetrant method: (a) undetected surface crack; (b) applied surface dye seeps into crack; (c) surface is cleaned and chemical developer applied; (d) crack detected by residual dye.

this problem, it is often used to inspect aircraft components, particularly engine parts, which are susceptible to surface cracking.

23.2.8 Acoustic emission

Acoustic emission involves the detection of defects using sounds generated by the defects themselves. Figure 23.12 shows the operating principles of the acoustic emission test method. The component is subjected to an applied elastic stress; usually just below the design load limit. When cracks and voids exist in materials, the stress levels immediately ahead of the defect are several times higher than the surrounding material (as explained in chapter 18). This is because cracks act as a stress raiser. Any plastic yielding and microcracking that occurs ahead of the defect owing to the stress concentration effect can generate acoustic stress waves before any significant damage growth. The waves are generated by the transient release of strain energy owing to microcracking. The waves are detected using sensitive acoustic transducers located at the surface. The transducers are passive, that is, they only 'listen' for sounds and do not generate the acoustic waves that the transducers used

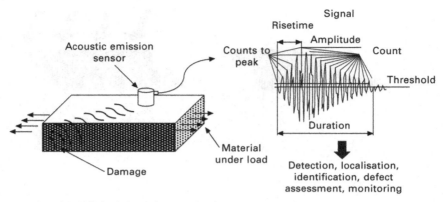

23.12 Principles of acoustic emission.

in ultrasonics do. Several transducers are placed over the test surface to determine the damage location. Certain defects have a characteristic sound frequency value and this is used to determine the type of damage present in the material. For example, delaminations in carbon–epoxy composites have a characteristic frequency of about 100 kHz whereas damaged fibres emit sound at around 400 kHz.

Acoustic emission has several advantages, including rapid inspection of large components and the capability to determine the location and type of damage. The method can be used to continuously monitor components while operating in-service, and in this respect it is a SHM technology. The downside is that the component must be further damaged to generate the acoustic emission signal.

23.3   Structural health monitoring (SHM)

23.3.1 Background to SHM

Conventional NDI poses several problems for the aerospace industry that are difficult to resolve using the standard test methods. NDI methods such as ultrasonics, radiography and magnetic particle inspection require point-by-point inspections which are slow and labour intensive. Wide-area NDI techniques such as thermography allow for rapid inspection, although it is difficult to detect damage in nonplanar and complex structures such as joints. Most NDI methods require the grounding and stand-down of aircraft, often for lengthy periods because the inspections are slow. Further problems with many NDI tests are difficulties in the detection of damage in physically inaccessible areas of the aircraft and the inability to continuously monitor the formation and growth of damage over the aircraft life. For these reasons, there is growing awareness in the aviation industry that the real-time, continuous monitoring of in-service aircraft requires the use of SHM.

SHM involves the continuous measurement and assessment of in-service structures with little or no human intervention. The information provided by SHM systems about the physical condition of an aircraft structure is used by airline companies and aerospace engineers to make continuous lifecycle management decisions. Continuous assessment can significantly reduce aircraft downtime because it minimises the need for routine ground-based inspections for damage that may not be present. The early detection of damage using SHM improves the structural reliability and safety of aircraft. Early damage detection also offers the possibility of reducing design safety factors applied for damage tolerance, which translates into more lightweight aircraft structures, as well as minimising the repair process.

SHM systems are classified as passive or active. A passive system relies on taking measurements during normal in-service operation or detecting individual damage events, such as bird strike. In contrast, active systems stimulate the structure with an input and measure the output response of the structure. Data collected from the sensors is analysed using intelligent software processors for the determination of structural health. The processors are stored on the aircraft, acting much like a black-box flight recorder, or located at a ground-based facility with the signal transmitted direct from the aircraft.

SHM uses surface-mounted or embedded sensors for the real-time structural monitoring of damage initiation and growth. Various sensor types are available, including the following: fibre-optic sensors, piezoelectric transducers, dielectric sensors, and comparative vacuum galleries.

There are several similarities between the SHM of aircraft and the nervous system of humans. The idea of a SHM 'nervous' system installed in a commercial airliner is shown in Fig. 23.13. SHM sensors are distributed throughout an aircraft, in much the same way as nerves are spread throughout the body. The sensors detect changes to the 'health condition' of the structure in real-time, again like the nervous system, and this information is sent to a central processor. The processor provides immediate information about the damage during aircraft operations. Although it is possible to distribute sensors at many locations throughout an aircraft to assess the overall structural condition, this is usually unnecessary. It is only necessary to locate sensors at critical and highly loaded structures, such as the wing box, wing–fuselage connections and landing gear, and at structures prone to environmental or impact damage, such as leading edges susceptible to bird strike.

SHM systems are designed for either local or wide-area monitoring. A local system is concerned only with pre-determined structural 'hot spots' such as joints, leading edges and door frames. SHM systems for local monitoring include Bragg grating optical fibre systems and comparative

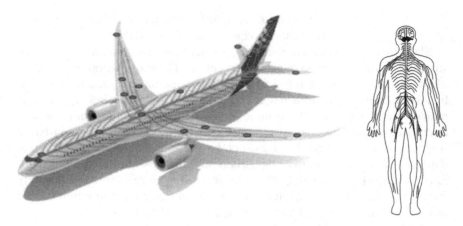

23.13 SHM and the human nervous system work on similar principles (from S. Black, Structural health monitoring: composites get smart, in *High-performance composites*, 2008, p. 47–52).

vacuum monitoring. Global systems are concerned with damage detection and identification over a much larger area and generally aim to detect larger major structural damage, and examples include acousto-ultrasonics and random decrement analysis based on vibrational responses.

Unlike conventional NDI, SHM for aircraft damage is not a fully mature technology, and SHM is currently not widely used in aircraft or helicopters. There are several military aircraft and, to a lesser extent, civil airliners fitted with SHM sensors. However, SHM is not a mainstream technology for damage monitoring, and the aviation industry is still heavily reliant on NDI technology. As the industry gains a better understanding of SHM then it is likely that sensors will be used increasingly in aircraft. Short descriptions of several promising SHM technologies follow – i.e. optical fibres, piezoelectric sensors and comparative vacuum monitoring – to demonstrate their potential aerospace applications.

23.3.2 SHM systems using optical fibre sensors

Optical fibre sensor systems are one of the more mature SHM techniques. The optical fibre sensor consists of a central silica core surrounded by an annular silica cladding with a protective coating (Fig. 23.14). Sensors are also made of translucent plastic materials. The fibres are long and thin (50–250 μm in diameter) and are surface mounted on metal and composite structures. Sensors can be embedded within fibre–polymer composites and along the bond-line of structural joints.

The method of damage detection involves shining monochromatic light

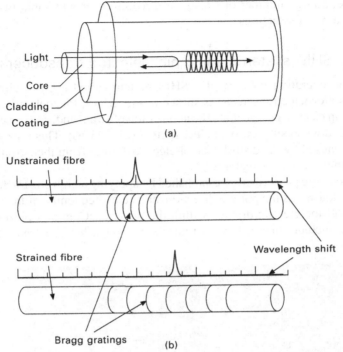

Light
Core
Cladding
Coating

(a)

Unstrained fibre

Wavelength shift

Strained fibre

Bragg gratings   (b)

23.14 Bragg grating sensor: (a) schematic showing core and cladding; (b) operating principles.

along the fibre. The fibre core has a higher refractive index than the cladding, which allows the light to be confined within the core with minimal loss over long distances. The core is inscribed with Bragg gratings, which are lengths in which the grating lines lead to changed reflection. The light is partly reflected back at the Bragg gratings. The spacing distance between neighbouring gratings is measured by the wavelength of the reflected light. The spacing between Bragg gratings changes when the fibre is strained; the spacing increases under tension and contracts under compression. The strain level applied to the sensor (and therefore the structure to which it is attached) is measured from the change in reflected wavelength caused by the change in spacing of the Bragg gratings. Certain types of damage, such as delaminations and broken fibres in composites and corrosion in metals, can reduce the local stiffness of structures. Damage is detected using optical fibres by an unexpected change in the measured strain (or stiffness) compared with the strain of the defect-free structure. Optical fibres are small and robust which makes them suitable for installation on aircraft structures. However, they can only detect damage in the vicinity of the Bragg gratings and, therefore, a large number of sensors are required for wide-area inspection of large aircraft structures.

23.3.3 SHM systems using piezoelectric transducers

The basic operating concept of a SHM system called acousto-ultrasonics that uses piezoelectric sensors is shown in Fig. 23.15. Piezoelectric sensors produce an electric charge upon the application of strain and, conversely, they can expand when subject to an electric field (Fig. 23.16). This piezoelectric effect is caused by the disturbance of electric dipoles from their equilibrium state within the sensor material.

Piezoelectric properties are established by applying a high electric field in a direction known as the polling direction, at an elevated temperature, in order to align all the electric dipoles within the material. Common piezoelectric materials include quartz, barium titanate ($BaTiO_3$), lead zirconate titanate

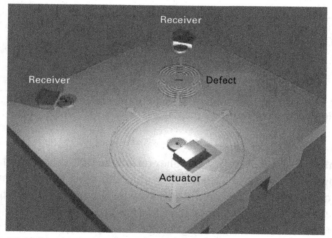

23.15 Piezoelectric structural health monitoring.

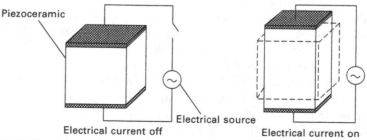

23.16 Piezoelectric effect in sensors.

[Pb(ZrTi)O$_3$] and polyvinylidene fluoride. Piezoelectric materials come in various shapes and forms, including wafers, plates, strips or fibres, which are attached to the structural component. The sensors are often thin and small (less than ~15 mm), and are bonded directly to the structure.

The system shown in Fig. 23.15 has three piezoelectric devices: an actuator that releases elastic stress waves when activated by an alternating electric charge and two receiving sensors that measure the strength and frequency of the waves by changes to their electrical properties. When an aircraft structure fitted with a piezoelectric SHM system is damage-free, the signal recorded by the receiving sensor has a characteristic wave intensity and frequency. Damage that changes the elastic properties of the structure, such as corrosion, fatigue or impact cracking, is detected by a change in the signal wave properties. It is feasible to use many piezoelectric actuators and sensors distributed over a large structure to perform wide-area, real-time and continuous inspections. The potential application of piezoelectric devices for the structural monitoring of aircraft has been demonstrated.

23.3.4 SHM systems using comparative vacuum monitoring

Comparative vacuum monitoring (CVM) is an emerging SHM system for use in large metallic and composite aircraft structures. The operating principle of CVM is simple; the system consists of a thin polymer sensor containing a series of thin holes that are connected to sensing and recording equipment (Fig. 23.17). The sensor is bonded to the aircraft structure, usually where damage is expected. The arrangement of the narrow holes, called 'galleries', inside the sensor is shown in Fig. 23.18. Air within every second gallery in a series is removed by a vacuum pump to place them in a state of low pressure (called vacuum galleries). Air is retained in the other galleries, and this creates a sensor system consisting of a series of parallel galleries that alternate between low pressure and ambient pressure.

CVM is based on the principle that a crack growing under the sensor links a low-pressure gallery and a neighbouring gallery at ambient pressure. Air flows through the crack into the neighbouring low-pressure gallery, and the pressure change is detected by sensing equipment. Using this approach, it is possible to determine the location and size of surface cracks caused by corrosion, fatigue or impact, provided of course the sensor is located at the defective location.

23.4   Summary

Aircraft structures and engine components must be nondestructively inspected after manufacturing and throughout their operational life for the presence of

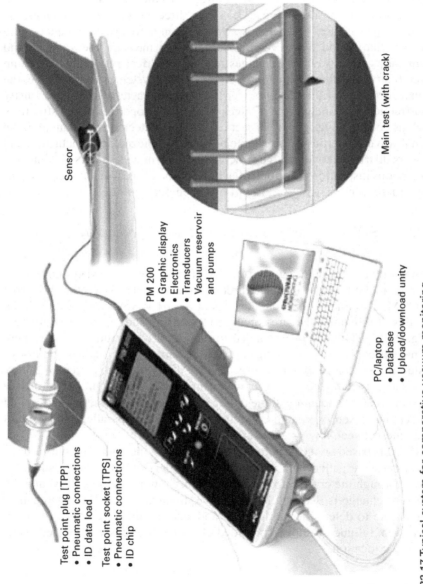

Sensor

Main test (with crack)

PM 200
• Graphic display
• Electronics
• Transducers
• Vacuum reservoir
  and pumps

PC/laptop
• Database
• Upload/download unity

Test point plug [TPP]
• Pneumatic connections
• ID data load

Test point socket [TPS]
• Pneumatic connections
• ID chip

23.17 Typical system for comparative vacuum monitoring.

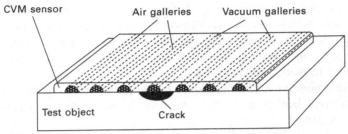

23.18 Comparative vacuum monitoring (CVM) sensor.

defects and damage. Most inspections are currently performed using NDT methods such as ultrasonics, radiography and thermography. Structural health monitoring (SHM) is emerging as an alternative to conventional NDI, in which sensor systems are used with little or no human invention to monitor aircraft for damage.

NDI methods have the capability to detect certain (but not all) types of damage in metals and composites. Ultrasonics, thermography and eddy current inspections are capable of detecting damage and cracks aligned parallel with the material surface whereas radiography is better suited to detecting cracks normal to the surface. It is often necessary to use two or more inspection methods to obtain a complete description of the type, amount and location of the damage.

Some NDI techniques can be used to inspect metals but not fibre–polymer composites. Of the NDI methods described in this chapter, damage in composites is difficult to detect using eddy current and magnetic particle owing to their low electromagnetic properties, and using liquid dye penetrant because most damage is internal (e.g. delaminations) and does not break the surface.

SHM has the potential to reduce aircraft downtime for routine inspections and reduce design safety factors for damage tolerance because of the early detection of damage. It is often only necessary to locate SHM sensors in components prone to damage (e.g. heavily-loaded parts, parts susceptible to impact damage), rather than covering the entire aircraft with a complex, integrated sensor network system.

SHM techniques are classified as local or global (wide-area). Examples of local health monitoring include Bragg grating optical fibre sensors and comparative vacuum monitoring, whereas wide-area monitoring techniques are acoustic emission and acousto-ultrasonics.

23.5   Terminology

Acoustic emission: NDE technique for detecting and analysing transient elastic waves released within material from the growth of damage under an externally applied stress.

Acoustic impedance: A frequency (f) dependent parameter used to describe the ease with which sound waves travel through solids, liquids or gases.

Acousto-ultrasonics: SHM technique that relies on the use of acoustic waves (usually Lamb waves) generated from piezoelectric sensors to detect damage in plate-like structures.

Active thermography: Thermographic technique that relies on the application of external heat (usually flash heating) to the material to generate an infrared signal.

Bragg grating: Periodically spaced zones in an optical fiber core with refractive indexes that are slightly higher than the core. The gratings selectively reflect a very narrow range of wavelengths while transmitting others. Used as an SHM sensor for the measurement of strain.

C-scan: Data presentation method (usually as a two-dimensional image) applied to ultrasonic techniques. Image shows the size and shape of the defective region, and by using time-of-flight data can reveal defect depth.

Comparative vacuum monitoring: SHM technique that uses a sensor to measure the differential pressure between fine galleries at a low vacuum alternating with galleries at atmosphere. If no flaw is present, the vacuum remains at a stable pressure. However, if a flaw develops, air flows through the passage created from the atmosphere to the vacuum galleries.

Eddy current: An NDE technique that uses an induced electric current formed within conductive materials which are exposed to a time varying magnetic field. Damage is recorded as a disturbance to the current.

Lamb waves: Elastic waves whose motion is along the plane of the plate. Lamb waves have lower frequencies than conventional ultrasonic waves, which allows them to travel longer distances along plates with a consistent wave pattern.

Laser ultrasonics: NDE technique that uses a laser to generate and detect ultrasonic waves for the inspection of damage in solids.

Liquid dye penetrant: An NDE technique for detecting surface porosity or cracks in metals. The part to be inspected is cleaned and coated with a dye that penetrates any flaws that may be present. The surface is wiped clean and coated with a chemical to absorb the dye retained in the surface defects indicating their location.

Magnetic particle: An NDE technique for determining defects in ferromagnetic materials. Finely divided magnetic particles, applied to the magnetised material, are attracted to and outline the pattern of the magnetic leakage fields created by the damage.

Passive thermography: Thermographic technique that relies on internally generated heat in the material to produce an infrared signal.

Piezoelectricity: The generation of electricity or electrical polarity in dielectric crystals subjected to mechanical stress, or the generation of stress in such crystals subjected to an applied voltage.

Pulse–echo ultrasonics: Ultrasonic technique that relies on the back-reflection of the acoustic beam from damage within a material to determine the location and size of the defective region.

Radiography: An NDE technique that relies on the transmission of radiation (usually x-rays) through a solid to produce an image produced on a radiosensitive surface, such as photographic film. Defective regions absorb the radiation at a different rate to the pristine material, and damage appears as a darker or brighter spot in the image.

Structural health monitoring (SHM): Process involving the observation of a structure over time using periodically sampled dynamic response measurements from an array of sensors, the extraction of damage-sensitive features from these measurements, and the statistical analysis of these features to determine the current state of system health.

Thermography: A non-destructive inspection technique in which an infrared camera is used to measure temperature variations on the surface of the body, producing images that reveal sites of damage.

Through-transmission ultrasonics: Ultrasonic technique that relies on the attenuation of the acoustic beam when passing through damage within a material to determine the location and size of the defective region.

Ultrasonics: NDE technique which relies on an ultrasonic beam passing through the material to detect the presence of damage by either back-reflection (pulse–echo) or attenuation (through–transmission) of the acoustic waves.

23.6   Further reading and research

Adams, D. E., *Health monitoring for structural materials and components: methods with applications*, John Wiley & Sons, 2007.

Chang, F. K., *Structural health monitoring: current status and perspectives*, Technomic Publishing, 1998.

Hellier, C. *Handbook of nondestructive evaluation*. McGraw–Hill, 2003.

Raj, B., Jayakumar, T. and Thavasimuthu, M., *Practical nondestructive testing (2nd edition)*, Alpha Science International Ltd., Pangbourne, UK, 2002.

Shull, P. J., *Nondestructive evaluation: theory, techniques, and applications*, Marcel Dekker Inc., 2002.

Disposal and recycling of aerospace
materials

24.1   Introduction

The disposal and recycling of aerospace materials is an important issue in
the whole-of-life management of aircraft. The use of sustainable materials
is becoming more important as the aerospace industry moves towards a
'cradle-to-beyond the grave' approach in the management of aircraft. Until
recently, the selection of materials for aircraft structures and engines was
based on cost considerations and performance requirements. Materials
are selected on economic considerations such as the costs of purchase,
manufacturing, assembly, and in-service maintenance. Materials are also
selected on performance requirements such as stiffness, strength, toughness,
fatigue life, corrosion resistance, maximum operating temperature and so
forth. The majority of the aerospace industry has previously given little
consideration to the materials beyond the end-of-life of the aircraft. In the
past, end-of-life meant the day the aircraft was taken out-of-service, never
to fly again.

The meaning of end-of-life for aerospace materials has recently changed
in an important way. There is a growing understanding in the aerospace
industry that end-of-life no longer means when the aircraft is taken out-
of-service, but extends beyond this point to include the management of the
aircraft after end-of-life (or 'beyond the grave'). Governments, environmental
organisations, and the wider public are placing greater demands on the
responsible management of products beyond the end-of-life when made
using nonrenewable resources. There are growing expectations that products
produced in large quantities can be recycled so their materials can be reused
rather than being disposed via landfill.

Recycling reduces the demand for the production of new metals, which
involves mining, extraction and refinement processes; all of which are
environmentally harmful. Recycling may also reduce the need for new
composite materials, which are produced using nonrenewable petroleum
products and use energy-intensive manufacturing processes. The other benefit
of recycling is the reduced demand on landfill and other hard waste disposal
methods.

Until recently, the recycling of aircraft materials was not a major
consideration for the aerospace industry. For decades, the majority of private,

558

civil and military aircraft ended their days in graveyard sites, such as the Mojave Desert in California which stores many thousands of retired aircraft and helicopters (Fig. 24.1). Aircraft are too large to bury as landfill, and are left in remote locations such as the Mojave Desert where the dry environment slows the destruction of the airframe, engines and avionics systems. Some of these old aircraft are used for ground training purposes, others are cannibalised for spare parts, and others are dismantled for recycling.

The aircraft recycling rate is currently about 60%, with the remainder representing aircraft that are left to decay. However, the pressure on these graveyard sites intensifies as greater numbers of passenger aircraft reach their end-of-life in coming years. Figure 24.2 show the retirement of aircraft per year between the years 1990 and 2012; over this period the number of retirements per year has increased by more than 500%. The number of retirements per year typically accounts for 1–3% of the entire fleet. Airbus estimates about 6400 airliners will retire before 2026.

The majority of aircraft in graveyard sites are constructed mostly of aluminium alloy. Most of the fuselage and wings of old civil and military aircraft are made using aluminium, which can be sold as scrap for recycling. With the greater use of composite materials in aircraft over the past ten to twenty years it is expected that the recycling of carbon fibre–epoxy will become increasingly important.

Growing global concerns about the environmental impact of retired aircraft as well as economic efficiencies are beginning to drive the aerospace industry

24.1 Aircraft graveyard in the Mojave Desert (USA).

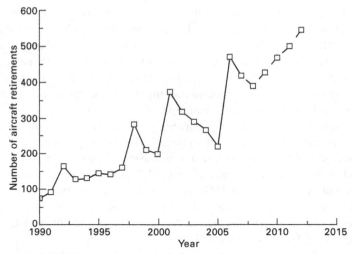

24.2 Passenger aircraft retirements, 1990–2012.

towards greater recycling. Several programmes and organisations have been established to manage the recycling of aircraft, such as the PAMELA project (Process for Advanced Management for End of Life Aircraft) by Airbus and the AFRA project (Aircraft Fleet Recycling Association) by Boeing. The aerospace industry is aiming to increase the recycling rate of aircraft materials from the current level of 60 to 80–90%. The increased recycling target is set in a future environment where greater numbers of aircraft are being retired. For this reason, an important factor in the design of new aircraft, which will eventually be taken out-of-service many years later, is the selection of sustainable materials that can be recycled. The aerospace industry has responsibility for considering the management of the aircraft beyond the end-of-life. This involves consideration in the design phase of the recycling of the entire aircraft, including the body, landing gear, cabin fittings, engines, avionic and hydraulic/control systems.

Sustainable engineering is a key issue in the design, manufacture and in-service support of aircraft and helicopters. An important component to sustainable engineering is the use of materials that can be recycled with little or no impact on the environment. As a minimum, recycling should be less harmful to the environment than the production of new material from nonrenewable resources. Table 24.1 shows the amount of energy consumed in the production of several metals used in aircraft from ore or recycled material, and the energy savings achieved with recycling are great. Table 24.2 shows the amount of carbon dioxide produced per kilogram of metal produced from the ore or recycled product. Producing metals from the processing of scrap consumes a lot less energy and generates less greenhouse gas than producing

Table 24.1 Energy needed to produce metal from ore or scrap

Metal	Energy (MJ kg^{-1})		Average energy ratio (ore/scrap)
	Primary from ore	Secondary from scrap	
Magnesium	350–400	8–10	42
Aluminium	200–240	18–20	12
Titanium	600–740	230–280	2.6
Nickel	135–150	34–38	3.9
Steel	32–38	9–11	3.6

Table 24.2 Carbon dioxide generation in the production of metal from ore or scrap

Metal	Carbon dioxide production (kg kg^{-1})		Average CO_2 ratio (ore/scrap)
	Primary from ore	Secondary from scrap	
Magnesium	22–25	1.8–2.0	12.5
Aluminium	11–13	1.1–1.2	10.5
Titanium	38–44	14–17	2.6
Nickel	7.9–9.2	2.0–2.3	4.0
Steel	2.0–2.3	0.6	3.6

from the ore. Extraction of metal from ore requires a large amount of energy because the ore must be mined, shipped great distances, freed from tailings, and reduced or smelted. The aerospace industry is also concerned with the economic impact of sustainable engineering. The cost of the recycling process, which includes the costs of removing the material from the retired aircraft, cleaning the stripped material, cutting and grinding the material into small pieces, transporting the scrap to the recycling processing plant, and the recycling process itself, should be at most equivalent to the cost of using new material.

In this chapter, we examine the recycling of aerospace materials. The recycling of metals and fibre–polymer composites used in aircraft structures and engine components is reviewed. The recycling methods described in this chapter are used not only to treat materials reclaimed from retired aircraft, but are also used to recycle waste material generated during the production of aircraft structures and engine components. The trimmings, swarf and other material removed in the casting, milling, machining and drilling of metal components can be recycled using the same methods as metal components removed from retired aircraft. The production of intricate metal components can require the removal of a large percentage of the original material (in some cases up to 90%), which should be recycled as part of the sustainable engineering approach to aircraft production. The manufacture of composite components also results in trimmed and dust materials, which can also be recycled using the same processing methods as used for composite parts removed from old aircraft.

24.2   Metal recycling

24.2.1 Aluminium recycling

Aluminium is the most common metal used in civil and military aircraft and, therefore, the ability to recycle this material using an economically viable and environmentally friendly process is essential to sustainable aerospace engineering. Aluminium is the second most recycled metal (after steel), and about one-third of all aluminium is extracted from scrap products. The most common source of scrap aluminium is general purpose items, such as beverage cans and household products. Aluminium reclaimed from aircraft is a small but growing source of scrap material.

There are several important benefits in the recycling of aluminium. Firstly, the quality of aluminium is not impaired by recycling; the metal can be recycled repeatedly without any adverse affect on the properties. Secondly, aluminium recycling is less expensive than the production of new aluminium from ore. Recycling of aluminium generally results in significant cost savings over the production of new aluminium, even when the costs of collection, separation and recycling are taken into account. As a result, it is financially viable to recycle scrap aluminium from aircraft, and based on current price the estimated value for reclaimed aluminium from the skin and airframe of a Boeing 747 is in the range \$200 000 – \$250 000. The process of aluminium recycling simply involves re-melting the metal, which is much cheaper and less energy intensive than producing new aluminium by electrolytic extraction (via the Bayer process) from bauxite ore. Recycling scrap aluminium requires about 5% of the energy needed to produce new aluminium. One of the challenges with recycling aluminium or any other metal is the unstable price of the recycled material. For instance, Fig. 24.3 shows the fluctuations in the price of recycled aluminium from 1990 to 2005, with the price adjusted for inflation. The price can vary greatly over relatively short periods of time, and this affects the financial profitability of the recycling process. Although the price of new metal also changes, it is almost always more expensive than recycled metal.

The third key benefit of aluminium recycling is that it is less polluting than producing new material. The environmental benefits of recycling aluminium are enormous. The electrolytic extraction of aluminium from bauxite is an energy-intensive process requiring large amounts of electricity. For example, the production of new aluminium in the United States consumes about 3% of the national energy requirements. The amount of nonrenewable resources

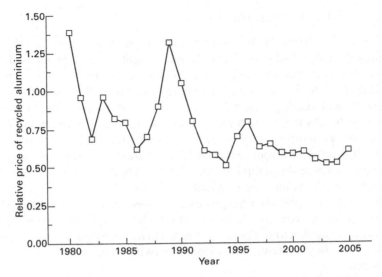

24.3 Variation in the price of recycled aluminium with the price normalised to the 1992 price and adjusted for inflation.

(e.g. coal, liquefied gas) needed to generate the electricity for the melting and refinement of scrap aluminium is much less than the production of new aluminium. Therefore, scrap aluminium has less environmental impact because less energy is needed to power the recycling process and therefore fewer greenhouse gases and other pollutants are generated. It is estimated that recycling 1 kg of aluminium saves up to 6 kg of bauxite, 4 kg of chemical products used in the electrolytic refinement process, and 14 kWh of electricity. Therefore, in the recycling of a mid-sized passenger aircraft, which contains about 20 tonne of aluminium, over 130 tonne of bauxite is saved along with 300 MWh of electricity needed to extract the metal from the ore. If brown coal is used to generate the electricity, then recycling also saves over 100 tonne of CO_2 and other pollutants.

Recycling of scrap aluminium from aircraft is a simple process. The aluminium components are removed from the aircraft, cut and shredded into small pieces, and then chemically treated to remove paint, oils, fuels and other contaminants. The aluminium pieces are compressed into blocks and then melted inside a furnace at 750 ± 100 °C. Refining chemicals (e.g. hexachloroethane, ammonium perchlorate) are added to the molten aluminium. The furnace is tapped to cast the refined aluminium into ingots, billets, rods or other product forms. The cast aluminium can be reused for any application, including the production of new aircraft parts.

24.2.2 Magnesium recycling

The use of magnesium in modern aircraft and helicopters is very small (typically under 1–2% of the airframe weight), although its high value makes recycling economically viable. The process of recycling magnesium is relatively simple. Scrap magnesium removed from the aircraft is stripped of paints and coatings, cleaned and then melted inside a steel crucible at 675–700 °C. As magnesium melts, there is the risk of ignition and burning owing to high-temperature oxidation reactions with air. To suppress burning, the melt is protected from oxidation by a covering of flux agents or an inert-gas stream. Once the scrap has completely melted and purified, the molten magnesium is cast into ingot moulds for reuse.

In addition to the melting process, magnesium is recycled by grinding the scrap into powder for steel production where it is used to remove sulfur impurities in molten iron. This use is limited to relatively pure magnesium scrap with a low alloy content, otherwise the alloying elements may contaminate the steel.

24.2.3 Titanium recycling

The aerospace industry is a large supplier of scrap titanium, which is removed from engines and structural components on aircraft and helicopters. Titanium is a valuable metal and therefore the strong economic incentive exists to recycle. The environmental impact of recycling titanium is less than producing new metal from the ore and, therefore, an environmental incentive for recycling also exists.

Recycling of titanium is more complicated than for aluminium because the metal is reactive at high temperature. Titanium is recycled by melting the scrap at high temperature (above 1700 °C). Liquid titanium reacts with nearly all refractory furnace linings as well as with oxygen and nitrogen in the atmosphere. Titanium cannot be melted in an open-air furnace because of oxidation. For this reason, the melting process is performed in a vacuum or inert atmosphere inside a furnace lined with nonreactive refractory material.

Vacuum-arc remelting (VAR), as the name implies, is a process whereby metal is melted under vacuum inside an electric arc furnace. The recycling process begins by forming the scrap titanium into a cylinder which serves as an electrode inside the arc furnace. The titanium electrode is placed just above a small amount of titanium resting at the bottom of a large crucible. An electric current flows into the upper electrode and creates an arc with the underlying titanium. The temperature generated by the electric arc causes the base of the titanium electrode to melt, with the liquid metal dripping into the crucible. The low pressure inside the furnace suppresses oxidation of the molten titanium and inhibits the formation of brittle titanium nitride precipitates by the reaction with atmospheric nitrogen. The vacuum also removes dissolved gases from the melt. After the electrode has completely melted, the titanium is cast into vacuum-sealed moulds for re-use in new aircraft components or nonaerospace applications (e.g. medical devices such as joint replacements and armour plates for military vehicles).

24.2.4 Steel recycling

Steel used in landing gear and other highly-loaded structures can be recycled. Steel is the most recycled of all the metals, with about 60% of steel products being recycled. The most common products for scrap steel are automobiles, food cans and appliances; scrap from aircraft represents a tiny fraction of the total amount of recycled steel. The recycling of steel involves melting the scrap metal at high temperature (1600–1700 °C) inside a furnace. Chemicals are added to the molten steel to remove carbon and other alloying elements as dross. The purified iron is then cast into ingots, billets or some other product for reuse.

There are strong environmental reasons for recycling the steel in aircraft. Every tonne of steel that is recycled makes the following environmental savings: 1.5 tonne of iron ore; 0.5 tonne of coal; about 250% energy saving compared with making new steel from iron ore; reduction of carbon dioxide and other gas emissions into the atmosphere by over 80%; reduction of slag and other solid waste products of 1.28 tonne. The only major problem with steel recycling for the aerospace industry is it is not profitable. The cost of removing steel components, cutting them into small pieces, transporting them to the recycling plant, and then recycling and casting the metal is greater than the sale value of the recycled steel. Table 24.3 gives the relative prices of scrap metals from aircraft, and steel is much less valuable than the other materials. The main incentive for the aerospace industry to recycle steel is environmental rather than economic.

Table 24.3 Relative prices (approximate) of recycled metals compared with steel

Metal	Price relative to steel
Steel	1
Aluminium	6.3 times higher
Magnesium	30 times higher
Nickel	50 times higher
Titanium	150 times higher

24.2.5 Nickel recycling

The recycling of superalloys from jet engines is attractive because of the relatively high price of nickel scrap. The recycling of superalloy can be divided into two categories: air-melted and vacuum-melted processes. Air melting, as the name implies, involves melting the scrap superalloy at about 1600 °C inside a furnace under normal atmospheric conditions. The molten scrap is refined by the removal of alloying elements, some of which are extremely valuable (e.g. W, Nb, Ta, Hf, V) and undergo further recycling and recovery. The purified nickel is then cast into ingots for reuse. Vacuum melting involves the melting and refinement of nickel scrap in a low-pressure furnace to eliminate dissolved gases and impurities. The low-pressure atmosphere is needed to produce high-purity nickel, which is virtually free of detrimental impurities. This process is used to recycle scrap in the production of superalloys for jet engine components, such as blades and discs.

24.3   Composite recycling

The disposal of composites in an environmentally friendly way is emerging as one of the most daunting challenges facing the aerospace industry. Carbon–epoxy composites are not sustainable materials because the thermoset polymer matrix cannot be recycled. The cross-linking of thermoset polymers is an irreversible process that cannot be undone when the material is ready for recycling. Another problem is that the cost of recycling composite material is not competitive with the price of using new material. The cost of recycling carbon fibre–epoxy is greater than the cost of new material. Furthermore, the mechanical properties of reprocessed composite are lower than the original material, and are usually too low to find application in high-performance structures requiring high stiffness and strength. For these reasons, the current practice for disposing of most composite products is landfill. Not only does this pose an environmental problem because of the many thousands of tonnes of waste material that occupy landfill, but the polymers and fibres are extremely durable and take many decades (or centuries) to break down in soil. There is approximately one million tonnes of composites manufactured each year (including materials for aerospace applications). Europe has introduced new regulations on the control of waste organic materials such as polymer composites. It is illegal to dispose of composites by landfill in many European countries whereas other countries have specified maximum limits which are well below the current amounts that need to be disposed. Other countries outside Europe are also enforcing stringent regulations on the disposal of waste composite.

Despite the challenges with recycling, various reprocessing techniques are available which are classified as regrinding, thermal or chemical processes.

Regrinding is the simplest and cheapest recycling process; it involves cutting, grinding or chipping the waste composite down to a suitable size to be used as filler material in new moulded composite products. The maximum particle size for most products is under several millimetres. Whereas regrinding is a simple process, the problem with using ground material in new products is that the continuous fibres have been broken down into small fragments, and thereby lost their ability to provide high stiffness and strength.

Thermal recycling involves the incineration of waste composite to burn off the polymer matrix and reclaim the fibres for reuse. Waste composite is incinerated above 500 °C in the absence of oxygen to break down the polymer matrix into oil/wax, char and gas. The process generates a large amount of greenhouse gas. The fibres are recovered for reuse after the matrix has been removed, but their mechanical strength is reduced by the high temperature needed to decompose the polymer. The strengths of both carbon and glass fibre decrease rapidly with increasing temperature above 300–400 °C, and the temperatures used to incinerate epoxy matrix composites (500–600 °C) result in a fibre strength loss of 80–95%. The large strength reduction means that recycled fibres are not suitable for use in high-performance structures. An added problem is that the cost of recycling composites by high-temperature incineration is often greater than the original cost of the material, and there is no financial incentive to reclaim fibres. Recycling by low-temperature incineration is currently under development to minimise the loss in fibre strength, but the process is not ready for large-scale processing.

Although reclaimed fibres cannot be used in aerospace applications, there is a potentially large market for low-grade carbon fibre in other industries. It is believed that the net profit from reclaiming carbon fibres from pyrolysis is about $5 kg^{-1}. This translates to $275 000 in combined scrap value for a B787-8, with similar figures for a A350. Considering both B737 and A320 may be superseded by high-composite replacements within the next ten years, the potential reclamation value is several billions of dollars. Clearly, there is a commercial case for advancing composite recycling technology.

Chemical processing is another approach to reclaim the fibres in composite materials. The process involves using strong acid (e.g. nitric acid, sulfuric acid) or base solvent (e.g. hydrogen peroxide) to dissolve the polymer matrix, leaving the fibres for recovery and reuse. Acid or base digestion processes are less harmful to carbon fibres than thermal recycling, with only a 5–10% loss in strength. However, the chemical dissolution of the polymer matrix is slow, much slower than incineration, and therefore large digestion facilities are required for commercial-scale recycling.

Problems exist with the regrinding, thermal and chemical processes for recycling composite waste. The composite industry is investing in the development of new, more environmentally friendly and cost-effective processes. At the moment, however, the recycling of composite components from aircraft is not environmentally friendly or economically viable.

24.4   Summary

Materials selection must consider whole-of-life management issues, including the selection of sustainable materials that can be recycled using processes that are cheaper and more environmentally friendly than the processes used to make new materials. The recycling of structural and engine materials is becoming more important to the aerospace industry with the increasing rate of aircraft retirements.

The aerospace industry is moving towards high targets (above 80%) in the recycling of structural materials. About 60% of the airframe is currently recycled, but the industry is aiming to increase this to 80%. Careful consideration of the selection of sustainable materials in the design phase of aircraft is essential to ensure high levels of recycling.

Recycling of metals is possible without any loss in mechanical performance, and these materials can be recycled and reused an infinite number of times without any detrimental effect on properties. The energy consumed in the recycling of metals is much less than the energy needed to extract metal from ore. The commercial incentive to recycle metals such as titanium, nickel, aluminium and magnesium is strong because of the high sale value of the scrap, whereas the value of steel is much less.

The recycling of aluminium and steel components is performed using standard melting, refinement and casting processes. More specialist recycling processes are needed for the other metals, such as vacuum melting of titanium and nickel alloys to avoid excessive oxidation and to eliminate trapped gases in the molten metal.

Recycling of fibre–polymer composites is difficult, particularly with thermoset matrix materials. Composites are recycled using grinding, incineration or chemical processes, although the cost of recycling is not competitive against the cost of new material. Furthermore, the fibres are weakened by grinding and thermal recovery processes, thus limiting their reuse in structural products requiring high strength.

24.5   Further reading and research

Ashby, M. F., *Materials and the environment: eco-informed material choice*, Butterworth–Heinemann, Oxford, 2009.
Fiskel, J. *Design for environment: a guide to sustainable product development*, McGraw–Hill, 2009.
Lund, H. F., *Recycling handbook (2nd edition)*, McGraw–Hill, 2001.

Materials selection for aerospace

25.1 Introduction

The process of selecting materials to be used in the airframe and engine is an important event in the design of aircraft. The material used can be just as important as the design itself; there is little point creating a well-designed structural or engine component if the material is unsuitable. The key objective of materials selection is to identify the material that is best suited to meet the design requirements of an aircraft component. Selecting the most suitable material involves seeking the best match between the design requirements of a component and the properties of the materials that are used in the component. There are an extraordinarily large number of requirements for the materials used in aircraft. For example, Fig. 25.1 shows just a few of the key requirements for different sections of an airliner. Each component must be carefully analysed for its main property requirements in order to select the best material.

Materials selection in aerospace involves one of two situations: the selection of either so-called revolutionary or evolutionary materials (as described in chapter 2). Revolutionary materials selection involves selecting a material that has not been used previously in aircraft, such as the first-time application of GLARE to the Airbus A380. The selection of revolutionary

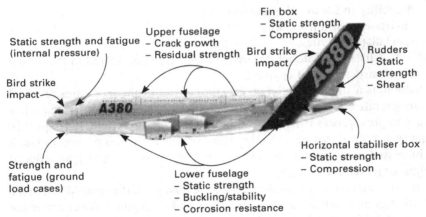

25.1 Examples of design requirements for the fuselage and empennage of the Airbus 380.

569

material may also involve choosing an existing material for a new application, such as the first-time use of carbon fibre–epoxy composite in the fuselage of the Boeing 787 airliner. Materials selection of an evolutionary material involves selecting an existing material for an application where it has been used before. It can also involve using a material that has been improved slightly from an existing type, such as a new aluminium alloy with a slightly modified alloy content or heat treatment compared with aluminium alloy used previously. The selection of a revolutionary material is usually done to achieve a large improvement in one or more aspects of the aircraft, such as a large reduction in cost or weight or a substantial improvement in fatigue life or damage resistance. Selecting an evolutionary material often results in a smaller improvement in performance, but using the material results in less risk. Regardless of whether new or existing materials are chosen, the process by which the best material is selected is the same.

The process of selecting a material that best meets the design requirements involves considering many factors such as cost, ease of manufacture, structural performance, and operating life. Other considerations can include the space and volume available for the component, the operating environment and temperature, and the number of parts to be produced. The most efficient design for an aircraft structure or engine is achieved by identifying early in the design phase those properties of the material that are most essential to achieving the design requirements. Materials are then selected based on their ability to meet these requirements.

The factors considered in materials selection can be roughly divided into the following categories:
- structural properties,
- economic and business factors,
- manufacturing issues,
- durability in the aviation environment,
- environmental impact, and
- specialist properties.

Table 25.1 lists the material properties most often considered in the design of aircraft structures and engines.

Seldom is a single material able to provide all of the properties required by an aircraft structural or engine component. The selection of materials is a complex process involving many considerations and it frequently requires compromises by accepting some disadvantageous properties (such as increased cost) in order to attain beneficial properties (such as reduced weight or increased strength).

In this chapter, we discuss the main factors and properties considered in the selection of materials for aircraft structures and jet engines. We examine the structural, business and economic, manufacturing, durability, environmental impact, and specialist properties that are considered in materials selection.

Table 25.1 Design-limiting properties used in materials selection

Class	Property
Structural	Density
	Elastic modulus (tension, shear, etc.)
	Strength (yield, ultimate, fracture)
	Impact damage resistance
	Hardness
	Fracture toughness (damage tolerance)
	Fatigue (life, strength)
	Creep resistance (creep rate, stress rupture life)
Economic and business	Cost (raw material, processing, maintenance)
	Availability
	Service life
	Regulatory issues
Manufacturing	Fabrication and casting (formability, machinability, welding)
	Dimensional (shape, surface finish, tolerances, flatness)
	Number of items
	Nondestructive inspection for quality assurance
Environmental durability	Corrosion rate
	Oxidation rate
	Moisture absorption rate
Environmental impact	Sustainability
	Greenhouse and other emissions during manufacture
	Recycling
	Waste disposal (hazardous)
	Health hazards (carcinogens, flammable)
Specialist	Thermal conductivity
	Electrical conductivity
	Thermal expansion
	Thermal shock resistance
	Stealth (electromagnetic absorbance/infrared)

In addition, there is an introduction to the process and methodology that aerospace engineers use to select the most appropriate materials for new designs or the modification of existing designs.

25.2   Materials selection in design

Selecting the right material for an aircraft component can seem an overwhelming task because there are over 100 000 materials available. There may appear to be too many choices from which to find the one material that is best suited to meet the design requirements of the component. However, materials selection is not a random, chaotic process in which the engineer is expected to 'stumble across' the best material among an immense variety of choices with little or no guidance. Materials selection is an ordered process by which engineers can systematically and rapidly eliminate unsuitable materials and identify the one or a small number of materials which are the most suitable.

The process of selecting materials and how they are then manufactured into an aerospace component is intimately linked to the design process. The design process involves several major stages performed in the sequence shown in Fig. 25.2. The process begins with a detailed assessment of the market need for a new aircraft or the modification of an existing aircraft type. Important issues are assessed in this stage to establish the economic viability for commercial aircraft and the war-fighting requirements for military aircraft. Answers are sought to key questions that must be addressed in the market assessment. What are the required range, speed, passenger capacity and payload for the commercial airliner? What is the required fuel economy of the aircraft, and are there environmental impact limits for greenhouse gas emissions and noise? What are the roles and functions of the new military aircraft? Does the aircraft require stealth and other covert capabilities? How many aircraft are required, and what is their anticipated operating life? Analysis of the market need is not performed by the engineer alone; a team is involved that includes aerospace design engineers, manufacturing engineers, materials engineers, financial analysts and possibly customer representatives. The team then proceeds onto the design of the aircraft using the information obtained in the market analysis.

Following the assessment of market need, the design process proceeds in three main stages: *concept design, embodiment design*, and (lastly) *detailed design*. Materials selection occurs in all three stages, but is fluid at the concept design stage where a large number of materials might be considered

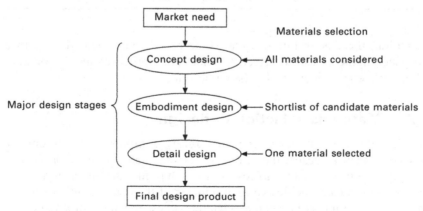

25.2 Major stages of design and their relationship to materials selection (adapted from G. E. Dieter 'Overview of the materials selection process', *ASM handbook volume 20: materials selection and design*. American Society of Materials International, Ohio, 1997).

as candidates, and as the design progresses towards completion the selection process becomes more focused towards a single material. As the materials selection process advances through the design stages it does not always lead to a 'correct' solution, although the choice of some materials is clearly better than others. Rarely does a single material meet all the design requirements, and often two or more materials are closely matched and would be equally suitable. In this respect, materials selection is different to other problems in aerospace engineering such as aerodynamics and structural analysis, which generally involve a single, correct answer.

Although there is no universally accepted definition of concept design, for the purpose of materials selection, it involves describing how a new product will be configured to meet its design objectives and performance requirements. Another useful definition of concept engineering is the process of translating customer needs (based on market research) to design features and measurable performance parameters. For complex designs, the basic configuration is determined at the concept stage, but all the minor design details may still remain unknown. The functional requirements of the design are determined at the concept stage, but the properties of the materials are not. The design requirements for a component specify what it should do but not what properties its materials should have nor how it will be made. Consider the wing design for a new type of stealth fighter; the functions would include the provision of lift and low radar visibility. These and other functions are determined at the concept design stage, but the materials and manufacturing process used to construct the wing are still not known. However, important questions related to the properties of the wing material are identified:

- Does the wing material require a radar absorption value above a specified limit?
- Will the wing experience heating during supersonic flight?
- Are any special properties required, such as ballistic protection or impact strength?

At the concept design stage, the options for materials are wide and essentially all types are considered. As the concept design develops, the choice is often made about the general class of material, such as whether it is to be metal alloy or composite, but the exact type of metal or composite is not known. If an innovative choice of material is to be made, such as constructing the wing using a new type of composite material with improved radar absorption properties, then it must occur during the concept design stage. Choosing an innovative material later in the design process is often too late because too many other decisions about the design have been made to allow for a radical change.

The embodiment stage of design involves determining the shape and approximate size of the product. The loads exerted on the component and environmental operating conditions are assessed in greater detail than during concept design. During embodiment design, the material properties important to the design are identified, such as cost, weight, strength and corrosion resistance. These properties are then ranked in order of importance. Once the design properties have been identified and ranked then a specific class of material is chosen; for example a variety of titanium alloys or steels. The properties of the candidate materials must be known to a high level of precision at this stage.

Once this is complete, the design process proceeds to detailed design which involves completing all the design details and then converting the design to specifications (e.g. dimensions, tolerances, materials, surface finish) and accompanying documentation. At this level, the decision is narrowed to a single type of material whose properties best match the design requirements.

25.3   Stages of materials selection

The design process from concept design to detailed design involves the progressive culling from a large number of material choices towards a single material, but it does not explain how the selection is made. The process of materials selection involves four main steps in the order: translation, screening, ranking and (finally) supporting information (Fig. 25.3).

25.3.1 Translation

Materials selection begins with translation, which involves examining the functions and objectives of the design. The functions define what the component is designed to do. For instance, any aerospace component used in the airframe or engine has one or more functions: to support a given stress level; to support a load at a given temperature; and so on. The objectives define what aspects of the design need to be maximised or minimised; such as maximum strength for minimum weight or greatest corrosion resistance for minimum cost.

The objectives are subject to a set of constraints, which are the conditions of the design that must be met and cannot be adjusted. As examples, the constraint may be that the component has to be within a certain size (e.g. aircraft landing gear); that the component must operate at high temperature (such as 800 °C) without softening and plastically deforming (e.g. turbine blade); that the component must withstand a specified number of loading (fatigue) cycles without cracking (e.g. helicopter rotor blade); and that the component must survive a bird impact of a defined weight and speed without causing damage (e.g. inlet blade to turbine engine). Examples of common objectives and constraints are given in Table 25.2.

The approach for selecting materials is shown by two examples in Fig. 25.4: an aircraft undercarriage and a gas turbine engine. The list of requirements

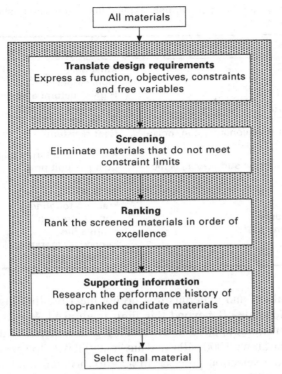

25.3 The four steps of materials selection: translation, screening, ranking, and supporting information (adapted from M. Ashby. *Materials selection in mechanical design*, Butterworth–Heinemann, Massachusetts, 1999).

that the material must meet are expressed as objectives and constraints. The objectives for the landing gear include minimum weight and volume and being structurally reliable. The objectives for the engine include a high thrust-to-weight ratio and the ability to produce low levels of greenhouse gases and operate using conventional jet fuel. The outcome of the translation step is a list of constraints expressed as design-limiting properties that must be met by the material. For example, the Young's modulus and strength must exceed limiting values whereas the rates of corrosion and creep must be under specified limits. Based on the objectives and constraints, the materials selection process moves to the next stage of screening.

25.3.2 Materials screening

Screening involves eliminating those materials whose properties do not meet the design constraints. The constraint defines an absolute upper or lower

Table 25.2 Common objectives and constraints applied in the translation stage of materials selection for aircraft

Common objectives	Common constraints
Minimise:	Must be:
Cost	Lightweight
Weight	High structural efficiency
Volume	Corrosion resistant
Maintenance and repair	Recycled
Environmental impact (fuel consumption, noise)	Heat resistant
Maximise:	Must meet a target value of:
Structural efficiency (specific stiffness, specific strength, etc.)	Cost per unit product
	Stiffness
Fatigue life	Strength
Damage tolerance	Fracture toughness
Impact damage resistance	Impact strength
Durability and operating life	Fatigue life
	Service temperature
	Corrosion rate

limit on property values, and materials that do not meet the limiting value are screened out. No trade-off beyond this limit is allowed. For example, in Fig. 25.4, one constraint applied to the landing gear material is that the yield strength must be above 1000 MPa, and any material with lower strength is eliminated from the selection process. In addition to constraints on the mechanical properties, other constraints may be applied related to economic/business, manufacturing and environmental factors as well as specialist properties. These can also be used to screen out materials. A description of the major factors and properties considered in materials selection is provided later in this chapter.

25.3.3 Material indices

Once the screening process is complete, the materials that pass are then ranked in the order that they surpass the design constraint limits. How well a material exceeds the constraint limit is quantified using a material index. In other words, a material index measures how well a candidate material that has passed the screening step can do the job required by the component. There are many material indices, each associated with maximising or minimising some property value, such as maximum strength per unit weight or minimum manufacturing cost per unit product.

Equations for calculating the index values for stiffness, strength and cost for different design shapes are given in Table 25.3. Other equations are used for calculating index values for properties such as thermal shock resistance, vibration damping and so on; some of which are given in

Objectives
  As light as possible
  As small as possible
  Safe and reliable
  Low maintenance cost

Constraints
  High modulus (>150 GPa)
  High strength (>1000 MPa)
  Damage tolerant (K_c >30 MPa m$^{1/2}$)
  Corrosion resistance (<10 μm year^{-1})

Material data
  Density
  Elastic modulus
  Strength (yield, fatigue)
  Impact toughness
  Fracture toughness
  Corrosion durability
  Dimensions

Selection process
  Screening
  Ranking
  Documentation (supporting information)

Final selection

(a)

25.4 Selecting materials for the (a) undercarriage of a commercial
airliner and (b) gas turbine engine. The objectives and constraints
are listed on the left side and the properties used to screen and rank
the materials are given on the right side.

Objectives
 As light as possible
 High thrust/weight ratio
 Safe and reliable
 Fuel efficient
 Low gas and noise emissions
 Low maintenance

Constraints
 High temperature modulus (>80 GPa)
 High temperature strength (>500 MPa)
 Damage tolerant (K_c > 30 MPa m$^{1/2}$)
 High creep resistance (<0.0001 yr^{-1})
 Corrosion and oxidation resistance
 (<10 μm year^{-1})

Material data
 Density
 Elastic modulus (800 °C)
 Strength (yield, fatigue) (800 °C)
 Creep resistance
 Fracture toughness
 Hot corrosion and oxidation durability
 Thermal expansion

Selection process
 Screening
 Ranking
 Documentation (supporting information)

Final selection
(b)

25.4 Continued

Table 25.3 Indices for stiffness and strength-limited design

Design objective	Design shape	Minimum volume index	Minimum mass index	Minimum cost index
Stiffness-limited design	Tie	$1/E$	ρ/E	$C_m\rho/E$
	Beam	$1/E^{1/2}$	$\rho/E^{1/2}$	$C_m\rho/E^{1/2}$
	Panel	$1/E^{1/3}$	$\rho/E^{1/3}$	$C_m\rho/E^{1/3}$
	Shaft	$1/G^{1/2}$	$\rho/G^{1/2}$	$C_m\rho/G^{1/2}$
	Cylinder	$1/E$	ρ/E	$C_m\rho/E$
Strength-limited design	Tie	$1/\sigma_y$	ρ/σ_y	$C_m\rho/\sigma_y$
	Beam	$1/\sigma_y^{2/3}$	$\rho/\sigma_y^{2/3}$	$C_m\rho/\sigma_y^{2/3}$
	Panel	$1/\sigma_y^{1/2}$	$\rho/\sigma_y^{1/2}$	$C_m\rho/\sigma_y^{1/2}$
	Shaft	$1/\tau_y^{2/3}$	$\rho/\tau_y^{2/3}$	$C_m\rho/\tau_y^{2/3}$
	Cylinder	$1/\sigma_y$	ρ/σ_y	$C_m\rho/\sigma_y$

E: Young's modulus; G: shear modulus; σ_y: yield strength; τ_y: yield shear strength; ρ: density; and C_m: cost per unit weight.

Table 25.4 Selected material indices

Design objective	Material index
Maximise thermal insulation	$1/k$
Minimise thermal distortion	k/α
Thermal shock resistance	$\sigma_f/E\alpha$
Maximise damage tolerance (beam, plate, etc.)	K_{Ic} and σ_f
Maximum pressure vessel strength	K_{Ic}/σ_f

E: Young's modulus; k: thermal conductivity; α: thermal expansion coefficient; σ_f: failure strength; and K_{Ic}: fracture toughness.

Table 25.4. Equations such as these are used to calculate the index value for each candidate material that passes the screening stage, and then the materials are ranked in order of excellence. It is also important to consider at this stage whether the materials can be fabricated into the component. There is no point ranking a material if it cannot be cost-effectively processed into the final product.

25.3.4 Supporting information and final selection

Once the shortlist of candidate materials are ranked in order of excellence using the index values, the final stage of the selection process is performed, and this involves the use of supporting information for a detailed profile of each material. Supporting information involves important factors other than the material properties that are relevant to the design, such as previous uses of the material in similar applications; the availability of the material; whether the company has prior manufacturing experience with the materials; certification issues associated with the material (i.e. has it been previously certified by aviation regulators); whether the material has any special handling requirements or poses occupational health and safety problems during manufacturing; whether the material can be recycled; and so on. Many sources, including databases and case histories, are used to collect as much information as possible about each material. The supporting information is analysed for each candidate material, and from this the final material is selected.

25.4   Materials property charts

The process of screening and ranking materials on their properties can be exhaustive when an extremely large number of materials are under consideration. With over 120 000 materials being available, the task of individually assessing each material against the objectives and constraints of the design is not practical. Material property charts are used to rapidly screen out large numbers of materials and to identify those materials that meet the property constraint. Material property charts plot two properties, as shown for example in Fig. 25.5 for Young's modulus against density and

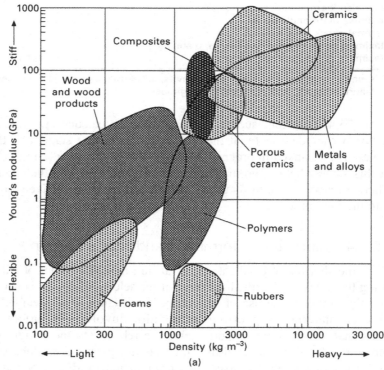

25.5 Examples of material property charts: (a) Young's modulus–density and (b) strength–toughness (provided courtesy of the Department of Engineering, University of Cambridge).

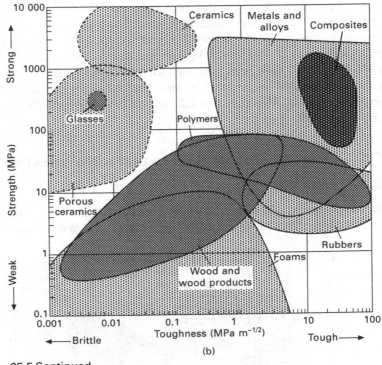

25.5 Continued

for strength against fracture toughness. These charts condense a large body of property data into a compact yet accessible and easy to understand form. The charts also reveal correlations between material properties, such as the relationship between strength and toughness or between elastic modulus and density capacity.

The properties of materials have a characteristic range of values, although for some properties this range can be large (covering up to five or more orders of magnitude). For instance, the Young's modulus for all metals is in the range of about 10 to 300 GPa and the density is between around 1300 and 20 000 kg cm^{-3}. With polymers, however, their modulus spans 0.08 to 10 GPa and density covers 800 to 2000 kg m^{-3}. When the property values for the various groups of materials are plotted they cluster together with, for example, the modulus–density combination of metal alloys forming a distinct grouping from ceramics, glasses, polymers and composites. A similar situation is found for strength–toughness and many other property combinations. A boundary is drawn around each group of materials and, for this reason, the graphs are also called bubble charts. These charts provide a global view of the relative performance of different classes of materials.

Material property charts are used to screen materials. The design constraint limits are plotted on the chart, and those materials that do not meet the limits can be eliminated from the selection process. For instance, Fig. 25.6 shows a material property chart whereby the limits of specific stiffness (Young's modulus/density) of 300 MN m kg^{-1} or specific strength (strength/density) of 100 kN m kg^{-1} have been imposed. All materials in the window defined by the limit labelled 'passed region' meet both these constraints, and can be considered further. Materials that fall outside this region can be eliminated from the selection process.

25.5   Structural properties in materials selection

25.5.1 Density

The use of lightweight materials together with optimised design has always been the most effective way of reducing the structural mass of aircraft. The

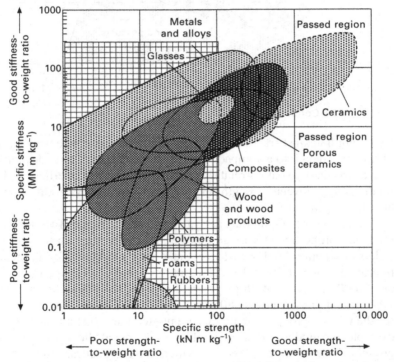

25.6 Example of the use of a materials selection chart for screening. Only materials that exceed the constraint limits of specific modulus > 300 MNm kg^{-1} or specific strength > 100 kNm kg^{-1} are considered, and all other materials (within the shaded region) are eliminated (modified from chart provided courtesy of the Department of Engineering, University of Cambridge).

use of low-density materials on their own does not necessarily provide a large weight saving; it must be combined with design methods for reducing mass to be fully effective. Similarly, lightweight design on its own does not reduce the mass significantly unless it includes the use of low-density materials.

The structural mass of most aircraft is within the range of 20 to 40% of the take-off gross weight (Fig. 3.1). Therefore, using low-density materials in the airframe translates to a large saving in the overall aircraft weight. For example, the structural mass of a B737-NG is about 22 500 kg (depending on the exact aircraft type), and substituting all of the aluminium alloy with a density of 2.7 g cm^{-3} with slightly lighter carbon fibre–epoxy composite with a density of 2.0 g cm^{-3} in principle provides a weight saving of around 3500 kg (or about 15% of the total mass). On a smaller scale, the use of composite material in the front fan case and fan blades of a gas turbine engine for a mid-sized airliner can reduce the mass by 100–200 kg.

As an approximation, reducing the structural mass by 1 kg on a mid-sized airliner provides about another 1 kg reduction in aircraft weight through the use of smaller engines to maintain the same airspeed as well as smaller wings to keep the same wing loading. A strong incentive exists to achieve even a modest reduction in the weight using light materials. It is estimated that, for every 1 kg saved on an averaged-sized airliner, the fuel consumption can be reduced by about 800 l year^{-1}. Therefore, a large reduction in structural weight translates to a substantial reduction in fuel burn with corresponding reductions in fuel cost, greenhouse and other gas emissions. Figure 25.7 shows the projected reductions in fuel consumption and gas emissions by

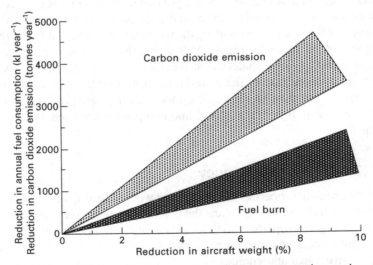

25.7 Approximate reductions in annual fuel consumption and carbon dioxide (greenhouse) gas emission with percentage reduction to the weight of a mid-sized passenger aircraft.

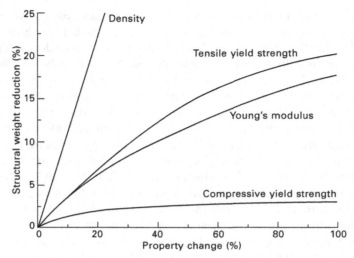

25.8 Effect of property improvement on the structural weight
(reproduced from F. H. Froes *et al.*, Proceedings of the International
Conference on Light Materials for Transportation Systems, Korea,
1993).

lowering the percentage structural mass of a typical mid-sized airliner through
the use of light materials.

Reducing the density of structural materials is recognised as the most
efficient way of reducing airframe weight and improving performance. It has
been estimated that reducing the material density is anywhere from 3 to 5
times more effective than increasing the tensile strength, modulus and fracture
toughness of the material. Figure 25.8 shows the effect of improvements to
various properties of a material on the structural weight change. A small
reduction in the density of the material is far more effective in reducing the
structural weight than increasing the mechanical properties, and this is the
main reason why light metal alloys and composites are used extensively in
aircraft.

25.5.2 Structural efficiency

Many mechanical properties are considered when selecting materials for their
structural efficiency, which means the mechanical performance of a material
per unit weight. Properties such as elastic modulus, strength, fatigue life
and fracture toughness are important in the selection of materials for both
airframe structures and engines.

Stiffness is an important design constraint for many aerospace structures
because of the need to avoid excessive deformation and buckling. Table 25.5

Table 25.5 Elastic modulus and structural efficiency of aerospace materials used in stiffness-critical applications

Material	Density (g cm^{-3})	Elastic modulus (GPa)	Structural efficiency		
			E/ρ (GPa m^3 kg^{-1})	E/ρ^2 (GPa m^6 kg^{-2}) $\times 10^{-6}$	E/ρ^3 (GPa m^9 kg^{-3}) $\times 10^{-9}$
Carbon–epoxy*	1.7	120	0.071	42	25
Magnesium	1.7	45	0.026	16	9.2
Glass–epoxy†	2.1	22	0.010	4.9	2.4
Aluminium	2.7	70	0.026	9.6	3.6
Titanium	4.5	110	0.024	5.4	1.2
Steel	7.8	210	0.026	3.5	0.4
Nickel	8.9	207	0.024	2.6	0.3

*Quasi-isotropic carbon-epoxy with 60% by volume of high modulus carbon fibres.
†Quasi-isotropic glass-epoxy with 60% by volume of E-glass fibres.

provides the elastic modulus and stiffness efficiency properties of various aerospace materials. It is usually the structural efficiency (also called specific property), expressed as the mechanical property normalised by the density of the material that is considered in aircraft materials selection rather than the absolute property of the material. This is because some lightweight materials may have relatively low stiffness and strength, but when these properties are normalised by the density they are superior to heavier materials with higher mechanical properties. For example, the elastic modulus of carbon fibre–epoxy composite used in stiffness-critical structures is about 120 GPa, which is less than the modulus of steel at 210 GPa. However, the composite material is around 3.5 times lighter than steel and, therefore, when the specific stiffness (E/ρ) of these materials is compared then the carbon–epoxy is nearly three times greater. The relative improvement in stiffness efficiency is even higher for beams (E/ρ^2) and plates (E/ρ^3) under bending loads.

The specific static strength is a key factor in materials selection for aerospace structures. Aircraft components are designed to withstand the maximum operating stress plus a safety factor, which is typically 1.5. The yield strength and strength efficiency of various aerospace materials are given in Table 25.6. The material with the highest strength is high-strength (maraging) steel, which is used in safety-critical components requiring high yield strength such as the undercarriage landing gear and the wing carry-through structure, although it does not have the highest strength efficiency. Titanium alloys and, in particular, carbon–fibre composites have high strength and structural efficiency and are also used in heavily-loaded aircraft components. Fibreglass composites also have high-strength efficiency, but their stiffness efficiency is relatively low.

Damage tolerance is another important property in materials selection, and

Table 25.6 Yield strength and structural efficiency of aerospace materials used in strength-critical applications

Material	Density (g cm^{-3})	Approximate yield strength (MPa)	Structural efficiency	
			σ/ρ (MPa m^3 kg^{-1})	σ/ρ^2 (MPa m^6 kg^{-2}) $\times 10^{-6}$
Magnesium alloys	1.7	230	0.14	80
Carbon-epoxy*	1.7	700	0.41	242
Glass-epoxy†	2.1	600	0.29	136
Aluminium alloys	2.7	525	0.22	72
Titanium alloys (Ti-6Al-4V)	4.5	1000	0.19	49
Steel (Maraging)	7.8	2000	0.25	33
Nickel alloys	8.9	600	0.07	8

*Quasi-isotropic carbon-epoxy with 60% by volume of high strength carbon fibres;
†Quasi-isotropic glass-epoxy with 60% by volume of E-glass fibres.

this defines the ability of a load-bearing structure to retain strength and resist crack growth when a defect or damage is present. Aerospace materials can contain small flaws such as processing defects (e.g. porosity, brittle inclusion particles) or in-service damage (e.g. impact, corrosion), and it is essential that these do not grow rapidly during aircraft operations otherwise it can lead to structural failure. Materials with the greatest damage tolerance generally possess high fracture toughness combined with excellent fatigue resistance defined by a slow rate of fatigue crack growth. Other mechanical properties considered in materials selection can include the structural efficiency (specific stiffness and specific strength) at high temperature, creep resistance, and fatigue performance (crack growth rate, life and residual strength).

25.6   Economic and business considerations in materials selection

An important consideration in the selection of materials is their whole-of-life cost. This cost includes all expenses associated with the material from initial manufacturing to final retirement of the aircraft, and consists of the costs of the raw material, processing and manufacturing, in-service maintenance, repair, and recycling and disposal. The decision on materials selection often comes down to a trade-off between performance and cost. Figure 25.9 shows the typical breakdown of the purchase and operating costs for a fighter aircraft, and the materials account for a small percentage of the total lifecycle cost. Therefore, using relatively expensive materials such as carbon-fibre composite or titanium instead of a cheaper material such as aluminium has little impact on the total lifecycle cost. Costs associated with the maintenance

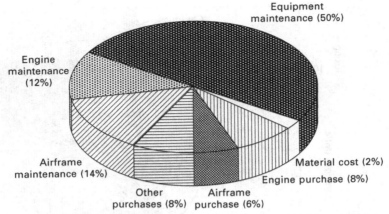

25.9 Relative distribution of costs for military aircraft. Material costs are a small fraction of the total cost.

of materials over the operating life of the aircraft are often at least two or three times greater than their initial purchase cost. Therefore, using materials that require less maintenance from in-service damage such as impact, fatigue or corrosion provides significant cost saving. The fuel cost is also a large operating expense, and a small reduction in aircraft weight by using lighter materials can also provide a substantial cost saving, as mentioned.

In the materials selection process, the cost is not considered in isolation from the other properties required from the material. The cost of the material is assessed against other important properties such as stiffness, strength and corrosion resistance, and a more expensive material may be selected because it has superior properties to a less expensive material. Sometimes, materials that are expensive are justified because they offer a unique property advantage or because they are cheaper to use than other lower-cost materials; for example, a design might be simplified and, thus, made at lower cost.

Figure 25.10 presents a materials selection chart of cost–strength for different groups of engineering materials. The cost of most materials, including the metal alloys and composites used in aircraft structures and engines, vary over a wide range depending on their composition and processing. Expensive metals such as titanium and nickel alloys are preferred over cheaper materials when high specific strength and creep resistance at elevated temperature are required. Similarly, carbon fibre composites are used in aircraft structures rather than the cheaper glass–fibre composites because of their superior stiffness and fatigue strength.

Another economic consideration in materials selection is that the purchase and maintenance costs of materials can change over time. The prices of raw materials are rarely stable and usually fluctuate up and down in response

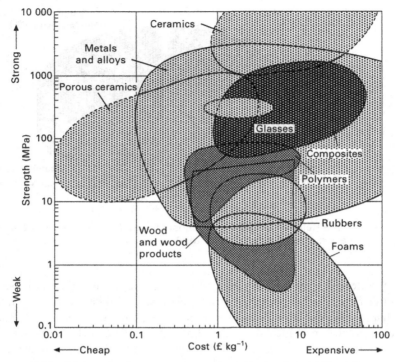

25.10 Materials selection chart for cost–strength. (Provided courtesy of the Department of Engineering, University of Cambridge.)

to supply and demand. Many aircraft types are sold by the aerospace manufacturers over a period of many years and, during this time, the price of raw materials can change considerably. As examples, the Boeing 747 has been in continuous production since 1969 and the production of the Airbus 300 commenced in 1974 and ceased 33 years later. Over these long periods, the price of the raw materials can rise and fall, which means that selecting a material because of its low cost does not always mean it remains cost competitive for the entire production period. Figure 25.11 shows the fluctuations to the price of titanium over a sixty-year period, during which the cost has varied more than 400%. Such large fluctuations in cost are difficult to predict, but must be recognised when selecting materials to be used in large quantities in aircraft construction.

The maintenance cost of the materials used in the airframe and engine can also change with time owing to deterioration from corrosion, fatigue and other factors. The cost of maintaining some types of materials (such as titanium and carbon fibre–epoxy) is usually less than other materials (such as aluminium or magnesium) owing to fewer problems with fatigue

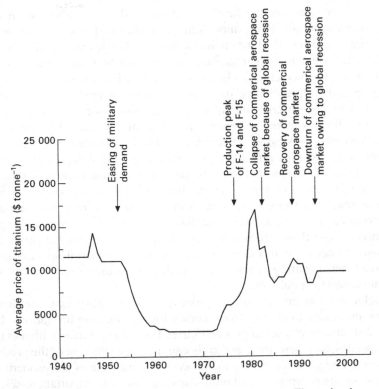

25.11 Average price of sponge titanium per tonne. The price is normalised to the cost of titanium in 1998.

or corrosion. The potential to reduce maintenance cost is a consideration in materials selection.

25.7   Manufacturing considerations in materials selection

Selecting the process by which raw materials are manufactured into airframe and engine components is an essential part of materials selection. Manufacturing includes the primary forming processes (e.g. casting and forging of metals and autoclave curing of composites), heat treatment (e.g. thermal ageing, stress-relief annealing), material removal processes (e.g. machining, drilling, trimming), finishing processes (e.g. surface coatings, anodising), joining processes (e.g. welding, fastening, adhesive bonding), and nondestructive inspection (e.g. ultrasonics, radiography).

Materials and manufacturing are closely linked, and it is impossible to select a material without considering how the material is to be manufactured into the final component. The choice of material is dependent on the choice of process by which it is formed, joined, finished and otherwise treated. The material properties affect the choice of process: ductile materials can be forged and rolled, whereas brittle materials such as composites must be processed in other ways. Conversely, the choice of process affects the material properties: forging and rolling change the grain structure, strength and toughness of metals whereas different manufacturing processes for composites result in different fibre contents and therefore different mechanical properties. The manufacturing process can change the material properties (beneficially or adversely), and thus affect the performance of the component in service. Some processes improve the properties, such as heat treatment raising the strength of metals or removing residual stresses, whereas other processes can degrade the properties, such as casting defects in metals or voids in composites. Selecting the best material for an airframe or engine component involves more than selecting a material that has the desired properties, it is also connected with the manufacturing of the material into the finished product and ensuring it meets the quality assurance requirements set by aviation safety regulators.

Selecting the manufacturing process is not an easy task for there are many processes to choose from, each with benefits and limitations. Figure 25.12 shows the classes of processes used for manufacturing with metals and composites, and there are many to choose from at each stage of the production process. The goal is to select the process that maximises the properties and quality of the component and minimises the cost. It is important to select the manufacturing process at an early stage in the materials selection process, otherwise the cost of changing the manufacturing route later can be costly.

The method of selecting the manufacturing process is shown in Fig. 25.13, and is similar in principle to the materials selection process. The starting point is that all processes are considered as possible candidates until proven otherwise. The sequential steps of translation, screening, ranking and search for supporting information are followed to eliminate unsuitable processes and to identify the best process. The translation step involves transforming the design requirements into constraint limits used in selecting the process. Constraints may include the size, shape, material type and processing temperature of the product. Limits are applied to the constraints, such as the process must be capable of making integrated products larger than 2 m or the process must heat treat the product in an inert atmosphere. Screening involves eliminating the processes which do not meet the constraint limits, and the shortlisted processes are then ranked in order of their ability to manufacture the product in terms of cost, batch size and so on. Supporting information is used to help identify the best process, such as the availability of the capital equipment or the technical complexity of the process operation.

An important consideration in the design and manufacture of an aerospace

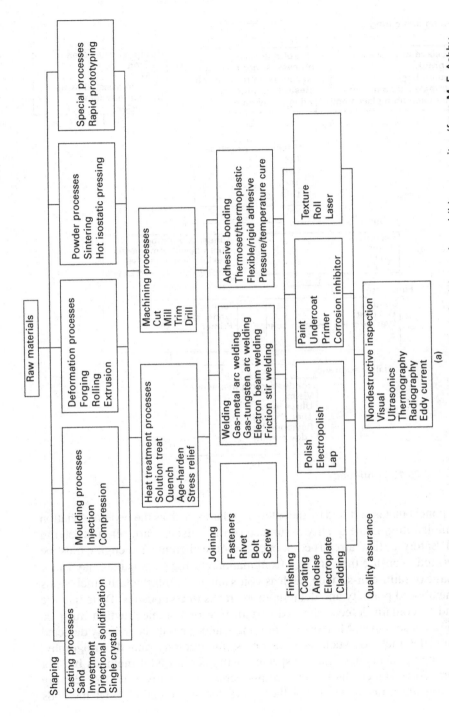

25.12 Classes of manufacturing process and process flowcharts for (a) metals and (b) composites (from M. F. Ashby. *Materials selection in mechanical design*, Butterworth–Heinemann, Massachusetts, 1999).

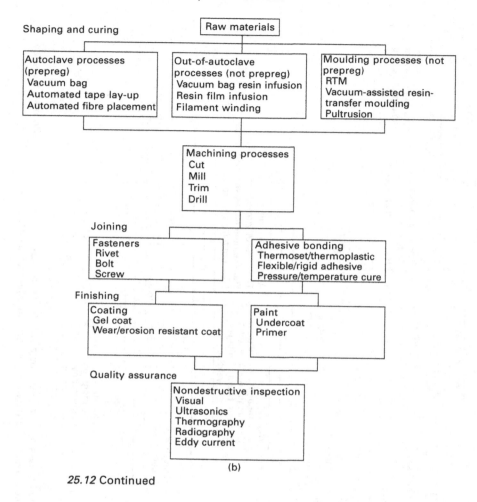

Shaping and curing

Raw materials

Autoclave processes (prepreg)
Vacuum bag
Automated tape lay-up
Automated fibre placement

Out-of-autoclave processes (not prepreg)
Vacuum bag resin infusion
Resin film infusion
Filament winding

Moulding processes (not prepreg)
RTM
Vacuum-assisted resin-transfer moulding
Pultrusion

Machining processes
Cut
Mill
Trim
Drill

Joining

Fasteners
Rivet
Bolt
Screw

Adhesive bonding
Thermoset/thermoplastic
Flexible/rigid adhesive
Pressure/temperature cure

Finishing

Coating
Gel coat
Wear/erosion resistant coat

Paint
Undercoat
Primer

Quality assurance

Nondestructive inspection
Visual
Ultrasonics
Thermography
Radiography
Eddy current

(b)

25.12 Continued

component that is certified by the safety regulators such as the Federal Aviation Administration is the quality assurance of the finished part, often involving NDI. Safety-critical aircraft components for the airframe and engine must be inspected to ensure they are free from manufacturing defects that may cause damage or failure in-service, such as voids and large intermetallic inclusions in metals and porosity and delamination cracks in composites. Parts that are found to contain defects above a certain size or volume fraction must be repaired or scrapped. Most defects are small, and can only be reliably detected using NDI methods such as ultrasonics, radiography and thermography. Some components are easily inspected using certain NDI methods but not others; for instance thick metal components can be inspected for casting voids using ultrasonics but not thermography. Also, certain NDI methods

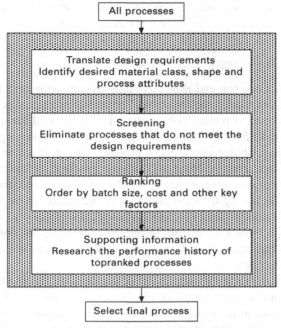

 does not reflect surrounding flowchart text; transcribing flowchart text below:

All processes

Translate design requirements
Identify desired material class, shape and
process attributes

Screening
Eliminate processes that do not meet the
design requirements

Ranking
Order by batch size, cost and other key
factors

Supporting information
Research the performance history of
topranked processes

Select final process

25.13 The four steps of manufacturing process selection: translation, screening, ranking, and supporting information (adapted from M. F. Ashby. *Materials selection in mechanical design*, Butterworth–Heinemann, Massachusetts, 1999).

are more suited to some materials than others; such as the application of eddy-current inspection to conducting materials, but it cannot be used on fibre–polymer composites. The shape, dimensions and material used in the component determines the type of NDI method that can be used to inspect for manufacturing quality, and this must be considered as part of the design process.

25.8   Durability considerations in materials selection

The durability of materials in the operating environment of the aircraft is an important consideration in minimising maintenance and extending the service life. The environment may be hot, humid, corrosive, abrasive or some other potentially damaging condition. Both metals and fibre–polymer composites are susceptible to environmental degradation during service, and selecting a material with the best durability is an important consideration. Figure 25.9 shows that maintenance of the airframe and engine accounts for a large percentage (~26%) of the whole-of-life cost, and a significant amount of this cost is the expense of inspecting materials for environmental damage and, when detected, repairing or replacing the component.

25.8.1 Corrosion in materials selection

Corrosion is the most common and expensive type of environmental damage to aircraft metals. Composites, polymers and ceramics are not susceptible to corrosion; the problem is mostly with metals. It is estimated the aviation industry spends around $2 billion per year on maintaining metal components damaged by corrosion. Water is the most common cause of corrosion to aircraft metals, although some acidic or alkaline solvents (e.g. paint strippers, cleaning agents) may also cause corrosion. Metals can be damaged in various ways by corrosion, including uniform attack over the entire surface of the material or localised attack causing surface pitting or cracking. Corrosion is an important consideration in materials selection because not only does it weaken aircraft components by removing material, but under stress it can also cause cracking.

The corrosion resistance of metals is dependent on many factors, including the type and concentration of the corrosive agent, temperature, external load, and material parameters such as the type of base metal, types and concentrations of alloying elements, microstructure (grain structure, precipitates), and residual stresses. Figure 25.14 provides an approximate ranking of the corrosion resistance of several materials in salt water, although such comparisons must be used with caution in materials selection. It is difficult to compare the corrosion resistance of candidate materials unless data is available from corrosion tests that are performed under conditions that closely replicate the in-service environment. There are two approaches used

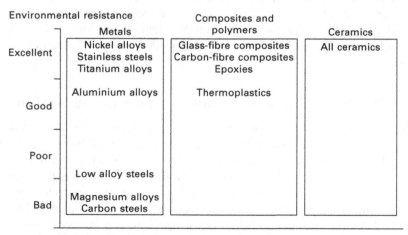

25.14 Ranking of the resistance of materials to attack by seawater.

to select aerospace materials for corrosion resistance: choose a material which has a high resistance to corrosion or protect the material using a corrosion resistant coating (such as anodised coating for aluminium or cladding for steel).

25.8.2 Oxidation in materials selection

Oxidation is another type of environmental damage that must be considered in materials selection, particularly materials for high-temperature applications, such as jet and rocket engines. Oxidation is a reaction process between the material and oxidising agents in the atmosphere, such as oxygen in air or sulfur dioxide in the combustion gas of jet fuel. The reaction damages the metal by forming a thick surface layer of brittle metal oxide. Oxidation is not usually a problem for aircraft materials unless they are used for high-temperature applications, when it is essential to select a material with high resistance to oxidation or a material that can be thermally insulated using an oxidation-resistant coating.

25.8.3 Moisture absorption in materials selection

Unlike metals, fibre–polymer composite materials are not susceptible to corrosion and are not used in hot environments where oxidation is a problem. However, composites are not immune to the environment and may be damaged by other ways. Composites are susceptible to environmental damage by absorbing water in the atmosphere. Water molecules are absorbed into the polymer matrix of composites where they cause softening and lower the glass transition temperature. With some types of composites, the absorption of water can cause delamination cracking, fibre/matrix debonding, and damage to the core material (with sandwich materials). Water can also be absorbed by organic fibres used in composites, such as aramid, which further weakens the material. The deterioration of composites by moisture absorption is a consideration in materials selection, particularly when the aircraft is required to operate in tropical regions where the atmosphere is hot and wet. The polymer matrix and organic fibres used in composites may also be degraded by long-term exposure to ultraviolet radiation in sunlight. Consideration of the environmental stability of composites is essential in materials selection, and numerous types of composites are available which are resistant to degradation by moisture and ultraviolet radiation.

25.8.4 Wear and erosion in materials selection

Damage by wear and erosion may be another consideration in assessing the durability properties of aerospace materials. Wear is not usually a serious problem except when materials are used in engines and other moving parts, when selecting a material with high wear resistance is important. Figure 25.15 presents a materials selection chart for the wear rate constant versus hardness. The wear rate constant k_a is a measure of sliding wear resistance: low k_a means high wear resistance at a given bearing pressure. With a single group of materials, such as metals or polymers, it is generally found that the wear rate constant decreases with increasing hardness. Therefore, selecting a material for high wear resistance is often based on the hardness and yield strength properties. Erosion is a specific type of wear involving the removal of material under impact from abrasive particles such as sand or dirt. Erosion is a consideration for materials used at the external surface of the main rotor blades for helicopters and propeller blades for aircraft. The erosion resistance of materials, like their sliding wear resistance, increases with the surface hardness and strength.

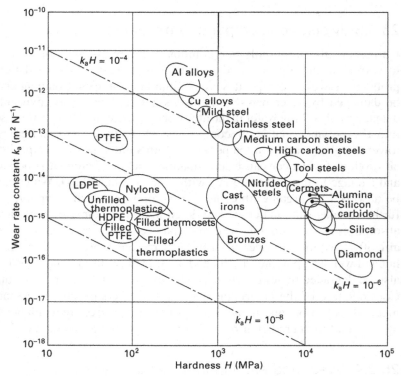

25.15 Materials selection chart for wear rate constant versus hardness. Reproduced from M. F. Ashby. *Materials selection in mechanical design*, Butterworth–Heinemann, Massachusetts, 1999.

25.9   Environmental considerations in materials selection

Consideration of the environmental impact of materials is fast becoming a key factor in materials selection for aircraft. Whenever possible, materials obtained from sustainable resources and which have low impact on the environment during their production, usage and disposal should be considered. The aerospace industry is keen to use sustainable materials to minimise the environmental impact of aircraft. Sustainability is defined as the ability of a material to be used infinitely, which involves recycling the material at the end-of-life for reuse in new aircraft or some other application. Recycling avoids the need to extract new material from a non-renewable resource such as ore for metals or petroleum products for carbon fibres and polymers. If some material is deemed 'sustainable' and cheap to recycle, this is a favourable selection factor, particularly for materials used in high tonnage such as aluminium. The ability to recycle varies considerably among the various aerospace materials, with aluminium and steel being relatively easy to recycle, titanium and magnesium being more difficult to recycle, and composite materials being extremely difficult (if not impossible) to fully recycle.

The aerospace industry is also keen to minimise the so-called 'carbon footprint' of materials, which is another consideration in materials selection. In the USA, Europe and many other places, emissions from manufacturing and recycling processes are significant factors to be dealt with. When selecting a material, consideration should be given to whether its use causes the emission of greenhouse gases and other pollutants to the atmosphere. For example, the production of 1 kg of aluminium, which includes the mining, refinement and smelting, requires about 284 MJ of energy and generates about 35 kg of carbon dioxide when the power source is grid electricity. In comparison, the production of 1 kg carbon–epoxy composite, which includes producing the fibres, polymer and manufacturing the material, requires about 40 MJ of energy resulting in about 5 kg of CO_2. Other waste by-products from manufacturing processes, such as effluent, should be considered.

25.10 Specialist properties in materials selection

Specialist properties are considered for aircraft materials used in components for a unique application. The specialist property may be the most important consideration in materials selection, and other properties such as the cost, ease of manufacture or mechanical performance could be of lesser importance. For example, resistance against cracking and spalling owing to rapid heating, known as thermal shock resistance, is an essential property for materials used in the exhaust casing of rocket engines. Listed below are several specialist properties which may be considered in materials selection:

- Electrical conductivity is an important property for materials used in the outer skin of aircraft. The material must have the ability to conduct an electrical charge in the event of lightning strike.
- Thermal conductivity is a consideration for materials used in high-temperature applications such as heat shields and engine components. Heat-shield materials require low thermal conductivity to protect the airframe structure from excessive heating.
- Thermal expansion is also a consideration for high-temperature materials. Materials with a low thermal expansion coefficient are often required to avoid excessive expansion and contraction during heating and cooling.
- Flammability is a consideration for materials where there is the risk of fire, such as aircraft cabins and jet engines. Flammability properties such as ignition temperature, flame spread rate and smoke may need to be considered.
- Stealth is an important property for materials used in the external surface of covert military aircraft. Materials with the capability to absorb radar waves and/or reduce the infrared visibility are important for stealth aircraft.

25.11 Summary

Materials selection is an important process in the design of aircraft. The objective of materials selection is to identify the material that is best suited to meet the design requirements. Selecting the material involves seeking the best match between the design requirements of the aircraft component and the properties of the materials.

Materials selection can be revolutionary or evolutionary. Revolutionary materials selection involves selecting a new material or a material that has not been used previously in an aircraft component. Evolutionary materials selection involves using an 'old' material or slightly modified version of an old material in a component where it has been used previously. Most choices for materials in aircraft follow the evolutionary path because it is less risky than using a revolutionary type of material.

Many factors must be considered in materials selection, which are classified as structural, economic/business, manufacturing, durability, environmental impact, and specialist properties for unique applications.

Following the assessment of market need, the design of aircraft involves several major stages in the order: concept design, embodiment design and detail design; followed by the final design solution. At the concept design stage, all materials are considered; then this is reduced to a shortlist of candidate materials in the embodiment design stage; and, finally, one material is chosen in the detailed design stage.

The process for selecting materials and the process for selecting the manufacturing process involves four main steps in the order: translation, screening, ranking and supporting information.

Materials selection charts are used to screen out materials which do not meet the design constraints, and materials indices are used for ranking the shortlisted materials in order of excellence.

The most effective way of reducing the structural mass of aircraft is using light-weight materials (together with optimised design). A reduction in material density is often more effective at reducing aircraft weight than using stiffer or stronger materials of higher weight.

The durability of materials in the aviation environment (e.g. heat, rain, humidity, erosive particles) is a key consideration in materials selection. Materials must be resistant to deterioration when used in service: metals must resist corrosion and oxidation; composites must be unaffected by moisture; and metals and composites must resist wear and erosion.

The environmental impact of using material is becoming an increasingly important consideration in materials selection. Sustainable materials obtained from renewal resources and which have minimal impact on the environment during their production and recycling are considered favourably.

25.12 Terminology

Concept design: Basic design of a product to meet the main functional objectives and performance requirements determined from market research. A large number of materials are considered at the concept design stage.

Constraint (design): Performance requirements that must be met by the product (and its materials).

Detailed design: Final stage of the design process involving all the detailed design work to complete the product. Also involves converting the design into specifications and documentation so the product can be produced. The material to be used in the product is selected in the detailed design stage.

Embodiment design: Determination of shape, size and other major design features of the product. A shortlist of candidate materials for the product are considered in the embodiment design stage.

Function (design): The purpose of the designed product.

Material indices: Quantitative measure of how well a material property (e.g. stiffness, strength, maximum operating temperature) exceeds the design constraint. Index values are used to rank shortlisted materials in order of excellence to exceed the design constraint limit.

Objective (design): The aims of the design, such as to minimise cost or maximise fatigue life.

Ranking: Listing shortlisted materials in their order of excellence to exceed the design constraint limit based on the index value.

Screening: Process of eliminating materials based on their inability to meet one or more of the design constraints.

Supporting information: Various resources (e.g. case histories, past experiences) used in the final stage of the materials selection process.

Translation: Process of translating the design into functions and objectives which can used to define the constraints for materials selection.

25.13   Further reading and research

Ashby, M. F., *Materials selection in mechanical design, 3rd edition*, Butterworth–Heinemann, Oxford, 2005.

Ashby, M. F. and Cebon, D., *Cambridge materials selector*, Granta Design Ltd., Cambridge MA, 1996.

Budinski, K. G. and Budinski, M. K., *Engineering materials: properties and selection*, Pearson Education Inc., Upper Saddle River, NJ, 2010.

Dieter, G. E. (editor), *ASM handbook volume 20: materials selection and design*, ASM International, Ohio, 1997.

Hinrichsen, J., 'The material down-selection process for A3XX', in *Around Glare*, edited by C. Vermeeren, Kluwer Academic Publishers, 2002, pp. 127–144.

Index

machining
  composites, 333–4
    drill bits used for carbon–epoxy and
      aramid fibre composites, 334
    peel-up and push down damage
      from incorrect drilling, 334
magnesium, 9–10, 24–5
  decline in use as aerospace material,
    25
  recycling, 564
magnesium alloys, 224–31
  classification system, 225–6
    ASTM lettering system, 226
  composition and properties, 227
    aerospace applications, 227
    pure Mg and its alloys, 227
  corrosion properties, 230–1
  helicopter gearbox casing, 225
  metallurgy, 225–31
  strengthening, 227–30
    effect of Al content on tensile
      properties, 229
magnetic particle, 545, 556
  inspection principles in surface cracks
    detection, 546
manual lay-up, 321–4
  autoclave for consolidation and curing
    of prepreg composites, 323
  plies orientations for quasi-isotropic
    cross-ply composite, 322
  vacuum begging operation, 323
manufacturing factors
  materials selection, 589–93
    classes of manufacturing process
      and flowcharts, 591–2
    four steps of manufacturing process
      selection, 593
'maraging', 244
maraging steels, 235, 244–6
  ageing temperature effect on strength
    and ductility, 246
martensite, 241–4
  body-centred-tetragonal structure, 243
  tempering temperature effect on
    medium-carbon steel, 244
mast, 52
material composition, 2
material density, 582–4
  annual fuel consumption and carbon

dioxide gas emission reduction,
  583
  effect of property improvement on the
    structural weight, 584
material indices, 576–9
  indices for stiffness and strength-
    limited design, 579
  selected material indices, 579
materials engineering see materials
  technology
materials property chart, 580–2
  illustration, 580–1
  screening use, 582
materials science, 2
materials screening, 575–6
materials selection
  aerospace, 569–600
    design, 571–4
    durability considerations, 593–6
    economic and business
      considerations, 586–9
    environmental considerations, 597
    manufacturing considerations,
      589–93
    materials property chart, 580–2
    specialist properties, 597–8
    stages, 574–80
    structural properties, 582–6
materials technology, 4
McDonnel-Douglas F-14 *Tomcat*, 203
  use of titanium alloys, 204
mechanical properties
  low through-thickness, 352
    elastic modulus, strength and
      fracture toughness properties of
      carbon–epoxy composite, 352
  microstructural defects effects on, 378
    void content on percentage tensile
      and interlaminar shear strengths,
      379
medium-carbon low-alloy steels, 235, 246
medium-carbon steels, 235
metal alloys
  casting, 134–48
    casting defects, 140–3
    processes, 143–8
    shape and ingot casting, 134
    solidification of castings, 134–5
    structure of castings, 135, 137–40

Printed in the United States
by Bookmasters

Printed in the United States
By Bookmasters